STRUCTURAL DESIGN WITH PLASTICS

STRUCTURAL DESIGN WITH PLASTICS

by

B. S. BENJAMIN

Professor of Architecture and
Architectural Engineering,
The University of Kansas

Second Edition
Revised and Enlarged

VNR VAN NOSTRAND REINHOLD COMPANY
NEW YORK CINCINNATI TORONTO LONDON MELBOURNE

Van Nostrand Reinhold Company Regional Offices:
New York Cincinnati

Van Nostrand Reinhold Company International Offices:
London Toronto Melbourne

Library of Congress Catalog Card Number: 80–29614
ISBN: 0–442–20167–2

Manufactured in the United States of America

Published by Van Nostrand Reinhold Company
135 West 50th Street, New York, N.Y. 10020

Published simultaneously in Canada by Van Nostrand Reinhold Ltd.

15 14 13 12 11 10 9 8 7 6 5 4 3 2 1

Library of Congress Cataloging in Publication Data

Benjamin, Bezaleel Solomon, 1938–
Structural design with plastics.

Bibliography: p.
Includes indexes.
1. Plastics in building. 2. Structural design.
I. Title.
TA668.B38 1981 693′.9923 80–29614
ISBN 0–442–20167–2

PREFACE TO THE SECOND EDITION

Since the publication of the first edition of this book in 1969, there has been considerable research devoted to structural plastics, particularly in the field of material properties, so essential to proper design. The high-performance plastics, SMC/BMC, and the reinforced thermoplastics are some areas in which tremendous progress has been made. I have tried to include all relevant work in as complete a manner as I possibly can. Due to limitations on the size of the book, however, I have been forced to be selective, and this has led to the exclusion of some work of considerable merit. For this I am truly sorry.

In the field of structural design, I have included solved problems on sandwich construction and plate theory, with new articles on the analysis and design of cylindrical barrel shells and hyperbolic paraboloids in plastics. A STRUDL Coding program for a folded-plate, conical dome in plastics has also been included.

In a work of this nature which brings together research done all over the world, it has not always been possible to have a consistent style. For example, a tensile strength of 8,000 psi is shown by some researchers as 8 on the graph, with the units defined as 10^3 psi. Other research workers show the same figure as 8 on the graph, with the units defined as 10^{-3} psi. There is little to choose between either method of representation, and I have allowed the wishes of the original researcher to prevail.

In the preparation of this work, over the last eleven years, I have been assisted by a number of my students, and I would like to acknowledge the help given by Christopher Atkins, M. W. Mustapha, G. Natkunan, Bill Huntress, Randy Comfort, Deena Koehn, Kent Rawson, Richard Musgrave, Satish Gadgil, Abdul Al-Hammad, and Peggy Sneegas.

I would also like to pay a tribute to the Society of the Plastics Industry and the Society of Plastics Engineers in the United States. These two fine institutions have done a great deal to publicize the nature, and encourage the use, of plastics materials, and their help is gratefully acknowledged. The

SPI Conference on Reinforced Plastics/Composites needs particular recognition. I have been a Chairman of the Building Construction Session and can vouch for the tremendous importance and usefulness of these conferences. I would like to add a word of thanks to Charles Condit, past Staff Vice-President of the SPI and Joe McDermott, presently of the SPI, for the consideration that they have always shown to me at the SPI Conference. I was an academic in an industrial world, but they made me feel comfortable, and their concern made the difference between just another conference and a great one.

Finally I am, as always, deeply grateful to my wife Nora and my children Ashley and Jennifer. Their unwavering support to me over all these years is very warmly appreciated.

B. S. Benjamin

PREFACE TO THE FIRST EDITION

This book treats all aspects of structural design in plastics materials. Various structural plastics, with particular reference to material properties, are discussed at great length. The approach is strictly an engineering one. The design of civil engineering structures such as folded plates, shells, stressed skin structures, etc., both in single skin as well as in sandwich construction, is considered. Modifications to conventional engineering design, to take into account the special properties of these materials (such as their viscoelastic nature), are presented. New concepts such as the structural and architectural life of the structure are also considered.

The effect of varying stress level on the long term life of the plastics laminate, and hence of the structure, remains as an unsolved problem. A prototype folded-plate, conical-dome roof structure built by the author and erected at Hatfield in Hertfordshire (United Kingdom) is yielding valuable information on the varying stress level to which the laminate can actually be subjected. Test results reveal that the stress level changes with every instant of time. Besides this fact, it varies all over the structure as well! Theoretically, in order to design correctly, one should be able to predict the long term behavior of the laminate under a randomly varying stress level that has been subjected to a statistical analysis. The difficulty in predicting the ultimate strength of the laminate under varying stress levels is also currently being considered by the author. Tests on several hundred samples of continuously laminated GRP performed by one of the author's students, Mr. A. C. Harold, indicate that the reduction in ultimate strength with time as a function of the stress level and of the percentage reinforcement, is better represented by a surface. Attempts are being made to define, in general terms, the mathematical equation of the surface into which appropriate constants can be fed in order to then define the surface, in specific terms, for particular materials and conditions. The problem is further complicated by the fact that though the life of the laminate reduces continuously with time, the ultimate strength does not. The ultimate strength remains fairly constant until a certain percentage of the life of the laminate has been exhausted, after which the laminate suffers rapid mechanical failure. The entire approach to the problem has necessarily to be a statistical one.

In the writing of this book I have received considerable help from many individuals, firms, and organizations, and I have attempted to acknowledge their assistance at appropriate points in the text. However, I want to express my gratitude here to the Plastics Institute and the British Plastics Federation of Great Britain, and the Society of Plastics Engineers and the Society of the Plastics Industry of the United States. In addition, I wish to thank Mr. Rassell of Cape Insulation Co., Mr. Brown of BTR industries, Mr. Halfnight of Halfnight Protective Systems, Dr. Mitchell of Mitchell and Smith, Mr. Salmon and Dr. Narracott of Shell Chemical Co., Mr. Kabelka of the Czechoslovak Academy of Sciences, Mr. Yescombe, chief librarian at the Northern Polytechnic, and finally, my former colleagues at the Structural Plastics Research Unit of the University of Surrey—Professor Makowski, Mr. Wajda, Mr. Davies, Dr. Gilkie, Mr. Hughes, Mr. Robak, and Mr. Zerning.

This book was, however, in an indirect way, made possible by members of my family, all of whom contributed to it in no small measure—my parents for a worthwhile and expensive education, and my wife and children for their patience in permitting me the long winter evenings to work on it. Their help is gratefully acknowledged.

B. S. Benjamin

Hatfield, Herts, England
April, 1969

CONTENTS

Preface to the second edition v
Preface to the first edition vii

1. Plastics as a structural material 1

1.1 Introduction 1
1.2 Plastics suitable for structural applications 2
1.3 The general mechanical behavior of plastics 3
1.4 The fire resistance of plastics 9
1.5 Design with plastics 11

2. Reinforced plastics 24

2.1 Reinforcing materials 24
2.2 Glass 24
2.3 Jute and sisal reinforcement 25
2.4 Asbestos reinforcement 26
2.5 Graphite or carbon fiber reinforcement 27
2.6 Boron fiber reinforcement 28
2.7 Aramid fiber reinforcement 28
2.8 Whiskers and microfibers 30
2.9 Mica, wollastonite, and other filler reinforcements 30
2.10 Manufacturing processes for reinforced plastics 31
2.11 Design recommendations for the choice of a process 38
2.12 Fiberglass reinforced polyester 38
2.13 Jute fiber reinforced polyester 82
2.14 Asbestos reinforced polyester 86
2.15 Glass fiber reinforced epoxy composites 88
2.16 Phenolic, melamine, and silicone resin composites 101
2.17 Reinforced vinyl ester composites 103
2.18 High-performance composites 105
2.19 SMC/BMC 122
2.20 Reinforced thermoplastics 130

3. Low density core materials 142

3.1 Introduction 142
3.2 Rigid polyurethane foam 143

3.3 Expanded polystyrene 158
3.4 Expanded polyvinylchloride (PVC) 159
3.5 Phenolic foam 160
3.6 Polyester foam 171
3.7 Polysulfone foam 172
3.8 Structural foams 173
3.9 Poisson's ratio of the low-density core materials 179
3.10 Reinforced plastics honeycomb cores 179

4. Sandwich construction 187

4.1 Introduction 187
4.2 Sandwich behavior under transverse loading 187
4.3 Sandwich failure under compressive end loads 195
4.4 Rectangular sandwich panel under transverse loading 197
4.5 Buckling of a rectangular sandwich panel under compressive end loads 201
4.6 Analysis of a rectangular sandwich panel in bending 205
4.7 Analysis of a rectangular sandwich panel in buckling 207

5. Architectural forms for use with plastics 209

5.1 Introduction 209
5.2 Folded plate structures 209
5.3 Shell structures 214
5.4 Stressed skin systems 218

6. Analysis of folded plate structures 221

6.1 Existing classical methods for the analysis of folded plates 221
6.2 A modified approach to the analysis of folded plate structures in plastics 222

7. Prismatic folded plate structures 226

7.1 The influence of connections 226
7.2 The analysis of an internal unit 226
7.3 The analysis of the external unit 235

8. Composite folded plate structures 247

8.1 The composite folded plate rigid frame 247
8.2 The composite folded plate barrel vault 247

8.3 The analysis of the barrel vault 248
8.4 Design and testing of a prototype composite folded plate barrel vault of sandwich construction 258

9. Shell structures 275

9.1 Shell structures in plastics 275
9.2 Design of a FRP dome of single skin construction 275
9.3 The barrel shell 295
9.4 The hyperbolic paraboloid (hypar) 300

10. Stressed skin structures 306

10.1 Skeleton analysis 306
10.2 Flat prismatic plates under axial compression 306
10.3 Angle sections under axial compression 308
10.4 Effect of variation of the angle of bend 310
10.5 Effect of material properties 311
10.6 Prismatic folded plates 311
10.7 Nonprismatic folded plates 314
10.8 Stiffness method of analysis 317
10.9 Finite elements method 331
10.10 STRUDL Coding and analysis for a folded plate, conical, FRP dome structure in single skin construction 332

11. Miscellaneous structures 346

11.1 Inflatables 346
11.2 Plastics radomes 351
11.3 Tensioned structures 353
11.4 Prefabricated low-cost housing 355

References 367

Author Index 379

Subject Index 383

[illegible]

References [illegible]

Author index [illegible]

Subject index 381

CHAPTER 1

PLASTICS AS A STRUCTURAL MATERIAL

1.1 INTRODUCTION

In the last decade, the use of plastics in building construction, in the automotive industry, and in aerospace technology has mushroomed beyond all expectations. There is virtually no industry in which plastics have not been able to compete with conventional materials for specific applications. In the home, the office, the factory or workshop, and in the recreation and leisure industries, plastics have contributed substantially to the development of better, cheaper, and more functional products. For structural applications, however, except in aerospace, where weight is the major consideration, progress has not been as dramatic. Plastics have some advantages—and some disadvantages—when used structurally, in comparison to conventional materials such as steel, concrete, or aluminum. Their disadvantages, apart from a somewhat higher cost, lie mainly in their properties, which is considered in greater detail in Section 1.3. It would be appropriate to discuss here, therefore, the advantages of those plastics suitable for structural applications.[1] These may be listed briefly as follows:

1. Structural plastics have higher strength/weight ratios than most other engineering materials.
2. They are attractive in appearance and can contribute substantially to the architectural beauty of the structure. They can easily be made translucent, or can be pigmented, to give desired tints or colors to the structure.
3. Many plastics have exceedingly good corrosion resistance, superior to that of many conventional engineering materials.
4. Plastics are very amenable to prefabrication. This property, together with their light weight, makes them very useful materials in systems

of industrialized building, and for low-cost housing for underdeveloped countries.

5. A high degree of built-in thermal insulation can be obtained by the use of suitable plastics.

In order to be able to design in plastics, it is first necessary to thoroughly understand the material itself. This may not be easy, because there are many types of plastics used for a variety of purposes. For structural purposes, however, there are only a small number of efficient plastics, and these will now be considered in greater detail.

1.2 PLASTICS SUITABLE FOR STRUCTURAL APPLICATIONS

Plastics can be basically classified into two main groups—thermoplastic materials and thermosetting materials. Thermoplastic materials soften with the application of heat. They exhibit considerable creep under load, particularly at elevated temperatures. Polyvinylchloride (PVC) and polymethylmethacrylate (acrylics) are examples of thermoplastics that can be used for light structural applications. Generally they are in sheet form and can be used with or without reinforcement. They are easily amenable to a number of fabrication processes such as molding or vacuum forming. By these processes the most intricate shapes can be created to suit architectural requirements. If unreinforced, they are isotropic and show very consistent properties. Acrylic domeshells built in Finland are an example of the structural use to which these materials can be put.[2] They have a diameter of 19 ft, 4 in. (5.89m) and a height of 6 ft (1.83m). They were designed for a snow load of 30.7 lb/ft^2 (150 kg/m^2) and a wind load of 20.5 lb/ft^2 (100 kg/m^2).

Thermosetting materials, once cured, do not soften upon application of heat. When reinforced they exhibit higher strengths and less tendency to creep than the thermoplastics. Polyester, epoxide, phenolic, and silicone resins are examples of thermosetting plastics. Of these, the polyesters are the most widely used. They are relatively inexpensive, with good properties in reinforced laminates. Epoxide resins have extremely good properties but are more expensive than polyesters. Phenolic resins have good properties that are maintained even at high temperatures. However, their applications are limited by their need of hot cures and high laminating pressures. High price and rigid processing requirements have also restricted the widespread use of silicone resins.

These thermosetting resins cannot be used alone structurally, and they require reinforcement with a substance such as glass, graphite, jute, cloth, paper or asbestos. Asbestos-reinforced phenolic composites are suitable for high-temperature work. The most common reinforcement is glass, but graphite, boron and aramid fibers are more suitable for high-strength applications. Research on sapphire whiskers is still under way. The main reinforcements are discussed in Chapter 2.

In single skin construction, the reinforced laminates can be used by themselves. For large spans, however, it is necessary to use sandwich construction in which the thin reinforced laminates are bonded to a low density core. The facings then take the bending moment and the axial forces, and the core is assumed to take all the shear. The cores can also be of low-density plastics, and the six main types are the following:[3,4,5]

1. Rigid polyurethane foam
2. Expanded polystyrene
3. Expanded polyvinylchloride (PVC)
4. Phenolic foam
5. Polyester foam
6. Polysulfone foam

A new development in this field has been the availability of structural foam,[6] which can be made from a variety of resins including ABS. This material is discussed in great detail in Section 3.8.

Of the first four types listed above, rigid polyurethane foam and phenolic foam are thermosetting plastics, the other two being thermoplastics. Rigid polyurethane foam has an advantage over the others because it can be foamed directly onto or in between laminates, thus making it particularly suitable for in situ work. Further, the polyester laminates can be directly "laid" on it. This cannot be done with expanded polystyrene, as it is chemically attacked by the polyester resin during the cure of the laminates. The cured laminates, therefore, must be fixed to the expanded polystyrene core by means of a separate adhesive layer. Expanded PVC and phenolic foam are not attacked. The surface has to be sealed, however, in order to prevent the penetration of the laminating resin into the foam.

The properties of some of these plastics, including polyester and polysulfone foams, is detailed in later chapters. The general mechanical behavior of plastics is considered below. Section 1.5 contains a more rigorous analysis of the viscoelastic nature of plastics.

1.3 THE GENERAL MECHANICAL BEHAVIOR OF PLASTICS

As has been seen, the types of plastics suitable for structural applications are rather limited. To be able to use these plastics efficiently, however, it is necessary to understand their mechanical behavior and the differences in such behavior between these materials and commonly used engineering materials. The following are most important points:[1]

1. The stress/strain curves of plastics are not usually linear up to yield. In some cases, there may be no yield at all.
2. The modulus of elasticity in tension, of plastics, is not necessarily the same as that in compression.
3. The modulus of elasticity of plastics is very low.
4. Plastics can exhibit anisotropic behavior.
5. The mechanical behavior of plastics is affected by the rate of straining of the material.
6. The mechanical behavior of plastics is affected by the temperature.
7. Plastics creep considerably under load with time.
8. Plastics show a reduction in ultimate strengths with time, even under static loading.
9. The properties of plastics can be affected by environmental conditions.

Each factor will now be considered in greater detail.

1.3.1 The Stress/Strain Curves of Plastics Are Not Usually Linear up to Yield—in Some Cases There May Be No Yield at All

The importance of this factor depends largely on the type of plastics used. For glass fiber reinforced plastics, the nonlinearity up to strains of about 0.3% is small, and the tangent modulus at the origin may be used as the short term modulus. Sometimes the yield and ultimate strengths are so close that they can be considered to be identical. The material can therefore be designed with reference to ultimate strength, with suitable factors of safety. However, due to the absence of plastic flow at yield, the material is incapable of relieving stress concentrations. In thermoplastics, the presence of yield and a zone of plastic flow in the stress/strain curve depend on the temperature and the rate of straining of the material. At lower temperatures and higher straining rates, the thermoplastics show brittle fracture, with absence of yield.

In general, it can be stated that plastics display partly elastic and partly viscous behavior, and that this viscoelastic behavior is both temperature and time dependent.

1.3.2 The Modulus of Elasticity in Tension is Not Necessarily the Same as That in Compression

This is true of most plastics—assuming that one can define the moduli accurately. For the reinforced plastics and the thermoplastics, the difference in the two moduli is small and may be neglected in the analysis. The difference, however, can be very great for the low-density core materials such as rigid polyurethane foam or expanded polystyrene.

1.3.3 The Modulus of Elasticity of Plastics is Very Low

This is an important factor to be kept in mind when designing with plastics. Even Fiber reinforced polyester (FRP) has a modulus of only about 1×10^6 psi (6.89 GPa). The thermoplastics have even lower values than this. For example, the 1 minute flexural modulus of acrylic at 20°C (68°F) is about 0.44×10^6 psi (3.03 GPa). Were these materials used in conventional structural forms, the deflections would be so large that the carrying capacity of the structure would be seriously limited. Therefore, it is essential to use those structural forms that provide added stiffnesss by virtue of their shape. Folded plates, singly and doubly curved shells, and stressed skin space structures offer the most efficient possibilities.[7–9] Large local deflections that could occur in single skin construction can be avoided by the use of sandwich construction. Large deflections in plate theory can complicate the analysis and are discussed in greater detail.

1.3.4 Plastics Can Exhibit Anisotropic Behavior

The amorphous thermoplastics such as acrylics and PVC show isotropy that simplifies the analysis. Fiber reinforced laminates, reinforced with chopped strand mat, also show reasonable isotropy. Composites reinforced with woven roving or cloth show orthotropy, the two directions being the warp and the weft. The properties of the composite at any given angle between these two directions can then be determined.[10] This property of reinforced plastics can be used to great advantage in filament winding, where desired directional strengths can easily be obtained.

1.3.5 The Mechanical Behavior of Plastics Can Be Affected by the Rate of Straining of the Material

Except for very high rates of strain (such as impact), the mechanical properties for purposes of design may be assumed to be unaffected by this factor. This conclusion is particularly realistic because, in civil engineering practice, it is very difficult to define the rate at which loading is imposed on the structure. The exact mechanical behavior of plastics, even if fully known over a large range of straining rates, would not be very useful. At very high straining rates, the thermoplastics tend to show brittle fracture.

1.3.6 The Mechanical Behavior of Plastics is Affected by the Temperature

To consider the effects of temperature, a sharp distinction must be made between the fiber-reinforced plastics using thermosetting resins, and the thermoplastics. Of these the former are far less sensitive to temperature than the latter. The upper temperature limits of thermoplastics are fixed by their melting points. The amorphous plastics such as acrylics or PVC show a dependence on temperature until the glass transition temperature, above which they become melts. The reduction[11] in dynamic modulus with temperature is shown in Fig. 1. The semicrystalline plastics such as nylon 6, though they show a reduction in modulus by their glass transitions, still retain sufficient strength and rigidity above that temperature to carry load until they reach their crystalline melting points. The thermosetting plastics do not melt. Their upper temperature limit is determined by thermal oxidation, which results in degradation. Below this high degradation temperature, thermosets can exhibit behavior with or without transition as shown[11] in Fig. 2. Phenolic resin, once fully cured, has no transition until degradation occurs at a fairly high temperature. Epoxy, on the other hand, like most other themosetting materials, shows a transition, but still remains load bearing until the degradation temperature is reached.

At very low temperatures, the thermoplastics show brittle fracture, and care should be exercised in their use.

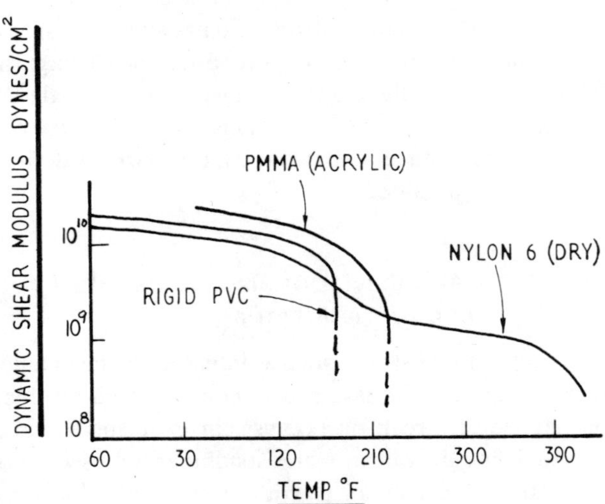

Fig. 1

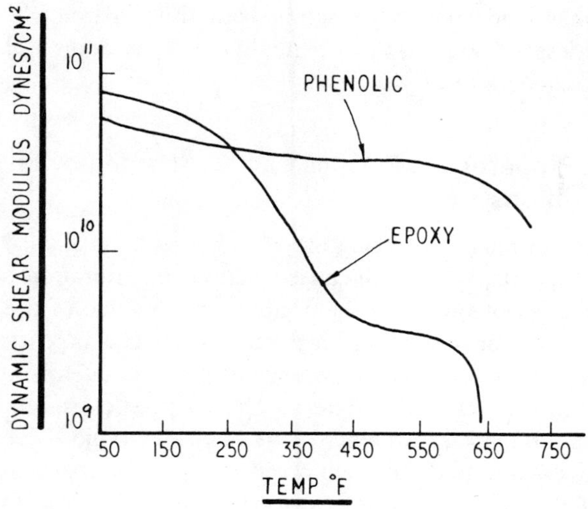

Fig. 2

1.3.7 Plastics Creep Considerably under Load with Time

Although other engineering materials, such as concrete, also creep with time, the creep in plastics is considerably greater in magnitude. It is also characterized by the fact that it is dependent upon the stress level and the temperature, and is wholly recoverable. As an example, an acrylic[12] that has a 1-minute flexural modulus of 0.44×10^6 psi (3.03 GPa) at 20°C (68°F) can have the value reduced to 0.27×10^6 psi (1.86 GPa) after 10,000 hours. Fiber reinforced plastics also creep with time, but at a slower rate than the thermoplastics. This factor of creep is so important in engineering design that it is treated in great detail in Chapter 2.

1.3.8 Plastics Show a Reduction in Ultimate Strengths with Time, Even under Static Loading

This property is also often defined as creep rupture. In effect, plastics exhibit a form of fatigue behavior, even under static loading. The thermoplastics show this behavior to an alarming degree. As an example,[12] an acrylic that has an ultimate tensile strength of 8600 psi (59.3 MPa) after 1 hour of loading at 20°C (68°F) has this value reduced to 5800 psi (40 MPa) after 1000 hours at 20°C (68°F). The fiber reinforced plastics initially show a slight increase in ultimate strength with time, due to the fact that a certain amount of curing occurs in the composite. After curing is complete, the ultimate strengths under load show a reduction with time in much the same way, although to a lesser degree, as the thermoplastics. This factor is also considered in much greater detail in Chapter 2.

1.3.9 The Properties of Plastics Can be Affected by Environmental Conditions

Some plastics, although not being attacked by acids or alkalis, are attacked by chemical solvents, which can cause a serious deterioration of the surface, if present in a gaseous form in the atmosphere. A table of the resistance of a particular plastics or resin to a wide range of chemicals is generally available. Such a table should be used with care, as some plastics suffer from environmental stress crazing or cracking. The unstressed plastics suffers no apparent damage, when immersed in the fluid, but can suffer rapid mechanical failure in a stressed condition. Fiber reinforced composites are even attacked by long immersions in water. This is caused by the slow deterioration of the bond between the resin and the fibers,[13] rather than by a direct attack of the resin.

Environmental conditions can also cause plastics to weather, though additives are now available to reduce the damaging effects of ultraviolet radiation. The weathering of plastics depends to a large extent on the formulation used and the control in manufacture. Pigments, fillers, reinforcements, stabilizers, and other additives profoundly affect weathering. The ultraviolet radiation in sunlight is a significant factor in the weathering of plastics. Heat and moisture also accelerate weathering. As a rough guide, it has been suggested[14] that a temperature increase of 10°C (18°F) doubles the rate of chemical reactions between the surface and its environment.

The weathering of plastics can be manifested in several ways—fading, yellowing, surface degradation, and erosion. In translucent materials there is, of course, loss of light transmission. Surprisingly, however, plastics weather better in industrially polluted atmospheres (free from those gases that specifically attack the surface) than in cleaner rural areas. The industrial atmospheres reduce the intensity of ultraviolet radiation and also cover the surface with a protective coating. In coastal areas, weathering is likely to be even more severe. Coatings can be used to protect the surface, but they are an additional expense. Gumerman and Mckay[15] carried out accelerated weathering tests on reinforced polyester panels coated with a thin film of polyvinylfluoride (PVF) as a finish. The panels exhibited an excellent degree of color and surface stability, not only by retention of gloss, but also by a complete lack of fiber exposure. Where light transmission is important, a life of 10 years has been suggested[14] for thermoplastic translucent roof sheeting. For structural purposes however, a life of 50 years or more can easily be expected for properly manufactured FRP sheets.

1.4 THE FIRE RESISTANCE OF PLASTICS

The most serious objection to the use of plastics for structural purposes involves their resistance to fire. It has delayed the even more widespread use of plastics for nonstructural purposes in building. Although there are no inexpensive, completely fire resistant plastics, there are fillers or additives that can make the plastics fire retardant or self-extinguishing. These additives, however, may reduce the weathering resistance and overall mechanical properties of the plastics.

The structural and nonstructural use of plastics in building, in any country, will be governed by the building regulations of that country. These regulations vary considerably from one country to another. In the United States, several codes govern these regulations, such as the Uniform Building Code, National Building Code, Southern Standard Building Code, Basic Building Code, and so on. In addition large cities may have their own regulations. Common to these codes is the method employed to control the use of materials, regardless of type, all materials are governed by performance standards. For instance, a plastics interior finish is treated just like any other interior finish. The "plastics" chapter that appears in the codes typically authorizes the approval of plastics for light transmitting purposes in applications such as skylights, panels in monitor and sawtooth roofs, wall panels, light diffusing systems in ceilings, light transmitting panels for partitions, awnings and canopies, and greenhouses. Also included are standards that regulate plastics when used as veneers, and for plastic elements if they form part of an assembly, which under provisions of the code is tested and accepted as a complete unit.

There are essentially six areas of interest. These are the classifications for the following: type of construction, occupancy, fire zones, interior wall and ceiling finishes, fire resistive standards, and the chapter on plastics. Depending on the type of occupancy, fire zone, and the like, the codes lay down required fire ratings for walls and floors. These range from 1 to 4 hours. The fire rating of walls can be obtained by tests that subject the wall assembly to a controlled fire on one side for the rated period, the temperature rise being noted on the other side of the wall. The heat by transmission must not result in a rise in temperature of more than 250°F (139°C). The wall also must not permit the passage of flame or gases. The wall should further be able to withstand water from a fire hose after half the rated period, or a maximum of 1 hour. If the wall is a load bearing wall, it must be able to carry a load of at least 80 lb/in.^{2} (56,240 kg/m^2) on the gross cross-sectional area of the wall. The fire rating of the wall assembly will therefore depend on the material and the thickness of the wall, and on the use of combustible or incombustible materials in wall finishing, wall insulation, and so on. In any case, it is the entire wall assembly that receives the rating.

It is easy to see that, when used alone, commercially available plastics cannot provide this measure of fire resistance in structural components. However, fire resistant materials are often used to give improved fire ratings to wall systems, such as wood stud walls or preformed metal wall panels, that might otherwise be unacceptable. Fire resistant gypsum wallboard and incombustible insulation can be used to give a wood stud wall a fire rating of 1 hour. Gypsum wallboard or gypsum plaster is often used to protect other materials with poor fire resistance, and this is to be recommended.

In addition to fire rating in hours for floor construction, the fire resistance of floor and ceiling materials is also evaluated by surface spread of flame, which is the rate at which flame will travel across the exposed surface. The test results in a "flame spread" index, which compares the rate for the material under test to noncombustible asbestos cement (classified as 0) and untreated red oak (classified as 100). There are three classifications: Class 25 (0–25), Class 75 (25–75) and class 200 (75–200). The lower the index, the more suitable the material, and Class 25 is therefore the most desirable. The fuel contributed and smoke developed are also rated in this test.

Roofing material is also classified for fire resistance, being rated A, B, or C, depending on the performance of the roofing in a "burning brand" test. This test measures the performance of the roofing, in its fire resistance, to ignition by flying burning brands, and surface spread of flame. In this case, however, the rating depends not only on the roofing itself, but also on the roof assembly. The provision of suitable underlayment can improve the class of rating.

The British Building Codes are very similar. They also classify the buildings into groups, based on occupancy and on the damage or consequences of a fire. In small residential rooms, for instance, damage to life is more important than damage to property, or to the structure itself.[16] In a warehouse, the ability of the structure to withstand the load for the duration of the fire is of prime importance. In built-up areas, the spread of fire between buildings is an important factor. British regulations also require all load-bearing external walls to have a certain minimum fire resistance in terms of hours, and even flame retardant plastics are unable to meet such stringent requirements. It is therefore more suitable to apply wall cladding panels to nonload-bearing walls, sufficiently remote from the boundaries.

The regulations pertaining to roofs are, likewise, more favorable to plastics materials. The performance of the roof is judged by its resistance to radiated and convected heat from the neighboring building, and to hot gases and flames from burning brands. The suitability of the material is classified by two letters—AA, AB, and so on. The first letter designates the resistance to penetration by fire, and the second designates the surface spread of flame. There are four categories, A to D. Thus material classified as AA would have a high resistance to penetration by fire, and a low surface spread of flame. The addition of the suffix X would denote that molten droplets formed during the test. The regulations specify that the roof may comprise any materials that achieve, at worst, an AC rating for all classifications of building occupancy.[17] Fire retardent grades of FRP can achieve AA, AB, or AC ratings and can therefore be used without restriction in the roof. Ceiling materials, which are classified as wall linings, are judged for their suitability by surface spread of flame tests. There are primarily four grades for ceiling materials. Class 1 materials have the highest degree of resistance to surface spread of flame, while Class 4 materials have the lowest. There is, in addition, a further Class 0, which is even higher than Class 1. The regulations require that ceilings of small residential buildings have no worse than Class 3 surface flame spread.

The generation of smoke and toxic gases is another problem that must be considered when evaluating the suitability of a particular plastics in a specific application. It should be realized that, in an actual fire, smoke and toxic gases and the depletion of oxygen kill more than actual flames, Asphyxiation is often the major cause of death. Attack by superheated air or gases and smoke inhalation come next in order of importance. Further smoke can often block the visibility of exits. Toxic products attack the lungs, irritate the eyes, and can dull the senses of the victim.

In some cases, efforts to make the materials fire retardent, by the use of additives, may increase the danger by resulting in the generation of heavy smoke and toxic gases. The tunnel test that measures the flame spread also determines the fuel contributed and the smoke generated. Here again the test makes comparisons to asbestos cement, which is rated as 0, and red oak, which is rated as 100.

1.5 DESIGN WITH PLASTICS

It has been seen that design with plastics, in order to be accurate, is inevitably complicated by the viscoelastic nature of the material. Engineering materials such as steel have properties that are time independent. Structures built of these materials, therefore, are designed by the ordinary elastic theory. This assumes that when the strains are small, stress is proportional to strain and this gives

$$\sigma = \epsilon E \tag{1}$$

where

σ = stress
ϵ = strain
E = modulus of elasticity

For viscoelastic materials, however, the above relationship is not valid, due to the time-dependent nature of the stress/strain relationship. In the most general form, this relationship can be expressed as[18]

$$\sigma = f(\epsilon, t, T. \ . \ . \ ) \tag{2}$$

where stress is a function of strain, time, temperature and other variables such as composition, wet or dry conditions, processes and conditions of manufacture, and so on. Equation 2 represents nonlinear viscoelastic behavior, under which category most plastics would be classified. It forms, however, a very complex basis for analytical treatment, and can be modified to

$$\sigma = \epsilon f(t) \tag{3}$$

Equation 3 neglects all effects except time. The effects of temperature can be considered by a time–temperature superposition. The other variables in Equation 2 are not very tangible quantities and cannot be subjected to strict mathematical analysis.

The time dependence of the mechanical properties of plastics can best be understood by the two phenomena of stress relaxation and creep.

If the strain is kept constant for a specimen under load, it is found that the stress decays as a function of time, as is shown in Fig. 3. The relaxation modulus E_r, now time dependent, may thus be calculated at any time t as

$$E_r(t) = \frac{\sigma(t)}{\epsilon} \tag{4}$$

If, alternatively, the stress is kept constant for specimens under load, the material creeps, as was pointed out in Section 1.3.7. The strain then continues to increase with time, as is shown in Fig. 4. Physicists define a creep compliance as

$$c(t) = \frac{\epsilon(t)}{\sigma} \tag{5}$$

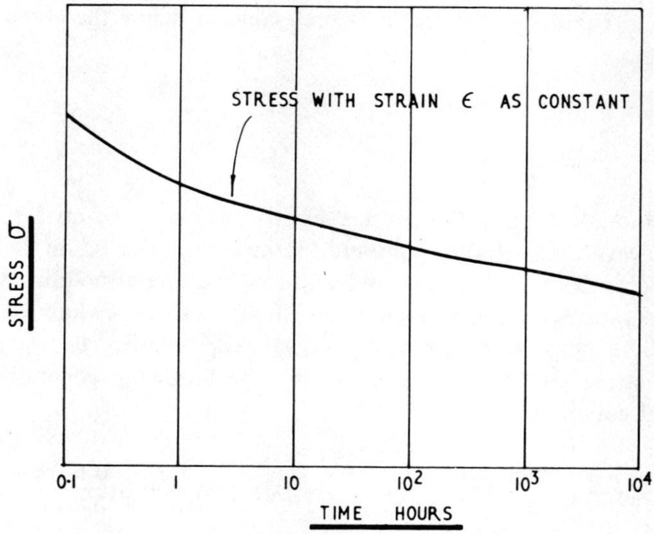
STRESS WITH STRAIN ϵ AS CONSTANT
STRESS σ
0·1
1
10
10²
10³
10⁴
TIME HOURS

Fig. 3

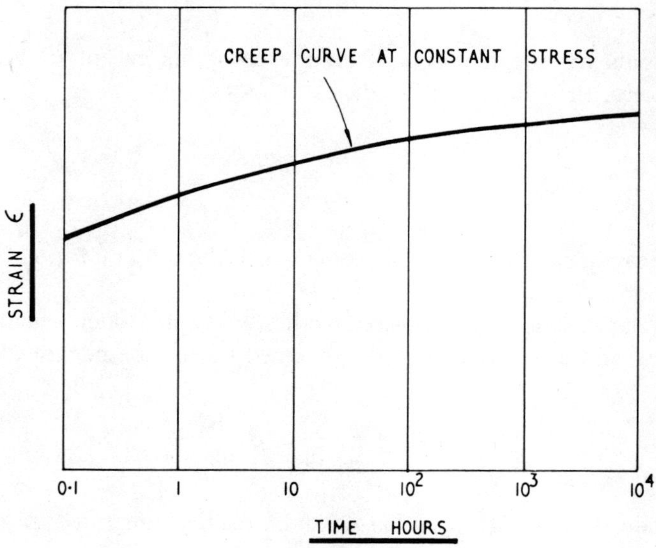
CREEP CURVE AT CONSTANT STRESS
STRAIN ϵ
0·1
1
10
10²
10³
10⁴
TIME HOURS

Fig. 4

However, in engineering terms, it is preferable to define the creep modulus as

$$E_{cr}(t) = \frac{\sigma}{\epsilon(t)} \tag{6}$$

The two moduli E_r and E_{cr} are not strictly the same. In civil engineering design, however, only the creep modulus is important, as load on the structure is assumed to be constant. The reduction of the creep modulus with time (and temperature) causes an increase in the deflections of the structure.

The relationship between relaxation and creep behavior can be expressed clearly if stress and strain are related by the following generalized linear differential equation:

$$\alpha_0\sigma + \alpha_1\frac{\partial^2\sigma}{\partial t} + \alpha_2\frac{\partial^2\sigma}{\partial t^2} + \cdots = \beta_0\epsilon + \beta_1\frac{\partial\epsilon}{\partial t} + \beta_2\frac{\partial^2\epsilon}{\partial t^2} + \cdots \tag{7}$$

where

$$\alpha_0, \alpha_1, \alpha_2 \cdots$$

and

$$\beta_0, \beta_1, \beta_2 \cdots$$

are constants for the material. If all the constants except α_0 and β_0 are equal to zero, then Equation 7 reduces to

$$\sigma = \frac{\beta_0}{\alpha_0}\epsilon = E\epsilon$$

Elastic behavior can therefore be looked upon as the simplest form of viscoelastic behavior.

Stress relaxation can be considered by the "Maxwell Model," which consists of a spring and a dashpot in series governed by the mathematical equation

$$\alpha_0\sigma + \alpha_1\frac{\partial\sigma}{\partial t} = \beta_1\frac{\partial\epsilon}{\partial t} \tag{8}$$

If the strain is constant, Equation 8 can be readily integrated to give

$$\sigma(t) = \sigma_0 e^{-t/\tau} \tag{9}$$

where

$$\sigma(t) = \text{stress at any time } t$$
$$\sigma_0 = \text{stress at time } t = 0$$

and $\tau = \alpha_1/\alpha_0$ is called the relaxation time.

Creep behavior can be considered by the "Voigt Model," which consists of a spring and a dashpot in parallel governed by the mathematical equation

$$\alpha_0\sigma = \beta_0\epsilon + \beta_1\frac{\partial\epsilon}{\partial t} \tag{10}$$

Since the stress is constant, Equation 10 may be readily integrated to give

$$\epsilon(t) = \frac{\sigma_0\alpha_0}{\beta_0}(1 - e^{-t/\tau}) \tag{11}$$

where

$$\epsilon(t) = \text{strain at time } t$$
$$\sigma_0 = \text{constant stress}$$

and $\tau = \beta_1/\beta_0$ is called the retardation time.
The relaxation modulus can be obtained from Equation 9 as

$$E_r(t) = \frac{\sigma(t)}{\epsilon_0} = \frac{\sigma_0}{\epsilon_0}e^{-t/\tau} \tag{12}$$

The creep modulus can be correspondingly obtained from Equation 11 as

$$E_{cr}(t) = \frac{\sigma_0}{\epsilon(t)} = \frac{\beta_0}{\alpha_0(1 - e^{-t/\tau})} \tag{13}$$

Equations 12 and 13 indicate that, with the mathematical assumptions made, stress relaxation and creep are simple exponential functions of time. For plastics materials, however, this is not so. A more accurate approach therefore is to consider a range of relaxation and retardation times. For stress relaxation, for instance, a number of Maxwell elements can be connected together in parallel, so that Equation 9 becomes

$$\sigma(t) = \sum_{i=1}^{n} \sigma_{0i}e^{-t/\tau i} \tag{14}$$

The stress relaxation modulus can then be obtained as,

$$E_r(t) = \sum_{i=1}^{n} \frac{\sigma_{0i}}{\epsilon_0} e^{-t/\tau i} \tag{15}$$

For creep, a number of Voigt elements can be considered in series, so that Equation 11 becomes

$$\epsilon(t) = \sum_{i=1}^{n} \frac{\sigma_{0i}\alpha_{0i}}{\beta_{0i}} (1 - e^{-t/\tau i}) \tag{16}$$

This gives the creep modulus under constant stress σ_0 as,

$$E_{cr}(t) = \sum_{i=1}^{n} \frac{\beta_{0i}}{\alpha_{0i}(1 - e^{-t/\tau i})} \tag{17}$$

Though the above moduli have been determined for a discrete range of relaxation and retardation times, they can also be determined in general terms, as integrals for a continuous range as well.

It should be realized, however, that from a design point of view the stress relaxation and creep curves for the material are already known. These curves should be supplied by the raw material manufacturer as part of the service he offers to encourage the use of his material. Of course it might not be possible to find the exact mathematical equation that fits these curves (and any suitable mathematical equation could have been used in the preceeding discussion), but it is possible to know the stress relaxation and creep modulus at any given time by the use of Equations 4 and 6.

Accurate prediction of the structure's behavior over a considerable period of time requires a knowledge of the past mechanical loading history of the material. The behavior of the structure at any instant in time can then be predicted by the use of the Boltzmann superposition principle. The principle can be applied to both stress relaxation and creep. Consider Equation 4, which gives

$$E_r(t) = \frac{\sigma(t)}{\epsilon}$$

At $t = 0$, let $\epsilon = \epsilon_0$ and stress $= \sigma_0$. Then, at time t,

$$\sigma(t_1) = E_r(t_1)\epsilon_0$$

At time t_1, let a strain ϵ_1 be added to the system. Then,

$$\sigma(t) = E_r(t)\epsilon_0 + E_r(t - t_1)\epsilon_1 \tag{18}$$

More generally, therefore,

$$\sigma(t) = \sum_{t_i = -\infty}^{t_i = t} E_r(t - t_i)\epsilon_i \tag{19}$$

For a continuous loading,

$$\sigma(t) = \int_{-\infty}^{t} E_r(t - t_i)\frac{d\epsilon}{dt_i}\,dt_i \tag{20}$$

Assuming that the lower limit of the integral is zero, and taking Laplace transforms of the equation, it can be shown that

$$\overline{\sigma}(p) = \overline{E}_r(p)\overline{\epsilon}(p) \tag{21}$$

where p is the transform variable.

For creep, consider Equation 6, which gives

$$E_{cr}(t) = \frac{\sigma}{\epsilon(t)}$$

If $\sigma = \sigma_0$ at $t = 0$, then strain at time t_1 is

$$\epsilon(t_1) = \frac{\sigma_0}{E_{cr}(t_1)}$$

Now let another stress σ be added to the system. Then the strain at time t is given by

$$\epsilon(t) = \frac{\sigma_0}{E_{cr}(t)} + \frac{\sigma}{E_{cr}(t - t_1)} \tag{22}$$

This can be generalized to

$$\epsilon(t) = \sum_{t_i = -\infty}^{t_i = t} \frac{\sigma_i}{E_{cr}(t - t_i)} \tag{23}$$

which for a continuous loading becomes

$$\epsilon(t) = \int_{-\infty}^{t} \frac{1}{E_{cr}(t - t_i)} \frac{d\sigma}{dt_i} dt_i \tag{24}$$

If the lower limit of the above integral is zero, it can be shown that

$$\bar{\epsilon}(p) = \bar{\sigma}(p) \frac{1}{\bar{E}_{cr}(p)} \tag{25}$$

by taking Laplace transforms of the equation. The analogy between Equations 21 and 25 and the usual elastic Equation 1 is unmistakable. From this we can conclude that Laplace transforms can be used to eliminate the time variable and to substitute it by the transform variable.[19,20] The ordinary elastic stress/strain relationships are still valid, provided that the ordinary elastic constants are replaced by their analogous viscoelastic counterparts. The ordinary elastic stress/strain relationships for two dimensional problems are

$$\begin{aligned} \epsilon_x &= \frac{1}{E}(\sigma_x - \nu\sigma_y) \\ \epsilon_y &= \frac{1}{E}(\sigma_y - \nu\sigma_x) \\ \epsilon_{xy} &= \frac{2(1 + \nu)}{E} \sigma_{xy} \end{aligned} \tag{26}$$

For linear viscoelastic analysis, these relationships can be modified to

$$\begin{aligned} \bar{\epsilon}_x(p) &= \frac{1}{\bar{E}(p)}[\bar{\sigma}_x(p) - \bar{\nu}(p)\bar{\sigma}_y(p)] \\ \bar{\epsilon}_y(p) &= \frac{1}{\bar{E}(p)}[\bar{\sigma}_y(p) - \bar{\nu}(p)\bar{\sigma}_x(p)] \\ \bar{\epsilon}_{xy}(p) &= \frac{2[1 + \bar{\nu}(p)]}{\bar{E}(p)} \bar{\sigma}_{xy}(p) \end{aligned} \tag{27}$$

Since the equilibrium and compatibility equations (which form the other foundations on which the elastic theory is based) are independent of material properties, it follows that linear or nonlinear elastic analysis can be used for the analysis of structures in plastics. This is provided the material is assumed to be linearly viscoelastic, and provided the elastic constants in the analysis are replaced by viscoelastic equivalents.

At this point, it is necessary to pause for an appreciation of the behavior of civil engineering structures in practice. Consider, for example, a simple roof structure. The only constant loading on the roof is its dead load, including permanent fittings and fixtures. Though this is a constant load, it is by no means the minimum load on the roof. The wind can throw the roof in suction, causing a decrease in the stresses due to dead load, or even an actual reversal of stress. The factors governing the action of the wind on the roof are very complex. They depend on the direction of the wind, its intensity, the shape of the roof, its slopes on the windward and leeward sides, the protection that the roof gets from neighboring buildings, and so on. The roof can also be subjected to snow loading, and an incidental live load. All these loads together form the maximum loads on the structure. It is hence impossible to determine, with any degree of accuracy, even the load that the roof carries at any particular instant of time, let alone its past history of loading.

Under these circumstances, it is not too inaccurate to substitute the transform variable p by the time variable t and use the time-dependent material constants directly in the elastic equations. Many researchers[20-22] have investigated the inaccuracies resulting from such a substitution and have come to the conclusion that they are not very significant. The error itself, of course, depends on how severely the properties are time dependent. The author[6] has investigated the problem experimentally by tests on models of polymethylmethacrylate as well as on a full scale barrel vault in FRP. The results of these tests are discussed in later chapters. They prove conclusively that the direct substitution of time-dependent quantities into the appropriate linear or nonlinear elastic solutions can predict the behavior of the structure with very fair accuracy.

Efficient design in plastics, therefore, requires a considerable knowledge of material properties, particularly how they vary with time and temperature. The stress/strain diagrams, for instance, plotted against time, or against log time as in Fig. 5, at a specific temperature, form a surface. From this surface[18,23] it is possible to determine the isochronous stress/strain curves at a specific temperature. These are merely constant time sections of the surface shown in Fig. 5. Although the linearity or nonlinearity of these isochronous stress/strain curves is important to the structural analysis of the system, the curves do not themselves directly produce data applicable to structural analysis and design. Of greater interest is the constant stress section of the surface shown in Fig. 5. These sections are, in fact, creep curves for the material. Of still greater interest are the stress relaxation and creep moduli, and their variation with time at specific temperatures. Because plastics are nonlinearly viscoelastic, the moduli are dependent on the strain itself. Since the initial strain in the creep experiment is lower than in the stress relaxation experiment, the creep modulus is initially slightly higher than the relaxation modulus as shown in Fig. 6. However, it approaches the relaxation modulus as the strain increases. If the strains are kept low, for purposes of design

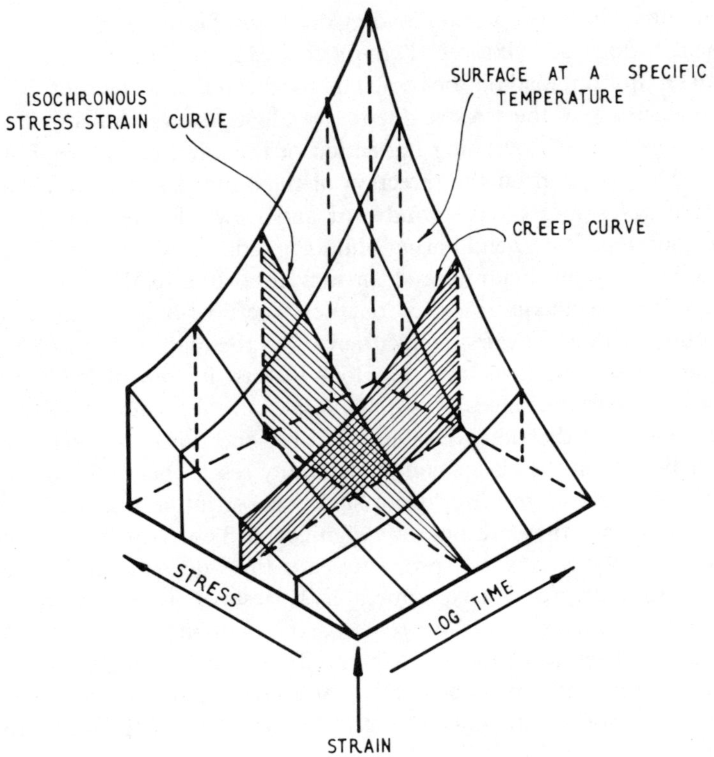

Fig. 5

the two moduli may be assumed to be equal. It may be noted here that it is not at all necessary to represent all these data as surfaces—although very elegant, they do not allow the designer to abstract values that are directly useful. Two-dimensional graphs—which are, in fact, sections through a more general surface—are far more valuable. It is then quite easy to interpolate between these curves for required values, as well as to extrapolate individual curves as necessary.

It has also been seen in Section 1.3.8 that the ultimate strengths of the material reduce with time. It becomes imperative, therefore, to define the long term properties of the material over the life of the structure. If the material is designed to work over a specific range of temperatures, the reduction in ultimate strength over that range of temperatures should be determined as well. This property should be available in the form of a curve as shown in Fig. 7. Such a curve, though indispensable to design, is based on the assumption that the stress in the material remains constant. The time to

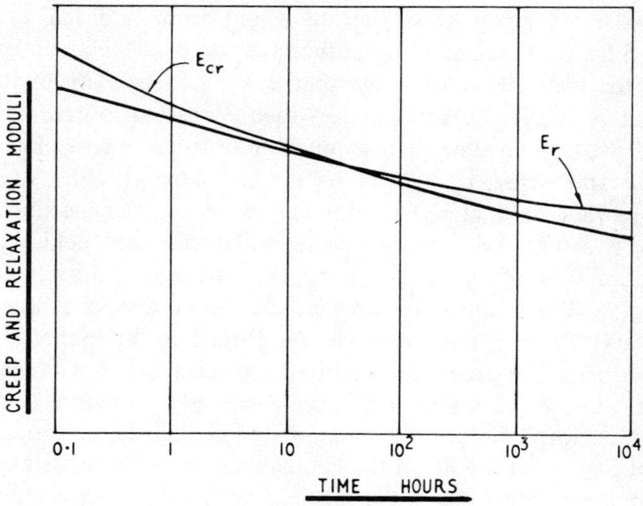

E_{cr}
E_r
CREEP AND RELAXATION MODULI
0·1
1
10
10^2
10^3
10^4
TIME HOURS

Fig. 6

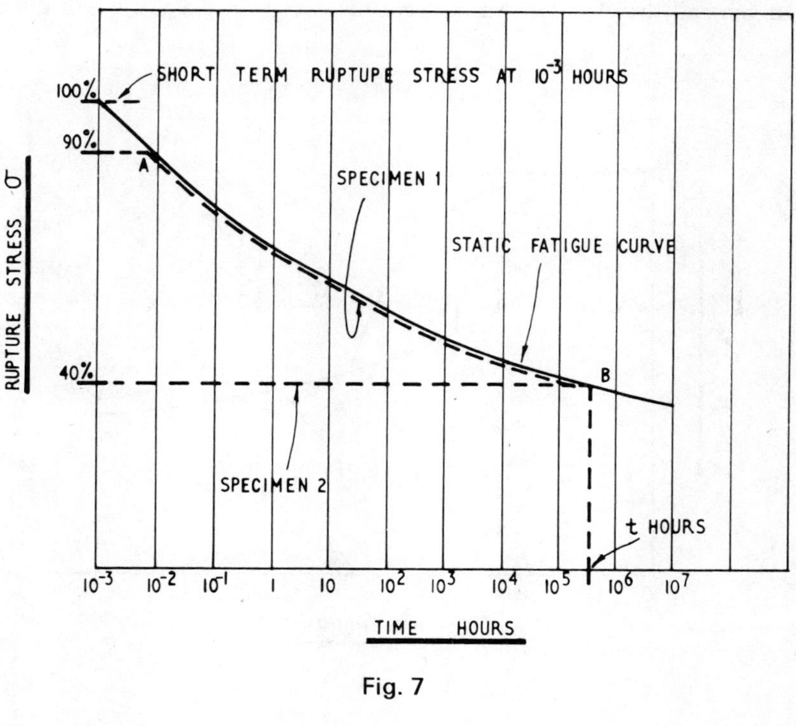

SHORT TERM RUPTUPE STRESS AT 10^{-3} HOURS
100%
90%
A
SPECIMEN 1
STATIC FATIGUE CURVE
40%
B
SPECIMEN 2
t HOURS
RUPTURE STRESS σ
10^{-3}
10^{-2}
10^{-1}
1
10
10^2
10^3
10^4
10^5
10^6
10^7
TIME HOURS

Fig. 7

rupture under a varying stress level, as would occur in actual practice, has not been fully investigated. The author's work is detailed in Section 2.4. Consider, for instance, a specimen that is stressed to 90% of its ultimate strength (at A in Fig. 7), and which is then permitted to gradually shed its stress such that at no time does it touch the rupture stress plot with time. At B, when the stress is 40% of its ultimate strength, the stress is held constant, so that the material reaches the curve and the specimen fractures at *t* hours. Now consider another specimen that has been held at a constant stress level of 40% of its ultimate strength. This is shown by the horizontal line in Fig. 7. The graph indicates that the second specimen must also fail after *t* hours. Yet the first specimen was loaded to far higher stress levels than the second. The problem is resolved, however, when it is realized that, though the life of the material reduces constantly with time, the ultimate strength does not. The ultimate strength can remain fairly constant until a substantial portion of the life of the laminate has been exhausted, after which time the material can suffer rapid mechanical failure. This is shown in Fig. 8. Plotting a range of stress levels at which the material can operate leads to a family of curves, the material switching from one curve to another as the stress level changes. This approach is considered in greater detail in Section 2.4.

In general, therefore, it can be seen that design in plastics is not difficult,

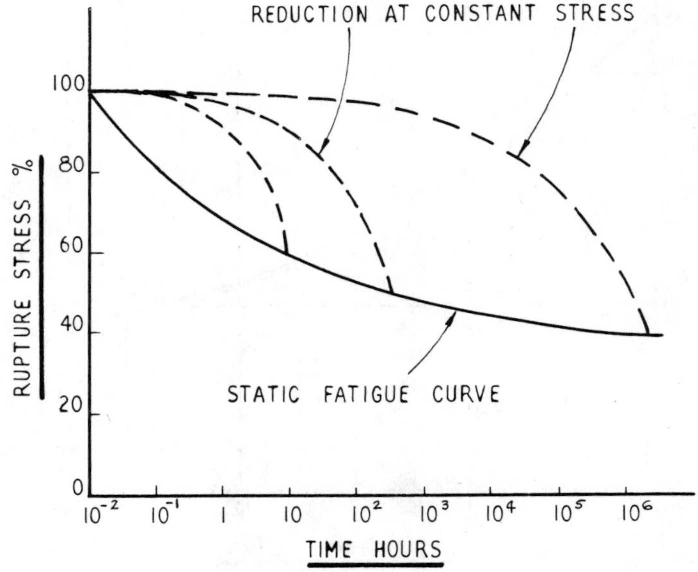

Fig. 8

provided certain new concepts are kept in mind.[24] First, it is essential to know the temperature of operation and the design life of the structure, as well as some details of its environment. The structure should therefore be designed to have a factor of safety based on ultimate strength values that are likely to occur, at that temperature at the end of that life, and not at the time of erection. However, since the factor of safety will automatically be higher at the time of erection, higher handling and erection stresses can be permitted. This can be of some advantage.

Secondly, particularly in single skin construction, deflections can be large in comparison to the thickness of the skin. In plastics, large deflections are compatible with a perfectly safe structure with stresses well below the permissible. Large deflections should be freely permitted, except in situations where they are likely to be positively uncomfortable or to lead to overall instability of the structure. Restricting the deflections for any other reason can only lead to uneconomical use of expensive materials.

Briefly, it may be concluded that, in order to design correctly in plastics, it is very essential to have a greater knowledge of both the material being used and the structure in question, than has hitherto been necessary.

CHAPTER 2

REINFORCED PLASTICS

2.1 REINFORCING MATERIALS

The thermosetting, as also the thermoplastic, resins can be reinforced to give strong engineering materials for use in a variety of applications in the automotive, aerospace, and construction industries. There is a wide range of materials that can be used as reinforcements or fillers for the resins, to impart specific properties. These include glass, asbestos, jute, mineral fibers, graphite, aramid, and boron, whiskers, and fillers such as mica flakes, Wollastonite, and even kaolin clay and sand. Some of these reinforcements are more suitable for the thermosetting resins, whereas others are better suited to reinforcing the thermoplastics. Some, such as graphite, aramid, or boron, may be used for high-strength applications, whereas others, such as jute, may be used in underdeveloped countries because of low cost and easy availability. In addition to strength, the reinforcements or fillers improve other characteristics, such as reduced shrinkage, improved abrasion resistance, increased toughness, dimensional stability, and so on. In some cases it is more suitable to use a combination of reinforcements in hybrid composites or laminates in order to offset the disadvantages associated with a particular reinforcement. The more common reinforcements to both thermosetting and the thermoplastic resins are considered below.

2.2 GLASS

Glass can be used in many different forms as reinforcement for both thermoplastic and thermosetting resins. The two most common types are A glass (high alkali content) and E glass (low alkali content). The most common form is fiber, but hollow glass spheres are also being used[25,26] as reinforcements in SMC/BMC compounds and as reinforcements to nylons and polyester resins. Glass fiber reinforcement is mainly available in three forms:[27] chopped strand mat, roving, and cloth.

Chopped strand mat is an important type of glass fiber reinforcement, and consists of randomly oriented glass fibers about 2 in. (51 mm) long. The resulting composite is essentially isotropic. The mat is available in various sizes ranging from about 1 oz/ft^2 (0.31 kg/m^2) to about 2½ oz/ft^2 (0.76 kg/m^2).

Roving consists of a definite number of continuous, parallel, bundled, but untwisted strands of glass. It is wound into cylindrical packages and supplied as cheeses. It is classified as 6, 12, 15, 60, 68, 120, or 150 end roving, depending on the number of strands bundled together. This type of reinforcement can impart very high directional strengths. For this reason it finds good application in filament wound structures such as pipes, tanks, rocket casings, and the like. Roving can also be woven to give unidirectional or bidirectional woven roving. In unidirectional roving, a heavy roving is used in the warp direction and a very light roving in the weft direction. In bidirectional woven roving, a heavy roving is used in both warp and weft directions, to give an orthotropic composite.

Cloths are woven fabrics of glass that give extremely strong composites. They are, however, expensive. Cloth reinforced laminates are anisotropic in their properties, the two principal directions being the warp and weft directions.

In addition to these three main types of glass reinforcement there are other types such as overlay mat, surface tissue, and needleloom mat. The first two types are very thin tissues of glass fiber that can absorb high percentages of resin. They are used where a resin rich surface is desired and to provide a fairly even surface on the inside, for hand lay-up laminates and moldings. Needleloom mat is made by needling chopped strands into a backing material such as cloth. This type of mat drapes better than ordinary chopped strand mat, and is hence more suitable for molded components.

Epoxy coated glass spheres[25,26] have also been used as reinforcement for nylon and polyester resins, in order to improve the physical properties and the moldability characteristics of these polymers. Molded parts were found to have a better surface appearance with glass spheres than with glass fibers. Glass microbeads have a particle diameter of up to 500μ (1μ = 10^{-4}cm). For reinforcement of thermoplastics, glass beads with a diameter of up to 100 μ are normally used. The properties of some of the reinforced thermoplastics containing such reinforcement is discussed in the article on reinforced thermoplastics in this chapter.

2.3 JUTE AND SISAL REINFORCEMENT

Jute reinforcement can be used either alone or in combination with glass fiber reinforcement. Unlike glass fiber reinforcement, however, jute reinforcement is available only in the form of a woven fabric, felt or cloth, and can be mainly of three types:[28,29]

1. Hessian, a plain woven fabric weighing about 1 oz/ft^2 (0.31 kg/m^2).
2. Heavy twill, which weighs up to 3 oz/ft^2 (0.92 kg/m^2) and can prove more economical than Hessian where bulk is required.
3. Unidirectional cloth, which weighs 1 oz/ft^2 (0.31 kg/m^2). The strength of the fabric is mainly in the warp direction, where it is 50% more than that of plain Hessian. It may also be used, like unidirectional woven glass roving, to obtain desired strengths in specific directions.

All these fabrics must be of the correct construction and surface finish to be suitable as a reinforcing material. Jute is a springy fiber, and jute cloth in its natural state is a hairy material. The cloth cannot therefore be used in a loom state because the surface hairs would tend to keep adjacent layers of reinforcement apart, resulting in resin rich intervening bands and leading to unsatisfactory composites. This can be overcome by mechanical pressing or flattening of the cloth. This type of finish is termed as a "calendered" finish and leads to satisfactory composites.

Sisal fiber[30] is a natural fiber, indigenous to India, which has low-cost housing applications. As the density of the sisal fiber is much less than that of glass, the specific strength of the fiber, taking the density into account, is much higher than glass.

2.4 ASBESTOS REINFORCEMENT

Asbestos is the term used to describe a group of fibrous minerals that occur naturally in rock formations. These are divided into two distinct classifications—serpentine and amphibole. Serpentine has only one fibrous form, called chrysotile, which, however, accounts for the major proportion of all asbestos fibers in commercial use. There are five naturally occurring fibrous forms of the amphibole group, the three significant varieties being crocidolite, amosite, and anthophyllite. All of the above four named varieties of asbestos are used as reinforcements for plastics.

The advantages of using asbestos reinforcement can be briefly listed as follows:[31]

1. High modulus of elasticity.
2. Excellent chemical resistance. Crocidolite, also called "blue" asbestos, because of its characteristic blue color, has the best general chemical resistance among inorganic reinforcing fibers.
3. High surface area offering an excellent fiber/resin bond.

4. High temperature resistance.
5. Fineness in fiber diameter giving improved porosity resistance.

The disadvantages of using asbestos reinforcements are their danger to health and the regulations that must be observed for reasons of safety. The high refractive index produces opaque moldings and restricts color possibilities in reinforced compounds and composites.

Asbestos reinforcement is available as a mat for normal contact molding. It can be used by itself or in combination with glass. It is also available as a tape for filament winding processes and can be used with bulk molding compounds utilizing polyester, epoxy, and phenolic resins.

The properties of chrysotile, crocidolite, amosite, and anthophyllite asbestos fibers[32-34] are given in Table 1. The comparison of the tensile strength of chrysotile asbestos fibers to other fibrous reinforcements[34] is given in Table 2.

2.5 GRAPHITE OR CARBON FIBER REINFORCEMENT

Graphite or carbon fiber reinforcement is eminently suited for use in high-performance composites. The outstanding properties of such fibers are high strength and modulus, low density, and resistance to environmental conditions

Table 1

Property	Chrysotile	Crocidolite	Amosite	Anthophyllite
Fiber length, in.	1/32–1	1/8–3	1/8–6	—
Fiber diameter, microns	0.02	0.1	0.1	0.1
Density, (g/cc)	2.4–2.6	3.2–3.3	3.1–3.5	—
Tensile strength, psi	824,000	876,000	287,000	353,000
Tensile modulus, psi	23.2×10^6	27.1×10^6	23.6×10^6	22.5×10^6

Table 2

Type of Fiber	Tensile Strength (psi)
Chrysotile asbestos	281,000 to 436,000
Fibrous glass	250,000 to 350,000
Cotton fibers	73,000 to 89,000
Graphite filaments	500,000
Graphite whiskers	3,000,000

and chemicals. The fibers are available as rovings, mat, and tow. The tow is composed of several thousand filament strands with zero twist, which are then impregnated with a resin such as epoxy, to form a ribbon prepreg.[35] The prepreg can be wound on to storage spools or braider packages for use in moldings.

Graphite fiber is also available as a milled fiber and in granulated form for use with injection molding compounds. Graphite can be used with any of the high-performance resins such as epoxy or polyimide, and the properties of these composites are considered in Section 2.18. The properties[36] of carbon or graphite fibers for use in such applications are given in Table 3. It may be seen that there are two types of fibers—high tensile strength and high modulus, and the properties of both of these have been given in Table 3.

Graphite fibers can also be used as reinforcements for nylon 66, polysulfone, polycarbonate, polypropylene, and the like, to give reinforced compounds. The properties of these materials are considered in Section 2.20.

Finally it should be noted that graphite fiber can be used in combination with other high-strength fibers such as aramid, boron, and the like, to tailor the hybrid matrix for specific requirements.

2.6 BORON FIBER REINFORCEMENT

Boron, like graphite, is another fiber reinforcement for high-performance applications. Boron has very high strength and modulus, higher even than graphite, but the density of boron fiber is also much higher than graphite. The comparative properties of boron[36] and other fibers is shown in Table 3.

Boron can also be used to form hybrid filaments[37] and the properties of composites using boron reinforcement are given in Section 2.18.

2.7 ARAMID FIBER REINFORCEMENT

Aramid is an oriented organic fiber with tensile strength about the same as that of high-strength graphite fibers and a modulus of elasticity between

Table 3

Fiber	Modulus (10^6 psi)	Strength (10^3 psi)	Density (g/cc)
HT-S	38.0	380	1.75
HM-S	50.0	320	1.90
Boron	60.0	500	2.63
S-Glass	12.5	500	2.49

graphite and glass. It has a low density, however, and this gives it better strength-weight properties than other continuous high-performance fibers. The properties of aramid fibers[38] (Kevlar 29 and Kevlar 49 marketed by Dupont) are given in Table 4.

The fiber is very tough and has particularly good application in hybrid composites with graphite and boron to improve impact resistance and lower costs. The toughness of the fiber itself makes it a little difficult to cut, and so the availability of the fiber in chopped or milled form would make it very suitable for use as reinforcement to thermoplastics and even structural foams. Milewski[39] has studied the comparative properties of the high-performance fibers and his results are shown in Fig. 9.

Aramid fibers[39] are available as yarns, rovings, woven rovings, and fabrics. Nonwoven fabrics and felts and heavy pultrusion rovings are under development.

Table 4

Property	Kevlar 49	Kevlar 29
Specific gravity	1.45	1.45
Tensile strength, psi	525,000	525,000
Tensile modulus, psi	19×10^6	9×10^6
Elongation to failure, %	2.8	3–4

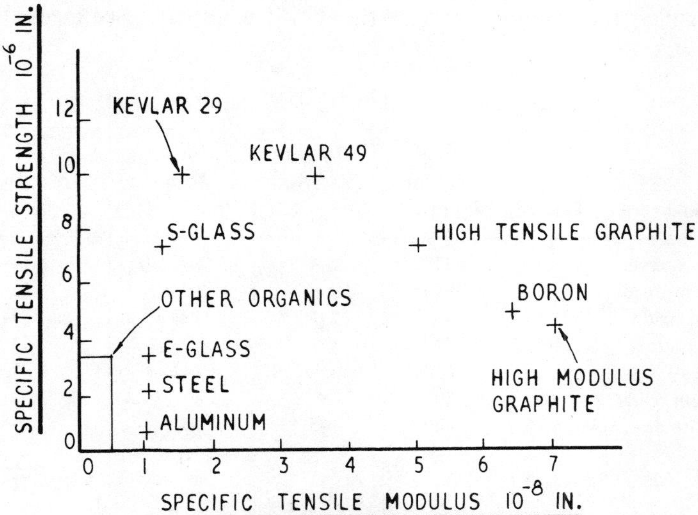

Fig. 9

2.8 WHISKERS AND MICROFIBERS

The growing of whiskers as extremely high strength reinforcement for use in reinforced plastics has not yet reached its full potential. These are filaments of single crystal structure formed from a variety of materials such as graphite, boron carbide, beryllium oxide, silicon carbide, and so on. Milewski[39] has compared the properties of some of these whisker reinforcements, and this is shown in Table 5. The silicon carbide (beta) whiskers have an average diameter of only 0.5 μ, and a modulus of elasticity that is extremely high, as shown in Table 5. The small diameter is particularly useful in the processing requirements for thermoplastic compounds. The whisker is expected to be commercially available.

Because of the very high cost of whiskers,[39] an attempt was made to develop microfibers. Microfibers differ from whiskers in that they are fine polycrystalline fiber bundles grown quite rapidly from a liquid matrix. Microfibers do have small diameter, but they do not have the extremely high strengths of the whiskers.

2.9 MICA, WOLLASTONITE, AND OTHER FILLER REINFORCEMENTS

Mica,[39-41] when suitably delaminated into flakes, is suitable for use as a reinforcing filler for both thermoplastic as well as thermosetting resins. Mica has several advantages as a reinforcement. It can impart a high degree of stiffness and improved strength, improve warp resistance and creep, and reduce shrinkage. It is not irritating to the skin. The impact resistance is lowered,

Table 5

Whisker	Density	Melting-Point (°F)	Tensile Strength (psi × 10^{-6})	Young's Modulus (psi × 10^{-6})	Strength/Density (in. × 10^{-6})
Aluminum oxide	3.9	3780	2–4	70–150	14–28
Aluminum nitride	3.3	3990	2–3	50	13–21
Beryllium oxide	1.8	4620	2.0–2.8	100	31–43
Boron carbide	2.5	4440	1	65	11
Graphite	2.25	6500	3	142	37
Magnesium oxide	3.6	5070	3.5	45	27
Silicon carbide (alpha)	3.15	4200	1–5	70	9–44
Silicon carbide (beta)	3.15	4200	1–5	80–120	9–44
Silicon nitride	3.2	3450	0.5–1.5	55	4.2–13

but this can be improved by using small amounts of glass fibers in a hybrid composite. The heat deflection temperature is improved as also reduced flammability.

Mica flake reinforcement can be used for resins such as ABS and Nylon 6/6 for injection moldable reinforced thermoplastics. It is in the form of platelets. The mica platelets have a strength of 300,000–450,000 psi (2.07–3.10 GPa) and an E value of 2.5×10^6 psi (17.24 GPa).

Wollastonite[39] is a pure white, commercially available, mineral fiber. It too is nonhazardous. It is low cost, and that is its prime advantage. It does not degrade the strengths and properties of the matrix.

Mineral fiber[39] can also be processed from blast furnace slag. It is an inert material, and can be used as a replacement for asbestos as a fibrous filler.

Clay and sand[42] can also be used as filler reinforcements. Sand is very suitable for use in reinforced plastics mortar pipes (RPM), and the properties of such pipes for sewer and irrigation applications are considered in Section 2.12.

2.10 MANUFACTURING PROCESSES FOR REINFORCED PLASTICS

In order to fully appreciate specific applications of reinforced plastics, it is necessary to understand briefly the various processes available for the manufacture of reinforced plastics components. These are the following:

1. Contact molding
2. Vacuum bag
3. Pressure bag
4. Autoclave
5. Spray-up
6. Encapsulation
7. Filament winding
8. Centrifugal casting
9. Continuous pultrusion
10. Directed fiber
11. Plenum chamber
12. Water slurry
13. Premix/molding compound
14. Injection molding
15. Continuous laminating.

It may be noted that the first six processes listed above are open mold processes, while the rest are closed mold processes. The directed fiber, plenum chamber, and water slurry processes are methods for forming preforms, which are then press molded. A preform is a mat of chopped strands bonded together in the shape of the end product. The injection molding process is for use with thermoplastic materials.

All of these processes have been described in an excellent manner by Bacon[43] and are summarized here by kind permission of the British Plastics Federation.

2.10.1 Contact Molding

This is the simplest method used for the manufacture of reinforced plastics components. The gel-coat resin is first laid-up against a highly polished and waxed mold. This forms the outer surface of the component after removal from the mold and may therefore have special properties for improved weathering, abrasion resistance, and the like, depending on the application to which the component is to be put. The laminating resin and suitable reinforcement is then laid on. The lay-up normally cures at room temperature, though heat can be used to accelerate the cure. The internal surface of the component, quite naturally, is rough, but a smoother surface may be obtained by wiping on cellophane.

This process has the advantage of using a minimum of equipment and low-cost molds that may be in any suitable material, such as wood, sheet metal or FRP itself. There are no size restrictions. It is a flexible process and design alterations can be readily made. However, the labor cost per unit is high. This may make it particularly suitable for use in underdeveloped countries to provide employment to large numbers of people in any low-cost housing application on a massive scale at the village level, using very low level technology. The quality of the component depends to a great extent on the skill of the operator.

2.10.2 Vacuum Bag

This process is similar to the contact molding process, except that, after lay-up, cellophane is placed over the component, the joints are sealed and vacuum is drawn. The resulting atmospheric pressure then eliminates voids and forces out entrapped air and excess resin. The advantages of this process over contact molding may be listed briefly as follows:

1. Higher reinforcement loadings are possible with less air voids.
2. The component has a better internal surface.
3. Better adhesion occurs in sandwich construction, due to the atmospheric pressure on the lay-up.

However, besides suffering from all the disadvantages of contact molding, it has the added disadvantage of needing additional equipment.

2.10.3 Pressure Bag

This process is also similar to the contact molding process. After lay-up, a tailored bag, normally of rubber sheeting, is placed against the component. The mold is then sealed with a pressure plate, and air or steam pressure is applied between the pressure bag and the plate, usually up to 50 psi (345 kPa).

This process retains all the advantages of contact molding. In addition dense, void free moldings are obtained. Higher reinforcement loadings and undercuts to the molding are also possible. It is particularly useful for sandwich construction. The process has, however, all the disadvantages of contact molding. Further, it has the added disadvantage of needing special equipment and being only applicable to female molds.

2.10.4 Autoclave

This is a modification of the pressure bag method. After lay-up, the entire assembly is placed in a steam autoclave at a pressure of about 50 psi (345 kPa) to 100 psi (689 kPa). The additional pressure achieves higher reinforcement loadings (in the case of glass reinforcement up to about 65%) and leads to very dense, void-free moldings. The autoclave is, however, expensive, and additional costs result from the labor involved in loading and unloading it. Further, the size of the autoclave imposes a limit on the size of the component.

2.10.5 Spray-up

In this process, particularly applicable to FRP, the glass fiber and the resin are simultaneously deposited in a mold. Glass roving is fed through a chopper and ejected into a resin stream which is then deposited on to the mold by either of two systems:

1. Two gun nozzles are used. One carries resin premixed with catalyst, while the other carries resin premixed with accelerator.
2. Only one gun nozzle is used. All the ingredients are fed into a single mixing chamber ahead of the spray nozzle.

By either method, the resin mix precoats the chopped strands, and the merged spray is directed onto the mold by the operator. The glass–resin mix is then rolled by hand to remove air, lay down the fibers, and smooth the surface.

The process has several advantages, which can be listed briefly as follows:

1. It uses portable inexpensive equipment.
2. It uses roving reinforcement, which is the cheapest form of glass reinforcement.
3. The labor costs for forming complex shapes is less than with a hand lay-up process.
4. Complex shapes even with reverse curves can be formed with a minimum of waste.
5. The process can be easily used for on-site fabrication.

The main disadvantage of the method is that, as a process, it is a little difficult to control. As compared to the other processes, there is a higher usage of materials. This is because it is difficult to control the composite thickness.

2.10.6 Encapsulation

In this process, milled fibers or short chopped strands are combined with catalyzed resin and poured into open molds to cast terminal blocks, electrical casings, electronic components, and so on. The fibers decrease shrinkage and crazing. They also increase the useful temperature range of the resin system. Curing is at room temperature. A post cure of 30 minutes at 200°F (93°C) is normal.

The process is simple, involves low tooling costs, and can be automated. Inserts can easily be provided. However, it is limited in application to small components, such as those mentioned above.

2.10.7 Filament Winding

This process uses continuous reinforcement to achieve efficient utilization of the strength of the reinforcement. Roving or tape is fed from a creel through a bath of resin and is wound onto a suitably designed mandrel. Preimpregnated reinforcement can also be used. Special lathes lay down the reinforcement in a predetermined pattern to give the maximum strength in the required directions. When the right number of layers has been applied, the wound mandrel is cured at room temperature in an oven. Filament winding has several advantages, which are listed below:

1. It gives the highest strength-to-weight ratio for the component. This is because it exercises the highest degree of control over the uniformity and the orientation of the reinforcement.
2. The process can be automated at high volumes.
3. Fusible or collapsible mandrels can produce complex shapes. Integral end closures can be wound in.
4. The manufactured components can be accurately machined.
5. It uses the cheapest form of reinforcement.

The only disadvantage of filament winding is that the shapes that can be easily wound are restricted to surfaces formed by revolution, such as round, oval, tapered, and so on.

2.10.8 Centrifugal Casting

Round objects, such as pipe, can be formed using this process. Chopped strand mat is positioned inside a hollow mandrel. The assembly is next placed inside an oven and then rapidly rotated. Resin mix is distributed uniformly through the reinforcement. Centrifugal action forces the resin and the reinforcement against the walls of the rotating mandrel prior to, and during, the cure. To accelerate cure, hot air is passed through the oven.

The process can be used to great advantage for the automatic production of shapes such as cylinders, without taper, and with uniform thickness. Though the equipment cost is high, the costs of the mandrel and the labor costs are relatively low. The process gives a good surface, both inside and outside, with a uniform, void-free wall thickness. It is possible to have threads, but on the external surface only.

2.10.9 Continuous Pultrusion

A continuous strand, in the form of roving or other form of reinforcement, is impregnated in a resin bath and drawn through a die that sets the shape of the stock and controls the resin content. Final cure is effected in an oven through which the stock is drawn by a suitable pulling device.

The process is suitable for small cross-sectional areas ranging in diameter from about ⅛ in. (3.2 mm) to 6 in. (152 mm). It can therefore be used for the production of unidirectional high-strength parts.

2.10.10 Directed Fiber

Roving is cut into 1 in. (25.4 mm) or 2 in. (51 mm) lengths of chopped strand that are then blown through a flexible hose onto a rotating preform screen. Suction holds them in place while a binder is sprayed on the preform. Curing takes place in an oven. One operator controls both the deposition of the chopped strands and the binder.

The directed fiber process is a preforming method; several advantages to using preform molding are listed below:

1. It is the most economical method for the production of complex shapes.
2. It can be easily automated.
3. It uses the cheapest form of reinforcement.
4. With this method it is possible to obtain close wall thickness tolerances, as the preform thickness can be varied over a continuous range.
5. Both decorative and functional embedded inserts are possible.

Preform molding has the disadvantage that the maximum molded wall thickness is limited. The size of the component is also necessarily limited by the size of the press in which the preform is molded.

2.10.11 Plenum Chamber

This is another of the performing methods. Roving is fed into the cutter at the top of the plenum chamber. The chopped strands are directed onto a spinning fiber distributor that separates the chopped strands and distributes them uniformly into a plenum chamber. The falling strands are sucked onto a preform screen. A resinous binder is then sprayed on. The preform is finally positioned in a curing oven. A new preform screen comes into position in the plenum chamber for a repeat cycle.

The advantages and disadvantages of preform molding have been discussed in Section 2.10.10.

2.10.12 Water Slurry

Chopped strands are preimpregnated with resin and blended with cellulosic fiber in a water slurry. The water is exhausted through a contoured and perforated screen, and the glass fibers and the cellulosic material are deposited on the surface of the screen. The wet preform is transferred to an oven where hot air is sucked through the preform. When dry, the preform is sufficiently strong to be handled and molded in a press.

2.10.13 Premix/molding Compound

Prior to molding, glass in the form of chopped spun roving, or another reinforcement, is thoroughly mixed with resin, pigment, fillers, and catalyst.

The premixed material can be extruded into a ropelike form for easy handling, or may be used in sheet or bulk form.

The premix is formed into accurately weighed charges and placed in a mold under heat and pressure. The pressure can vary from 100 psi (689 kPa) to 1500 psi (10,342 kPa). The cycling time depends on the cure temperature, the resin used, the reinforcement and the wall thickness. It can vary from 30 seconds to 5 minutes. Cure temperatures range from 225°F (105°C) to 300°F (149°C).

Transfer molding is a high-speed modification of compression molding (described above) using premix or molding compounds. The premix or molding compound is preheated in a chamber (called a pot) and is then forced into the hot mold cavity for curing. This process is used to mold small, complicated parts incorporating many delicate inserts.

This method of molding has many advantages over other methods. Any shape and cross section can be molded together with inserts and attachments, holes and threads, and so on. Although there is a high equipment cost, the labor costs are low, and the method leads to automated, low-cost, mass production of units. Molding, since it is in a press, can be to very close tolerances. The strengths achieved, however, are less than in preform molding.

2.10.14 Injection Molding

This high-production process is designed for use with reinforced or unreinforced thermoplastics. The glass and resin molding compound is softened in a heating chamber. The mass is then injected into a mold cavity that is kept at a temperature below the softening point of the resin. The component then cools and solidifies.

Injection molding is a fully automated process that can be used for high-production runs. Labor costs are very low. Very small, delicate, precision parts can be manufactured. Complex details can be easily incorporated. The method has the disadvantage of necessarily limiting the size of the component. The initial equipment cost, and the cost of the molds, is rather high.

2.10.15 Continuous Laminating

In continuous laminating, fabric or mat is passed through a resin dip and brought together between cellophane covering sheets. The lay-up is then passed through a heating zone and the resin is cured. The laminate thickness and the resin content are controlled by squeeze rolls as the various plies are brought together.

Continuous laminating is an automated process with low tooling costs ideally suited for the production of flat or corrugated panels of various cross sections. Though the widths are limited by the size of the rolls, there is no limitation to the length of the panels produced. A wide variety of surface textures are possible. The wall thickness is very uniform, though limited in size.

2.10.16 Other Plastics Processing Methods

In addition to the methods discussed, there are other processing methods[44] such as blow molding, calendering, casting, coating, extrusion molding, matched-die molding, rotational molding, slush molding, and thermoforming or vacuum forming.

In some cases, they are variations of the methods described. For instance, matched-die molding is a variation of the conventional compression molding described in Section 2.10.13. The process employs two metal molds, which seal in the plastic compound being molded, and allows trim of the reinforcement as well. The mat or preform reinforcement is positioned in the mold, and the mold is then closed and heated under pressures of 150 psi (1034 kPa) to 400 psi (2758 kPa). The mold is then opened and the part removed.

In another excellent paper Rosato[44] has described some of these and other processing methods together with their advantages and their limitations.

2.11 DESIGN RECOMMENDATIONS FOR THE CHOICE OF A PROCESS

In order to have a suitable and efficient end product, it is essential to choose the proper materials in addition to the most suitable manufacturing process. Bacon[43] has drawn a comparison between the various processes, and this is shown in Tables 6 and 7. Table 7 provides very useful design recommendations for engineers, regarding the choice of a suitable process.

Rosato[44] has also compared the processing methods as a function of part design, and his recommendations are given in Table 8.

2.12 FIBERGLASS REINFORCED POLYESTER

For most engineering applications, when viewed as a material independent of the manufacturing process, FRP can basically be divided into two types:

1. FRP randomly reinforced with chopped strand mat and, hence, essentially isotropic.
2. FRP reinforced with roving, woven roving or cloth fabric and, hence, anisotropic in character with varying directional properties.

Table 6

Process	Resin	Type of Glass Reinforcement	Normal % by Weight of Glass	Maximum Filler % by Weight	Moulding Temperature and Pressure	Moulding Cycles
Contact moulding	Polyester Epoxy	Mat, fabric, or woven roving	30–45	none	70–110°F, 0 lbf/in.²	30 minutes to 1 day
Vacuum bag	Polyester Epoxy	Mat, fabric, or woven roving	40–50	none	70–110°F, 12-14 lbf/in.²	As above
Pressure bag	Polyester Epoxy	Mat, fabric, or woven roving	45–60	none	70–220°F, 50 lbf/in.²	As above
Autoclave	Polyester Epoxy	Mat, fabric, or woven roving	45–60	none	70–250°F, 50–100 lbf/in.²	As above As above
Spray-up	Polyester Epoxy	Continuous roving	30	none	70–110°F, 0 lbf/in.²	As above
Encapsulation	Epoxy Polyester	Milled fibers	5	none	70–110°F, 0 lbf/in.²	120 minutes to 1 day
Filament winding	Epoxy Polyester	Continuous roving or narrow fabrics	50–75	none	70–150°F, wrapping tension	10–400 lbs per hour
Centrifugal casting	Polyester	Mat or continuous roving	30	none	180–200°F, 20–40 lbf/in.²	10–20 minutes
Continuous pultrusion	Polyester Epoxy	Mat, fabric, or Continuous roving	30–60	none	180–220°F, 0 lbf/in.²	up to 5–6 ft./minute
Matched die moulding with						
a) Preform	Polyester	Chopped strands	30	40	225–300°F, 100–300 lbf/in.²	1–5 minutes
b) Mat	Polyester Phenolic Melamine Silicone Epoxy	Mat	30	40	225–350°F, 100–3000 lbf/in.²	1–10 minutes
c) Fabric	Polyester Phenolic Melamine Silicone Epoxy	Preimpregnated woven and nonwoven fabrics	60	none	225–350°F, 100–3000 lbf/in.²	1–30 minutes
Compression moulding of premix/ moulding compound	Polyester Epoxy Other moulding compounds	Spun roving, continuous roving	25	65	250–350°F, 250–2000 lbf/in.²	0.5–5 minutes
Transfer moulding of premix/ moulding compound	Polyester Epoxy Other moulding compounds	Chopped strands, chopped spun roving	25	65	250–350°F, 1000–5000 lbf/in.²	0.5–5 minutes
Injection moulding	Thermoplastics	Continuous roving or milled fibers	10–40	5	275–750°F, 10000–25000 lbf/in.²	0.25–1.5 minutes
Continuous laminating	Polyester Acrylic	Mat or fabric	25–35	none	190–250°F, 5–20 lbf/in.²	40 ft.² to 50 ft.² per minute

Table 7

	Contact Moulding, Spray-up	Pressure Bag	Filament Winding	Continuous Pultrusion	Premix/ Moulding Compound	Matched Die Moulding with Preform or Mat
Minimum inside radius in inches	¼	½	⅛	N.A.	1/32	⅛
Moulded-in holes	Large	Large	N.R.	N.A.	Yes	Yes
Trimmed-in mould	No	No	Yes	Yes	Yes	Yes
Built-in cores	Yes	Yes	Yes	N.A.	Yes	Yes
Undercuts	Yes	Yes	No	No	Yes	No
Minimum practical thickness in inches	0.060	0.060	0.010	0.037	0.060	0.030
Maximum practical thickness in inches	0.50	1	3	1	1	0.25
Normal thickness variation in inches	±0.020	±0.020	±0.010	±0.005	±0.002	±0.008
Maximum build up of thickness	As desired	As desired	As desired	N.A.	As desired	2 to 1 maximum
Corrugated sections	Yes	Yes	Circumferential only	In the longitudinal direction	Yes	Yes
Metal inserts	Yes	Yes	Yes	No	Yes	Yes
Surfacing mat	Yes	Yes	Yes	Yes	No	Yes
Limiting size factor	Mould size	Bag size	Lathe bed length and swing	Pull capacity	Press capacity	Press dimensions
Metal edge stiffeners	Yes	N.R.	Yes	No	Yes	Yes
Bosses	Yes	N.R.	No	No	Yes	Yes
Fins	Yes	Yes	No	N.R.	Yes	N.R.
Moulded-in labels	Yes	Yes	Yes	Yes	No	Yes
Raised numbers	Yes	Yes	No	No	Yes	Yes
Gel coat surface	Yes	Yes	Yes	No	No	Yes
Shape limitations	None	Flexibility of the bag	Surface of revolution	Constant cross-section	Mouldable	Mouldable
Translucency	Yes	Yes	Yes	Yes	No	Yes
Finished surfaces	One	One	One	Two	Two	Two
Strength orientation	Random	Orientation of ply	Depends on the wind	Directional	Random	Random
Typical glass % by weight	30–45	45–60	50–75	30–60	25	30

Note: N.A.—Not applicable
N.R.—Not recommended

Table 8

Part Design	Blow Molding	Casting	Compression	Extrusion	Filament Winding	Injection	Matched-Die Molding	Rotational	Thermoforming	Transfer/ Compression	Wet Lay-up (Contact Molding)
Major shape characteristics	Hollow bodies	Simple configurations	Moldable in one plane	Constant cross section profile	Structure with surfaces of revolution	Few limitations	Moldable in one plane	Hollow bodies	Moldable in one plane	Simple configurations	Moldable in one plane
Limiting size factor	Material	Material	Equipment	Material	Equipment	Equipment	Equipment	Material	Material	Equipment	Mold Size
Maximum thickness, in.	> 0.25	None	0.5	6	3	6	2	0.5	3	6	0.5
Minimum inside radius, in.	0.125	0.01–0.125	0.125	0.01–0.125	0.125	0.01–0.125	0.06	0.01–0.125	0.125	0.01–0.125	0.25
Minimum draft, degrees	0	0–1	>1	NA[b]	2–3	<1	1	1	1	1	0
Minimum thickness, in.	0.01	0.01–0.125	0.01–0.125	0.001	0.015	0.005	0.03	0.02	0.002	0.01–0.125	0.06
Threads	Yes	Yes	Yes	No	No	Yes	No	Yes	No	Yes	No
Undercuts	Yes	Yes[a]	NR[b]	Yes	NR[b]	Yes[a]	NR[b]	Yes[c]	Yes[a]	NR[b]	Yes
Insert	Yes	Yes	Yes	Yes	Yes	Yes	Yes	Yes	NR[b]	Yes	Yes
Built in cores	Yes	Yes	No	Yes	Yes	Yes	Yes	Yes	Yes	Yes	Yes
Molded-in holes	Yes	Yes	Yes	Yes[d]	Yes	Yes	Yes	Yes	No	Yes	Yes
Bosses	Yes	Yes	Yes	Yes	No	Yes	Yes	Yes	Yes	Yes	Yes
Fins or ribs	Yes	Yes	Yes	Yes	No[e]	Yes	No[f]	Yes	Yes	Yes	Yes
Molded in designs and nos.	Yes	Yes	Yes	No	No	Yes	Yes	Yes	Yes	Yes	Yes
Surface finish[g]	1–2	2	1–2	1–2	5	1	4–5	2–3	1–3	1–2	4–5
Overall dimensional tolerance, in./in., plus or minus	0.01	0.001	0.001	0.005	0.005	0.001	0.005	0.01	0.01	0.001	0.02

[a] Special mold required. [b] Not recommended. [c] Only flexible material. [d] Only direction of extrusion. [e] Possible with special techniques. [f] Fusing premix/yes. [g] Rated. 1 = very smooth. 5 = rough.

Of these, the first is by far the most commonly used for hand lay-up processes. For small jobs it is to be recommended. From the designer's point of view there are disadvantages in using chopped strand mat. The resulting composite can be of indeterminate quality. Tests on specimens of FRP reinforced with chopped strand mat show large variations. This may partly be explained by the fact that it may not be possible to impregnate the mat very evenly, since the amount of resin used and its distribution within the laminate depends on the skill of the operator. All the figures for strengths, moduli, and so on, quoted for FRP reinforced with chopped strand mat, and made by a hand lay-up process, indicate a range of values with a variation of as much as 20%. FRP reinforced with woven roving or cloth fabric is a much more reliable material whose properties can be predicted with fair precision provided proper care has been taken in its manufacture. The thickness of the composite can also be better controlled, and the material has better properties than FRP reinforced with chopped strand mat.

The properties of FRP are completely defined by the following time- and temperature-dependent quantities:

1. Ultimate tensile strength
2. Tensile modulus of elasticity
3. Compressive strength
4. Compressive modulus of elasticity
5. Flexural strength
6. Flexural modulus
7. Shear strength
8. Shear modulus of rigidity
9. Poisson's ratio

Not all of these properties are necessarily used in structural design. In sandwich construction, for instance, the composites are in either tension or compression, and the flexural properties of the composite are relatively unimportant. In single skin construction it may be necessary to take bending moments into account, and the flexural properties are important. It must be realized, however, that not all of these properties are available, even from resin manufacturers. Again, most resin manufacturers can specify short-term properties of the composites, but little may be known about their long-term behavior. In this case, reference must then be made to the general body of scientific work in this field so that, in light of that work, the short-term properties can be modified.

The properties of FRP composites are largely dependent on the resin/glass fiber ratio or the glass fiber content. With chopped strand mat the glass fiber content is generally low. The higher the glass fiber content the stronger the laminate. However, too low a resin content may result in insufficient impregnation and poorer bonding. A low resin content also results in poorer weathering characteristics. For FRP reinforced with chopped strand mat, the glass fiber content by weight generally varies from 25% to 33%. For FRP reinforced with cloth, the glass fiber content can vary from about 30% to 50% and more. The upper limit depends on the molding pressure used.

The variation of the flexural and tensile modulus of elasticity with percentage reinforcement for FRP reinforced with chopped strand mat is shown in Figs. 10 and 11. Trampenau and Wilson[45] give average values of marine laminates using chopped strand mat, a composite of chopped strand mat and woven roving, and, finally, woven roving, with different reinforcement contents; these are shown in Table 9.

Scott and Sommella[46] give the average properties of FRP laminates with higher reinforcement contents and these are shown in Table 10.

It must be realized that the properties will depend not only on the type of reinforcement (chopped strand mat, woven roving, etc.) and its percentage,

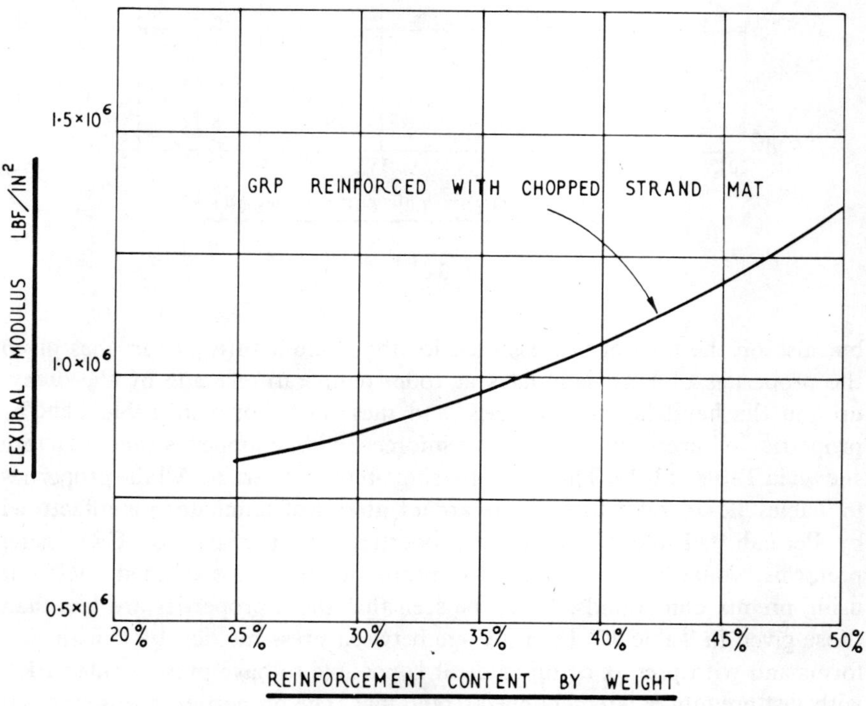

Fig. 10

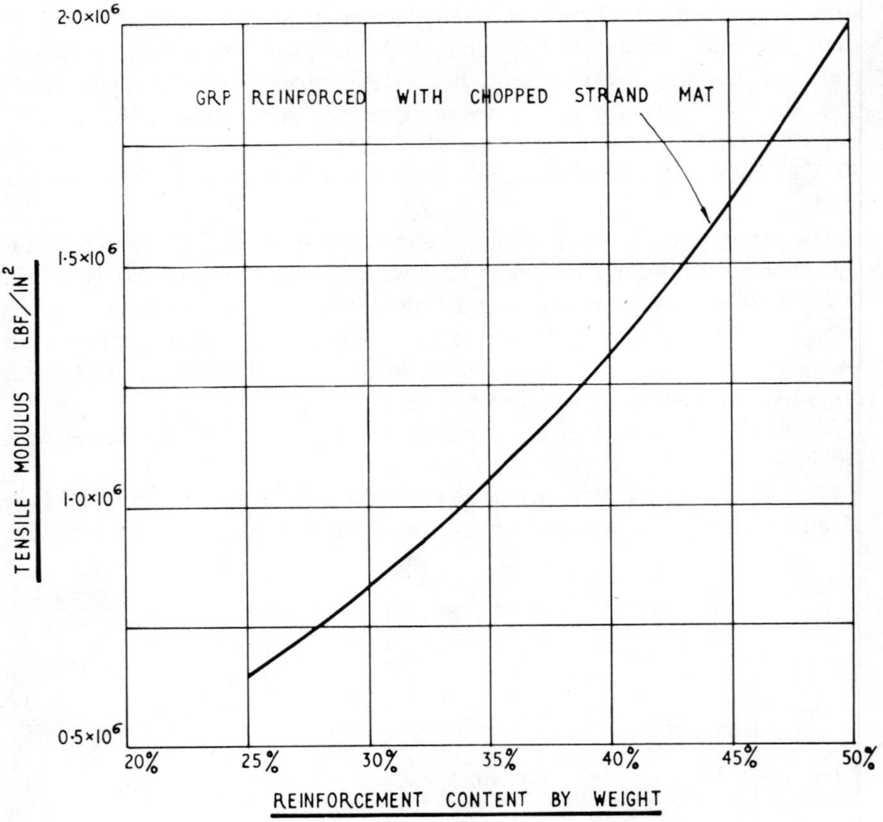

Fig. 11

but also on the type of process used for the manufacture. Desai[47] has given the properties of FRP laminates at room temperature made by the spray-up and the hand lay-up processes, and these are shown in Table 11. The properties of press molded FRP reinforced with chopped strand mat are shown in Tables 12–14. These are also short-term properties. All the properties in Tables 12–14 refer to flame retardant grades of laminates manufactured by Permali.[48] Table 12 gives the properties for press molded FRP using preforms, while Table 13 gives the properties for press molded FRP but using premix compounds. It can be seen that these properties are less than those given in Table 12. Intermediate between press molded FRP with pre-forms and with premix compounds, it is possible to have press molded FRP with just preimpregnated chopped strand mat. The properties of this material are given in Table 14.

Table 9

Property	Chopped Strand Mat Laminate (Low Glass Content)	Composite Laminate (Medium Glass Content)	Woven Roving Laminate (High Glass Content)
% Glass by weight	25–30	30–40	40–55
Specific gravity	1.40–1.50	1.50–1.65	1.65–1.80
Flexural strength, psi $\times 10^3$	18–25	25–30	30–35
Flexural modulus, psi $\times 10^6$	0.8–1.2	1.1–1.5	1.5–2.2
Tensile strength, psi $\times 10^3$	11–15	18–25	28–32
Tensile modulus, psi $\times 10^6$	0.9–1.2	1.0–1.4	1.5–2.0
Compressive strength, psi $\times 10^3$	17–21	17–21	17–22
Compressive modulus, psi $\times 10^6$	0.9–1.3	1.0–1.6	1.7–2.4
Shear strength, perpendicular, psi $\times 10^3$	10–13	11–14	13–15
Shear strength, parallel, psi $\times 10^3$	10–12	9–12	8–11
Shear modulus, parallel, psi $\times 10^6$	0.4	0.45	0.5

Notes:
1. These are average values, for guidance only.
2. Properties are from short term loading tests, wet condition. Composite and woven roving values are for warp direction.
3. The composite laminate above, is based on typical alternate plies of 2 oz/ft² mat and 24 oz/yd² woven roving.

Table 10

Property	Woven Roving Laminate	Unidirectional Laminate	Woven Roving-Unidirectional Composite
% Glass by weight	50	60–65	60
Specific gravity	1.72	1.9	1.8
Flexural strength, psi	32,500	114,000	65,000
Flexural modulus, psi	1.85×10^6	4.1×10^6	2.9×10^6
Tensile strength, psi	30,000	110,000	65,000
Tensile modulus, psi	1.75×10^6	3.9×10^6	2.9×10^6
Compressive strength, psi	19,500	100,000	60,000
Compressive modulus, psi	2.1×10^6	3.9×10^6	3.0×10^6

Notes:
1. These are average values, in the warp direction, and are for guidance only.
2. The woven roving-unidirectional composite assumes alternate plies of 50% woven roving and 50% unidirectional reinforcement.

Table 11

Material	Flexural Modulus (psi × 10^6)	Flexural Strength (psi)
FRP Spray-up/mat polyester resin (25% glass)	0.8	16,000
FRP Lay-up/mat polyester resin (25% glass)	0.8	18,000
FRP Lay-up/cloth polyester resin (45% glass)	1.7	37,000
FRP Lay-up/woven roving polyester resin (50% glass)	2.0	40,000

It will be noted that, in several of the tables, only the flexural modulus has been given. It has been found by some experimentalists that the modulus of elasticity in compression is not the same as that in tension. Kabelka,[49] as will be seen in the discussion on FRP reinforced with glass cloth fabric later in this chapter, reached the same conclusion. But this conclusion has been drawn from results obtained by the usual methods of testing in tension and in compression. Abbott and Broutman,[50] however, determined the modulus of elasticity of laminates using the stress wave technique. They reached the conclusion that the two moduli of elasticity are the same and are independent of the stress level.

Table 12

Specific gravity	1.52–1.68
Glass content	About 25%
Flexural strength, lbf/in.^{2}	27000–30000
Flexural modulus, lbf/in.^{2}	$1.6 \times 10^6 - 2.0 \times 10^6$
Tensile strength, lbf/in.^{2}	14000–21000
Compressive strength, lbf/in.^{2}	36000–42000
Shear strength, lbf/in.^{2}	12000–14000
Impact strength, ft. lbf/½in. notch	9–10

Note: The actual properties of the material will depend on the resin used and the exact glass content.

Table 13

Specific gravity	1.7
Glass content	About 25%
Flexural strength, lbf/in.^{2}	9000–20000
Flexural modulus, lbf/in.^{2}	1.2×10^6
Tensile strength, lbf/in.^{2}	6000–9000
Compressive strength, lbf/in.^{2}	18000–30000
Shear strength, lbf/in.^{2}	6000–10000
Impact strength, ft lbf/½ in. notch	4–10

Note: The actual properties of the material will depend on the resin used and the exact glass content.

Table 14

Specific gravity	1.78
Glass content	About 25%
Flexural Strength, lbf/in.^{2}	25000
Flexural modulus, lbf/in.^{2}	1.8×10^6
Tensile strength, lbf/in.^{2}	11000
Compressive strength, lbf/in.^{2}	33000
Shear strength, lbf/in.^{2}	10000
Impact strength, ft lbf/½ in. notch	3

Note: The actual properties of the material will depend on the resin used and the exact glass content.

The chemical resistance of FRP laminates reinforced with chopped strand mat has been studied by Rassell,[31] and his results have been shown in Figs. 12–18. The polyester resin used was Atlas Chemical Co. Atlac 382. Figures 12–18 give the percentage retention of flexural strength over periods of immersion, at a continuous boil, in various chemical solutions and in water. The extent of chemical attack depends to a large degree on the finish used. Green and Guevin[51] have reported on polyurethane finishes for FRP as being very satisfactory, as regards both chemical resistance and weathering. In all cases, part of the reduction in flexural strength is due to the increased temperature. These curves, therefore, define the ultimate deterioration that can take place. The deterioration at ambient temperatures would be much less. Pritchard and Taneja[52] have reported on the water damage to FRP composites. They tested various resins with glass reinforcements and with differing gel coat tissue, and the retention of flexural strength and modulus after boiling at 100°C (212°F) for 50 days is shown in Table 15. Hot water attacks the FRP composite by hydrolysis of the resin, fibers, and interfacial bond, causing swelling and shrinkage stresses in the matrix. Debonding and resin cracking then occurs. The cure system used for the polyester resin can also affect its

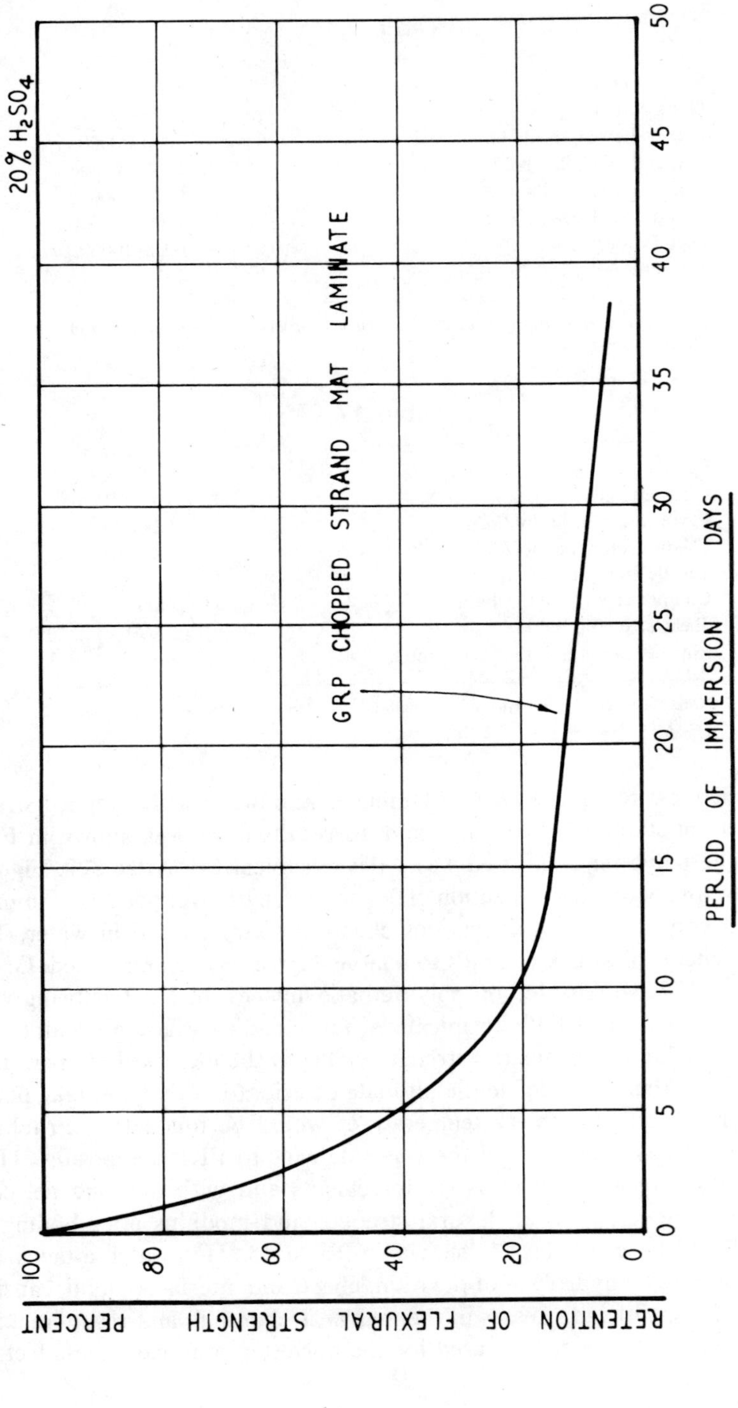

$20\%\ H_2SO_4$
GRP CHOPPED STRAND MAT LAMINATE
RETENTION OF FLEXURAL STRENGTH PERCENT
100
80
60
40
20
0
PERIOD OF IMMERSION DAYS
0
5
10
15
20
25
30
35
40
45
50

Fig. 12

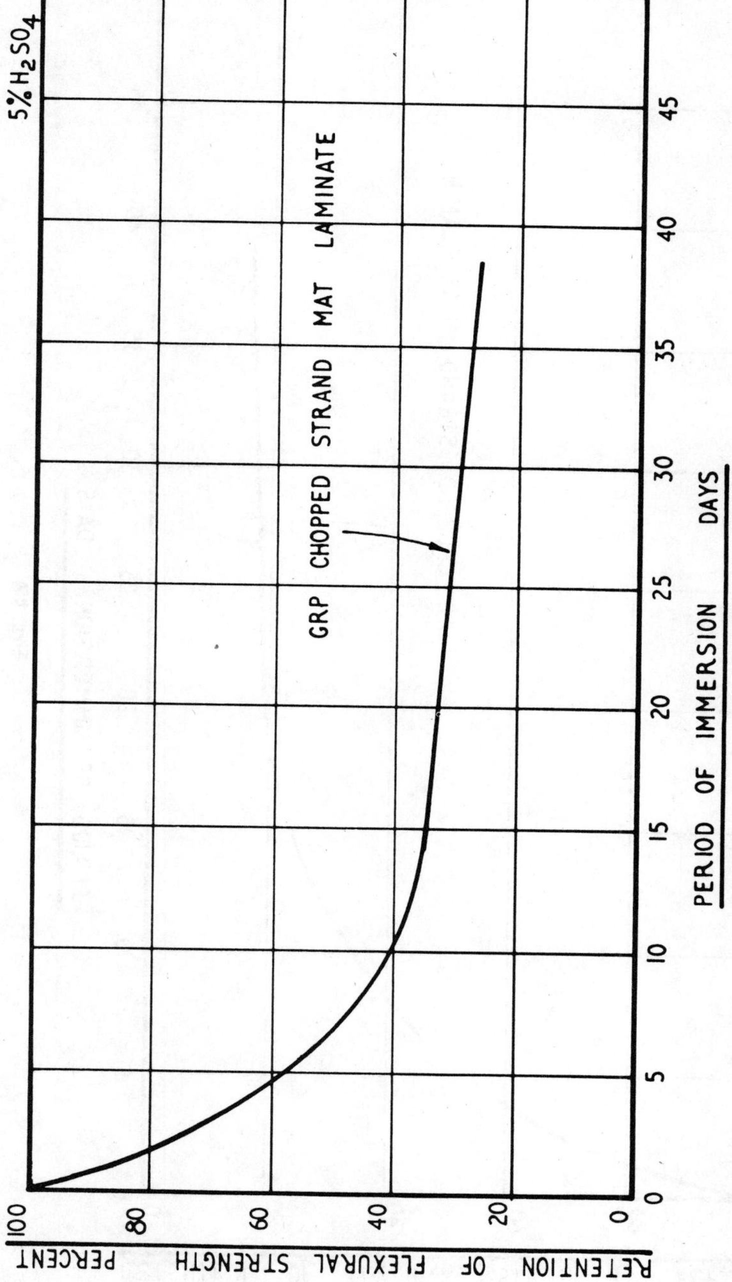
5% H_2SO_4
GRP CHOPPED STRAND MAT LAMINATE
RETENTION OF FLEXURAL STRENGTH PERCENT
100
80
60
40
20
0
0
5
10
15
20
25
30
35
40
45
50
PERIOD OF IMMERSION DAYS

Fig. 13

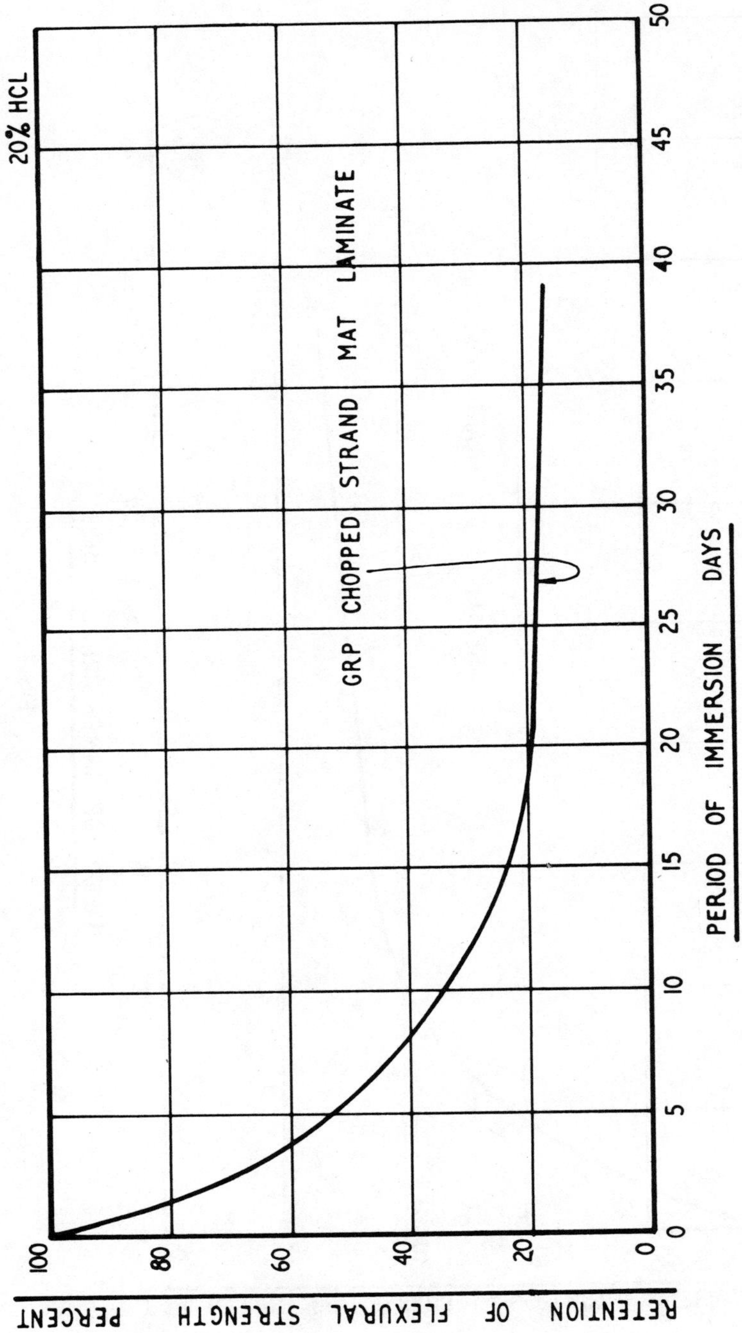
20% HCL
GRP CHOPPED STRAND MAT LAMINATE
PERIOD OF IMMERSION DAYS
0
5
10
15
20
25
30
35
40
45
50
PERCENT RETENTION OF FLEXURAL STRENGTH
0
20
40
60
80
100

Fig. 14

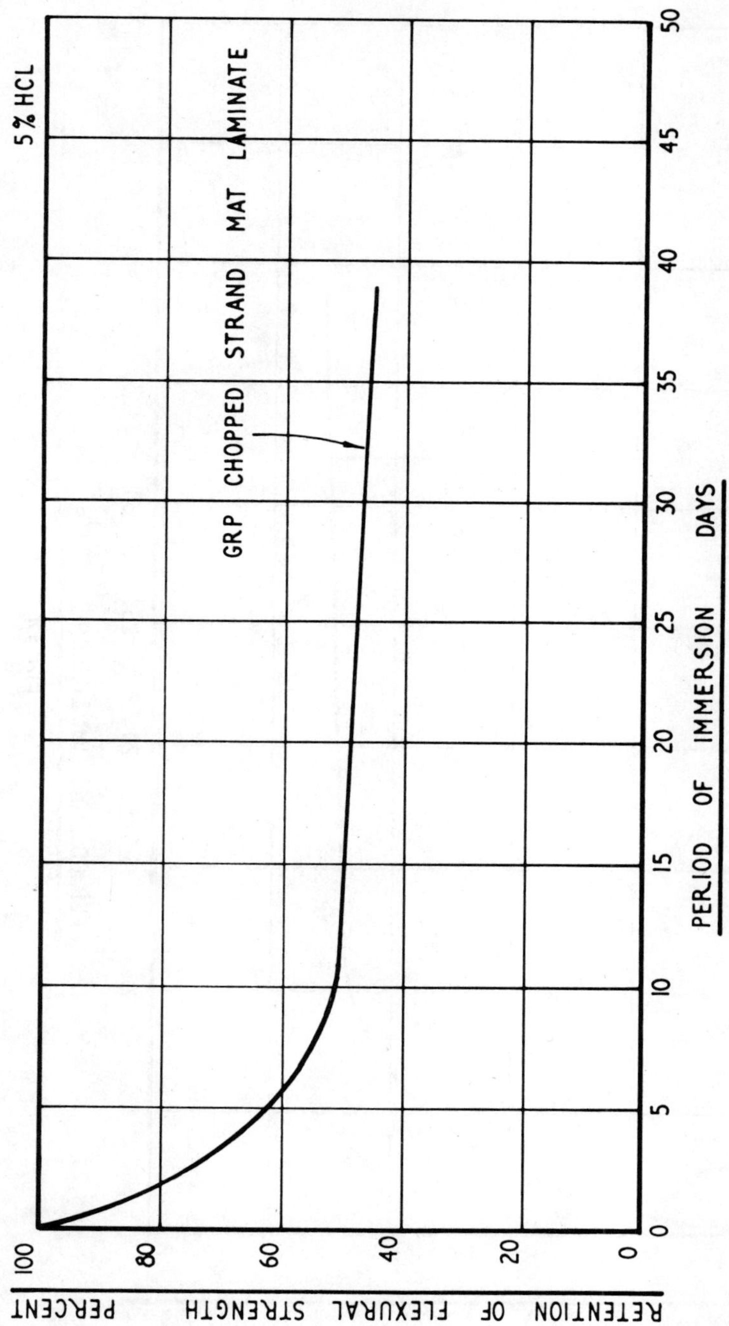
5% HCL
GRP CHOPPED STRAND MAT LAMINATE
PERIOD OF IMMERSION DAYS
0
5
10
15
20
25
30
35
40
45
50
RETENTION OF FLEXURAL STRENGTH PERCENT
0
20
40
60
80
100

Fig. 15

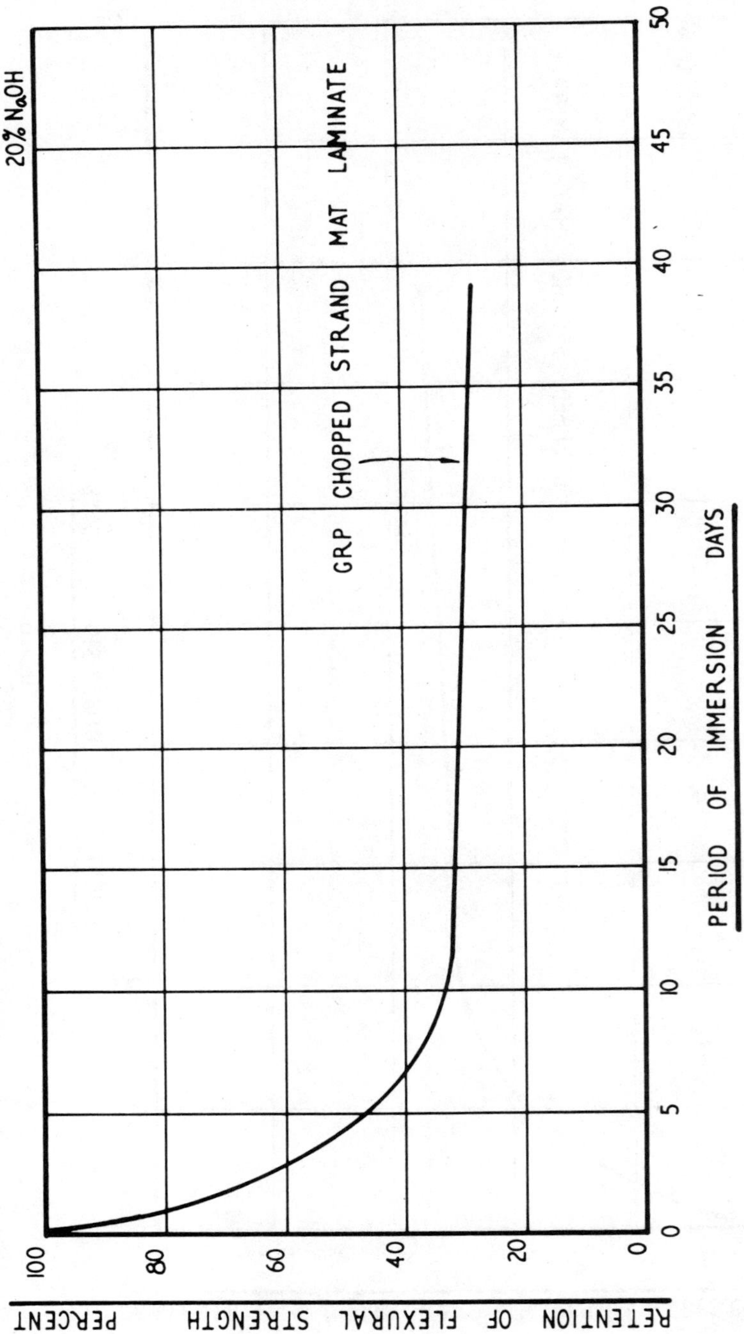
20% NaOH
GRP CHOPPED STRAND MAT LAMINATE
0
5
10
15
20
25
30
35
40
45
50
PERIOD OF IMMERSION DAYS
0
20
40
60
80
100
RETENTION OF FLEXURAL STRENGTH PERCENT

Fig. 16

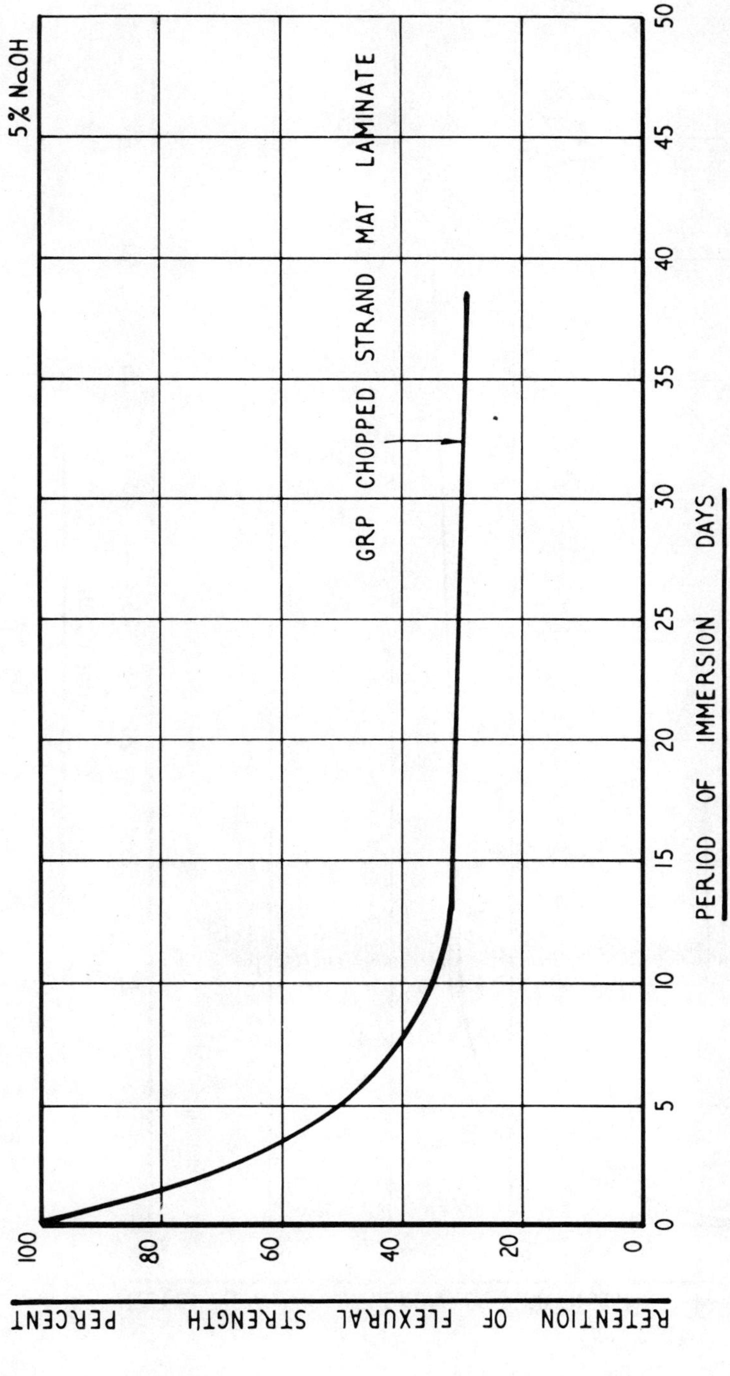
5% NaOH
GRP CHOPPED STRAND MAT LAMINATE
RETENTION OF FLEXURAL STRENGTH PERCENT
100
80
60
40
20
0
0
5
10
15
20
25
30
35
40
45
50
PERIOD OF IMMERSION DAYS

Fig. 17

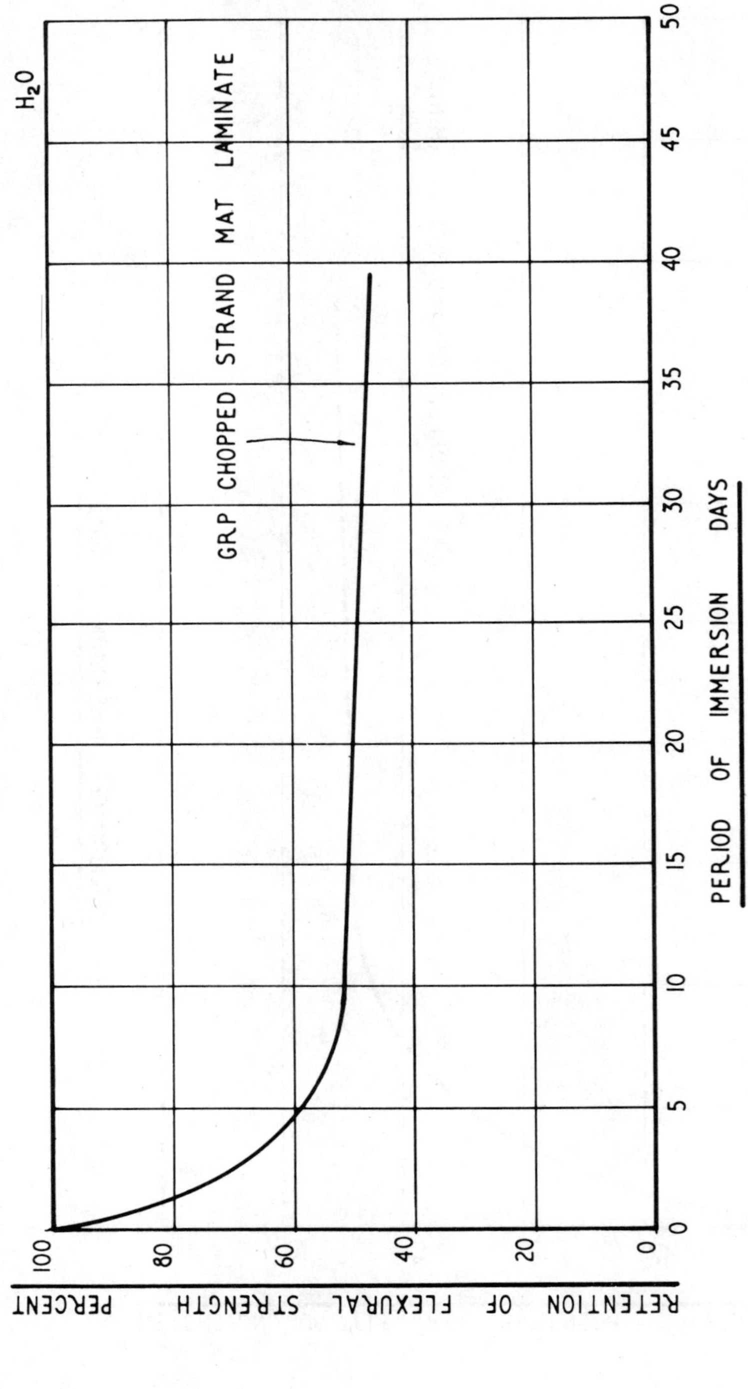
H_2O
GRP CHOPPED STRAND MAT LAMINATE
0
5
10
15
20
25
30
35
40
45
50
PERIOD OF IMMERSION DAYS
100
80
60
40
20
0
RETENTION OF FLEXURAL STRENGTH PERCENT

Fig. 18

Table 15

Laminate Type	Flexural Strength	% Retention Flexural Modulus	Barcol Hardness
Orthophthalic/glass tissue	37	58	65
Orthophthalic/carbon fiber	23	47	63
Isophthalic/carbon fiber	34	68	65
Isophthalic/glass tissue	41	72	81
Isophthalic/acrylic tissue	35	66	81
Vinylester/glass tissue	53	77	85
Bisphenol/glass tissue	49	97	83

chemical resistance and Dixon, Longenecker, and Ferrarini[53] give the chemical resistance of laminates exposed to various chemical solutions. Two cure systems were used. The percentage retention of flexural strength was considered by them with period of exposure in months, and the general behavior is not unlike that determined by Rassell.[31]

While an elevation in temperature causes a deterioration in the behavior of the laminate, the behavior of composites at cryogenic temperatures is much more promising. Brink[54] comes to the conclusion that the mechanical behavior of reinforced plastics increases as the temperature decreases regardless of the loading conditions. Hertz[55] concludes that for almost all design purposes, composites at cryogenic temperatures would be as good or better than at room temperatures.

It has been seen that FRP reinforced with cloth or woven roving is anisotropic in its properties. Some of the properties listed before must, therefore, be obtained in the directions of the warp and the weft. The properties of the composite in any direction can then be determined.

Let the principal axes of the composite be as shown in Fig. 19. Consider two axes ζ and η inclined at an angle α to the original axes as shown in Fig. 19. It can then be shown that

$$\frac{1}{E_\zeta} = \frac{\cos^4\alpha}{E_x} + \frac{\sin^4\alpha}{E_y} + \left[\frac{1}{G_{xy}} - \frac{2}{E_y}\mu_{xy}\right]\sin^2\alpha\cos^2\alpha$$

$$\frac{1}{E_\eta} = \frac{\sin^4\alpha}{E_x} + \frac{\cos^4\alpha}{E_y} + \left[\frac{1}{G_{xy}} - \frac{2}{E_y}\mu_{xy}\right]\sin^2\alpha\cos^2\alpha \qquad (28)$$

$$\frac{1}{G_{\zeta\eta}} = \left[\frac{1}{E_x} + \frac{1}{E_y} + \frac{2}{E_y}\mu_{xy}\right]\sin^2 2\alpha + \frac{\cos^2 2\alpha}{G_{xy}} \qquad (29)$$

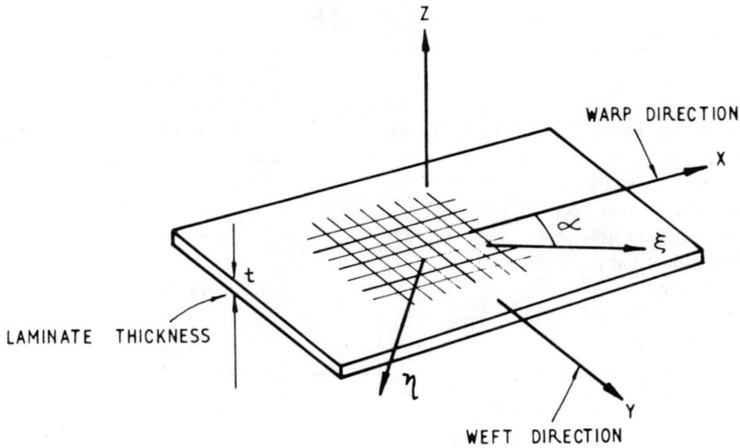

Fig. 19

$$\mu_{\zeta\eta} = \frac{\dfrac{\mu_{xy}}{E_y} - \left[\dfrac{1}{E_x} + \dfrac{1}{E_y} - \dfrac{1}{G_{xy}} + \dfrac{2}{E_y}\mu_{xy}\right]\dfrac{\sin^2 2\alpha}{4}}{\dfrac{\sin^4 \alpha}{E_x} + \dfrac{\cos^4 \alpha}{E_y} + \left[\dfrac{1}{G_{xy}} - \dfrac{2}{E_y}\mu_{xy}\right]\sin^2 \alpha \cos^2 \alpha} \tag{30}$$

The derivation of the above equations has been well documented in the literature.[56-58] Certain conclusions can be drawn from the above equations. First it can be seen that the value of $G_{\zeta\eta}$ at $\alpha = 45°$ is only a function of E_x, E_y, and μ_{xy} but not G_{xy}. This follows from Equation 29, which, at $\alpha =$ 45°, reduces to

$$\frac{1}{G_{\zeta\eta}} = \left[\frac{1}{E_x} + \frac{1}{E_y} + \frac{2}{E_y}\mu_{xy}\right]$$

Secondly, it has also been found that large variations in the Poisson's ratio do not affect the values significantly. A typical plot of E_ζ and E_η with α for specific values of the elastic constants along the X and Y axes is shown in Fig. 20. The problem, of course, becomes considerably more complicated when different plies are used to form the composite. This angle ply construction of composites is discussed for the high-performance composites considered in Section 2.18.

Kabelka,[49] at the Institute of Thermomechanics of the Czechoslovak Academy of Sciences, has performed considerable work on the properties of FRP reinforced with cloth fabric. His findings are reproduced here with the kind permission of the British Plastics Federation.

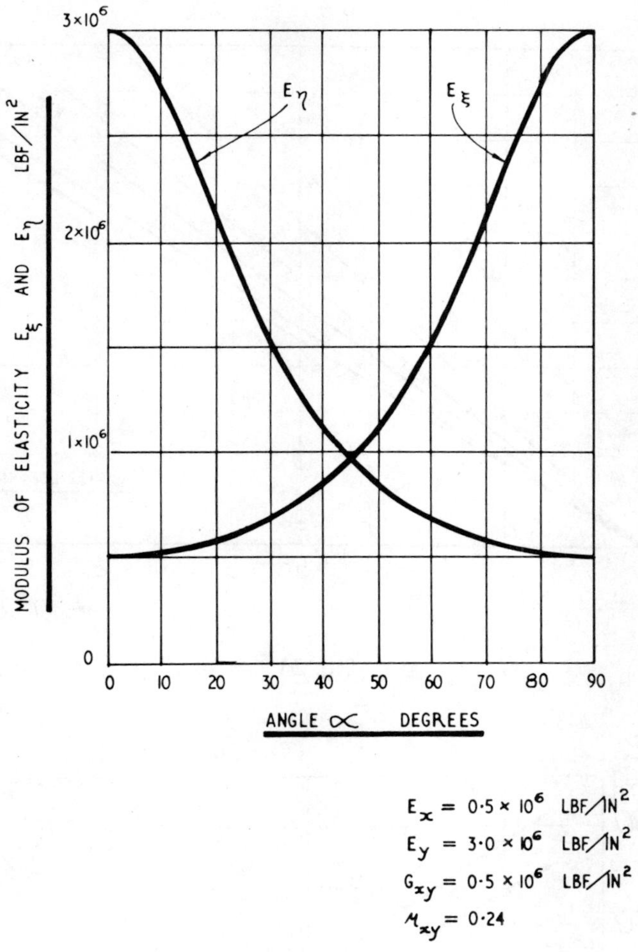

Fig. 20

All tests were carried out on FRP laminates reinforced with plain weave glass cloth fabric with a resin/glass ratio 1:1. The fabric weighed 0.387 kg/m² (1.27 oz/ft²) and was orthotropic, the properties in the warp and weft directions being almost equal.

The stress/strain relationships for the FRP composite in tension at 45°C (113°F) and at various times after loading are shown in Fig. 21. They are isochronous stress/strain curves as defined in Section 1.5. They are almost straight lines and, for purposes of design, can be assumed to be so.

The variations of E_{xc} and X_{xt} with temperature are shown in Fig. 22. At

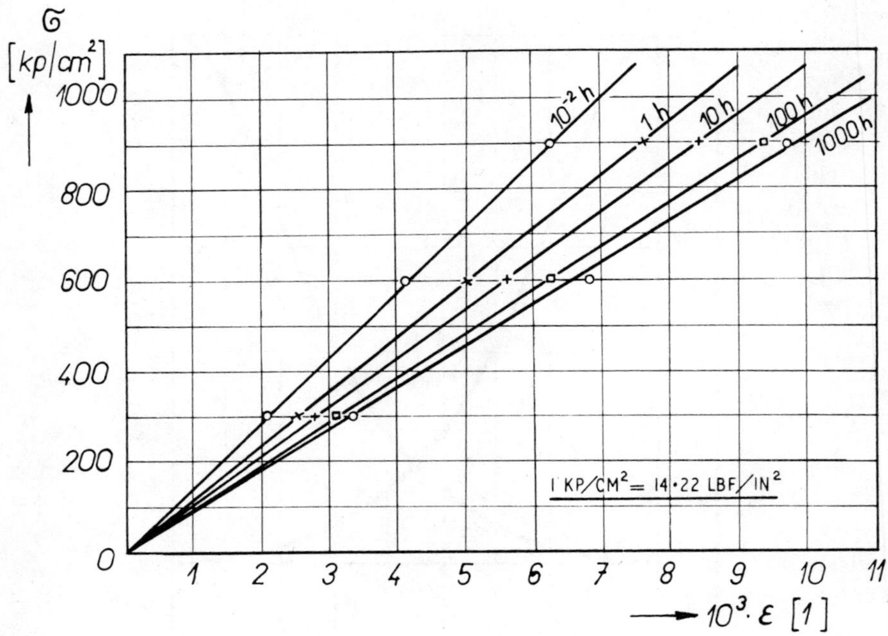
σ
[kp/cm²]
1000
800
600
400
200
0
1
2
3
4
5
6
7
8
9
10
11
10⁻² h
1 h
10 h
100 h
1000 h
1 KP/CM² = 14·22 LBF/IN²
10³·ε [1]

Fig. 21

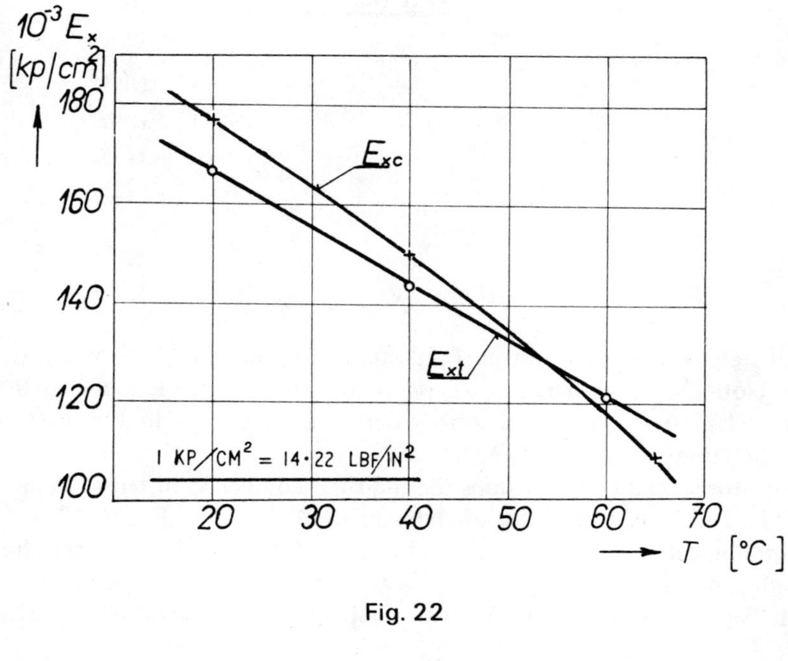
10⁻³ Eₓ
[kp/cm²]
180
160
140
120
100
20
30
40
50
60
70
Exc
Ext
1 KP/CM² = 14·22 LBF/IN²
T [°C]

Fig. 22

20°C (68°F), E_{xc} is higher than E_{xt} by about 7%. As has been mentioned earlier, this conflicts with the findings of Abbott and Broutman. The author feels that the method of testing composites in tension and in compression would play some part in the actual values obtained for the moduli, and, for purposes of design, $E_c = E_t$ is a very valid assumption.

The shear modulus of rigidity G_{xy} also varies with the temperature, and this variation is shown in Fig. 23. G_{xy} refers to the coordinate axes as defined by the warp and the weft directions. The shear modulus $G_{\zeta\eta}$ referring to the coordinate system of axes $\zeta\eta$ lying at an angle α to the warp and weft directions, can be determined by Equation 29, modified to

$$\frac{1}{G_{\zeta\eta}} = \frac{1}{E_x} + \frac{1}{E_y} + \frac{2\mu_{xy}}{E_x} - \left[\frac{1}{E_x} + \frac{2\mu_{xy}}{E_x} + \frac{1}{E_y} - \frac{1}{G_{xy}}\right] \cos^2 2\alpha \qquad (31)$$

The theoretical and experimental variation of $G_{\zeta\eta}$ with angle α is shown in Fig. 24.

Determination of the Poisson's ratio presents considerable difficulty, particularly where the strains have to be determined across the depth of the test piece, i.e., in the Z direction. The variation of the Poisson's ratio μ_{xy} and μ_{xz} with the temperature is shown in Fig. 25. The Poisson's ratio $\mu_{\zeta\eta}$ of

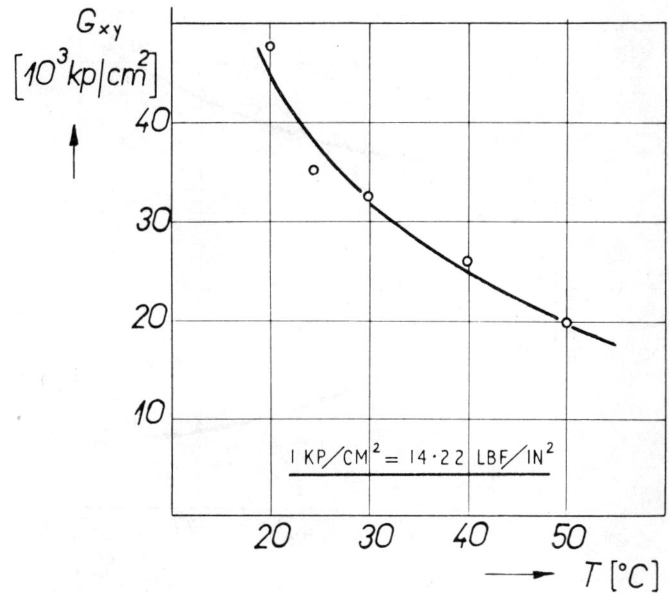

Fig. 23

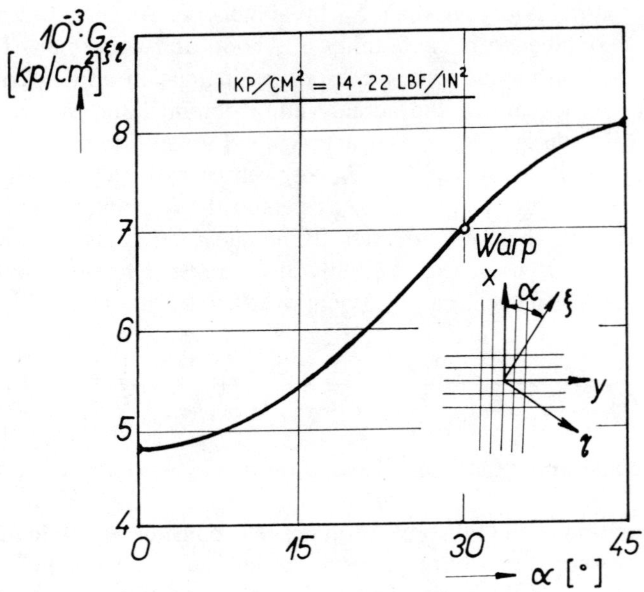

$10^{-3} \cdot G_{\xi\eta}$
[kp/cm²]
1 KP/CM² = 14·22 LBF/IN²
8
7
6
5
4
0
15
30
45
α [°]
Warp
x
α
ξ
y
η

Fig. 24

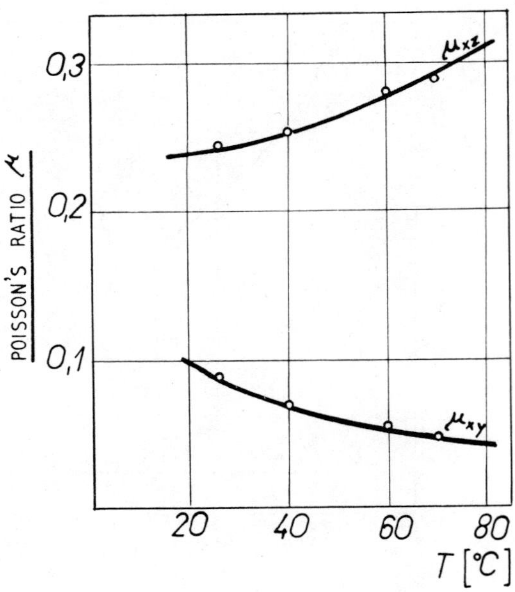

POISSON'S RATIO μ
0,3
0,2
0,1
20
40
60
80
T [°C]
μxz
μxy

Fig. 25

the material with reference to coordinate systems $\zeta\eta$ at an angle α to the warp and weft directions can be determined by the following equation:

$$\mu_{\zeta\eta} = \frac{\mu_{xy} - (1 - R)\sin^2 2\alpha}{1 + (R - 1)\sin^2 2\alpha} \tag{32}$$

where

$$R = \frac{E_x}{E_{45^\circ}}$$

The theoretical and experimental variation $\mu_{\zeta\eta}$ with angle α at 20°C (68°F) is shown in Fig. 26. It is interesting to note that the Poisson's ratio also depends on the resin/glass fiber ratio. At 20°C, (68°F), μ_{xy} can be determined from Fig. 26, when $\alpha = 0$, as 0.13. The resin/glass fiber ratio for these specimens was 1:1. In order to determine μ_{xz} it was necessary to form test prisms by compressing together several layers of resin and glass. This resulted in specimens with a glass fiber content of 74%. The corresponding value of the Poisson's ratio μ_{xy} from these specimens is 0.10, as can be seen in Fig. 25.

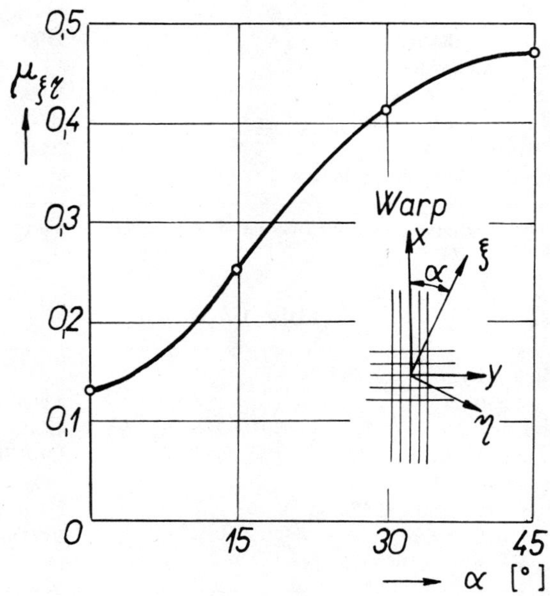

Fig. 26

Though an exact analysis of the warp and weft directions is essential for an exact design, there arise situations, such as a preliminary design or a cost estimate, in which it becomes necessary to treat a cloth laminate in the same way as an isotropic chopped strand mat laminate. In such cases, the properties listed in Tables 16 and 17 can be very useful. Table 16 gives the properties of a general purpose polyester resin/fine weave glass cloth composite. Table 17 gives the properties of composites using the same resin with 8 oz roving cloth reinforcement. In both cases the values refer to laminates ⅛ in. (3.2 mm) manufactured by Permali Limited.[48]

Of very considerable interest is the long-term behavior of FRP composites. The variation of compressive strength σ_{xc} with time at temperatures of 20°C (68°F) and 30°C (86°F) for the composites tested by Kabelka,[49] are shown in Fig. 27. The instantaneous short term compressive strengths were 1690 kp/cm² and 1550 kp/cm² at the respective temperatures. This in fact implies that the short term tests were carried out in about 10^{-3} hours, or about 3.6 seconds. From the graphs it can be seen that even after 10^6 hours (that is about 114 years), the compressive strength can be expected to be at least 40% of the instantaneous short term compressive strength at 20°C. This

Table 16

Specific gravity	1.83
Glass content	About 50%
Flexural strength, lbf/in.²	55000
Flexural modulus, lbf/in.²	2.2×10^6
Tensile strength, lbf/in.²	35000
Compressive strength, lbf/in.²	50000
Shear strength, lbf/in.²	20000
Impact strength, ft. lbf/½ in. notch	6

Note: The actual properties of the laminate will depend on the resin used and the exact glass content.

Table 17

Specific gravity	1.89
Glass content	About 50%
Flexural strength, lbf/in.²	40000
Flexural modulus, lbf/in.²	1.9×10^6
Tensile strength, lbf/in.²	34000
Compressive strength, lbf/in.²	50000
Shear strength, lbf/in.²	20000
Impact strength, ft. lbf/½ in. notch	16

Note: The actual properties of the laminate will depend on the resin used and the exact glass content.

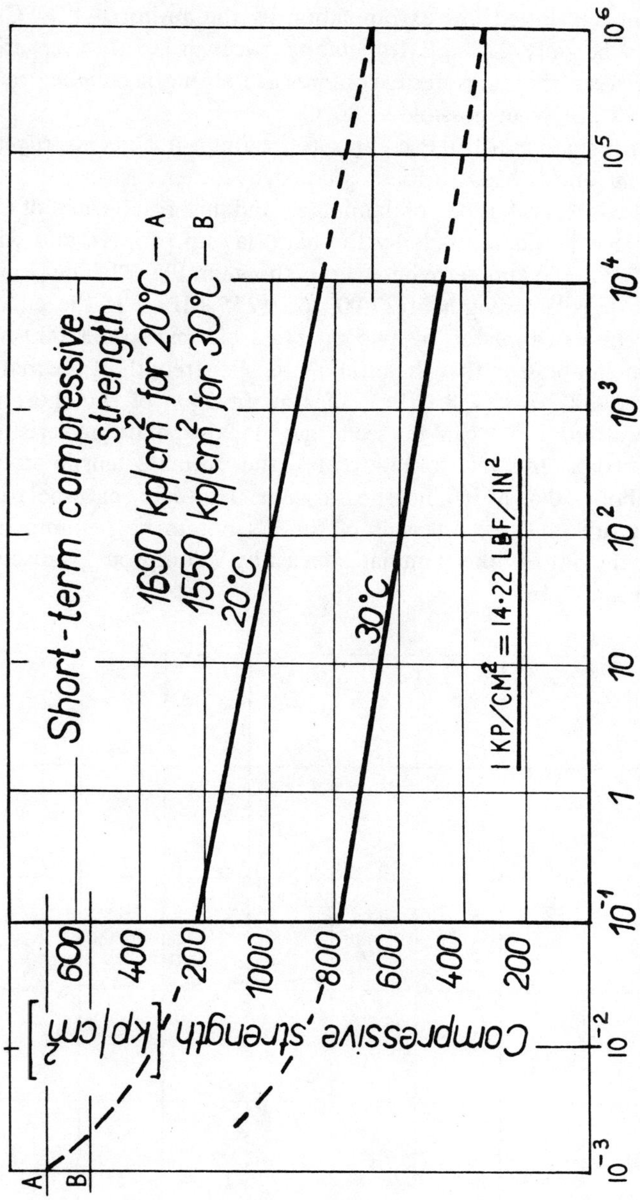
Short-term compressive strength
1690 kp/cm² for 20°C — A
1550 kp/cm² for 30°C — B
20°C
30°C
1 KP/CM² = 14·22 LBF/IN²
Compressive strength [kp/cm²]
600
400
200
1000
800
600
400
200
A
B
10⁻³
10⁻²
10⁻¹
1
10
10²
10³
10⁴
10⁵
10⁶
time [hours]

Fig. 27

value is based on the dotted line extrapolation by the author. At 30°C, the value appears to be only 22%. Extrapolating backwards, however, it can be seen that whereas the short term compressive strength can be reached at 20°C, this does not seem possible at 30°C.

The long term behavior of FRP composites, though not necessarily reinforced with cloth fabric, has also been studied by other research workers. Thompson[59] tested several types of laminates and his results are of some interest. For a FRP laminate made by the hand lay-up process, and with a resin content of 51%, the long term behavior is shown in Fig. 28. The laminate had an ultimate tensile strength of 37,000 psi (255 MPa). If the curve in Fig. 28 were to be extrapolated by two intervals of time to 100,000 hours (or 11.4 years), it can be seen that the ultimate tensile strength of the material would be about 20,000 psi (138 MPa). This about 54% of the short term ultimate tensile strength. It can be seen that if extrapolation were to be made for one further interval (to 114 years), the ultimate tensile strength would still be about 50% of its short term value. However, extrapolating a curve of this nature to three intervals of time is not to be recommended. The long-term behavior of other laminates tested by Thompson[59] is discussed in Sections 2.15 and 2.16.

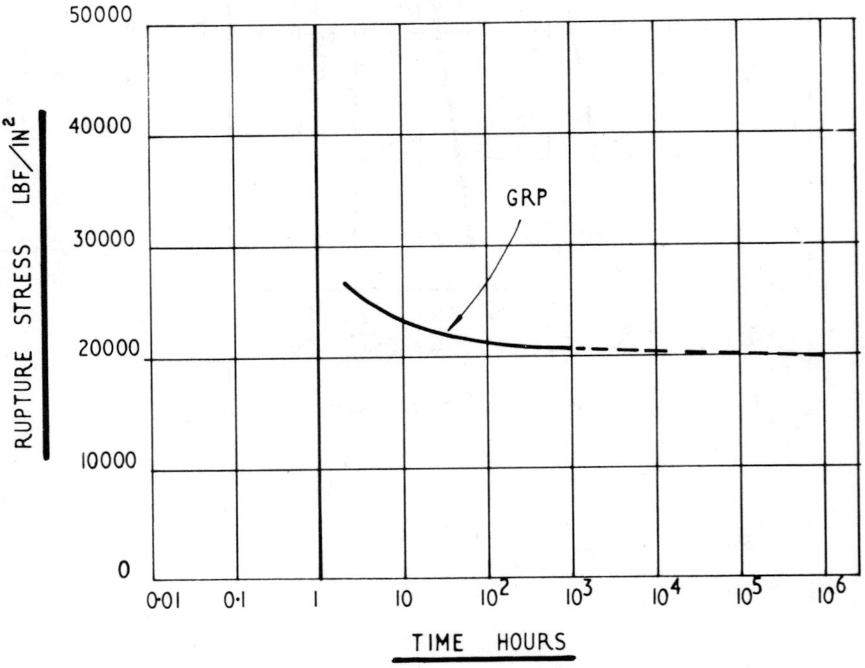

Fig. 28

Boller[60] has performed a considerable amount of work on the long-term properties of laminates under both dry and wet conditions. His tests, though only up to 10,000 hours, show quite conclusively that the percentage reduction of ultimate strength with time in the wet condition is far more severe than in the dry condition. He found that for polyester resin reinforced with woven roving, the percentage reduction in ultimate strength for laminates in the dry condition was to about 68.4% at 10,000 hours. This compared to a percentage reduction in ultimate strength for laminates in the wet condition to about 46.1% of the short term ultimate strength. Extrapolating his curves to 30 years produced the result that the ultimate strength of the laminate in the dry condition would be about 55% of the short term ultimate strength, and in the wet condition would be about 33% of the short term ultimate strength. It is easy to see that the extrapolation by one further interval to 114 years would not have produced any dramatic drop in the above percentages.

Yet other tests were carried out by Messers Sharpe Jones and Co. Limited, and quoted by Ainsworth.[61] These results are shown in Fig. 29. Both Figs. 28 and 29 are by kind permission of the British Plastics Federation. Figure

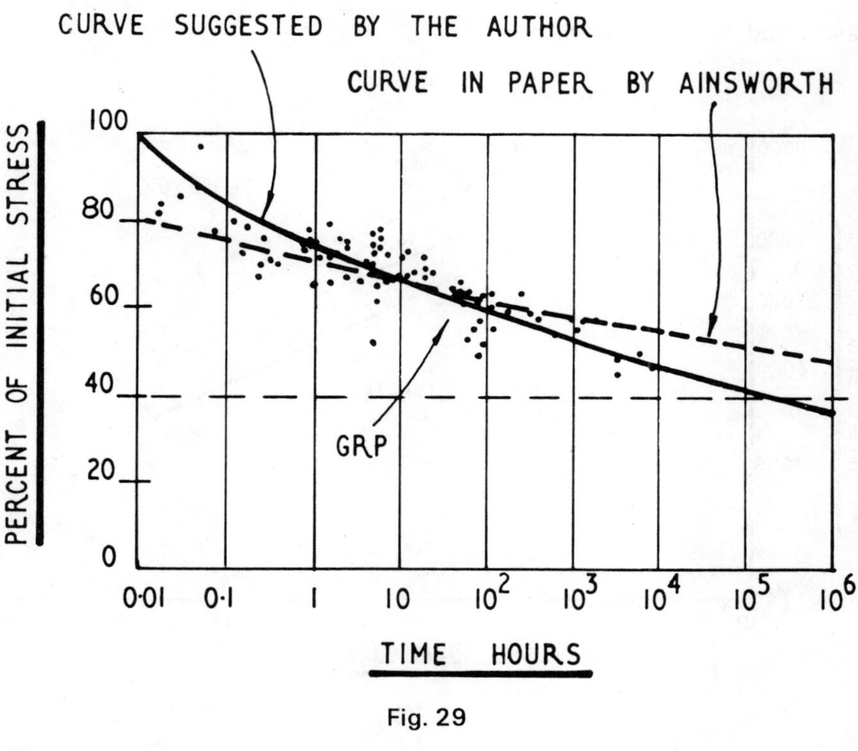

Fig. 29

29 shows the individual test results for specimens in tension. The resin was polyester resin SR19008 reinforced with "E" glass 12 end roving. If a curve is drawn through the points as in Fig. 29, it can be seen that the ultimate load that the material will carry at the end of 10^6 hours (or 114 years) is about 35% of its short term ultimate load. Ainsworth[61] suggests that 50% of the ultimate short term load will be sustained over a period of 30 years.

Bax[62] has reported tests on glass fiber reinforced polyester pipes, and these are shown in Fig. 30. The weeping of the pipes occurred at lower stress levels, but the results in Fig. 30 are for rupture.

Theoretical work on the long term rupture stresses in reinforced plastics, not only FRP, is of great interest. Some work has been performed by Goldfein.[63] He states that for a given stress, the time to rupture is related to the temperature by the equation

$$T(c + \log_{10} t) = \mathrm{K} \tag{33}$$

where

T = absolute temperature (460 + T°F),
t = time in hours,

and c and K are constants.

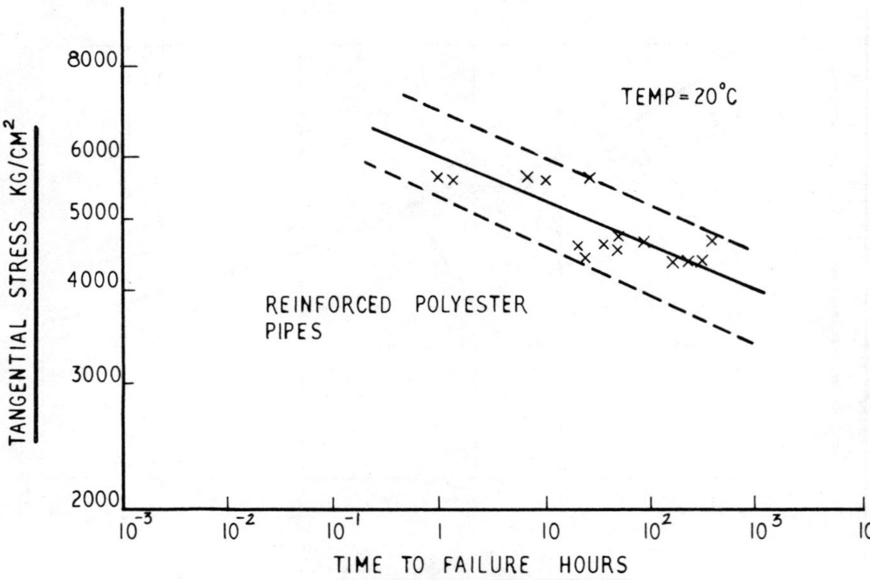

Fig. 30

He found that for all intents and purposes, the value of c is independent of the material or stress level and has been determined to be equal to 20. Equation 33 therefore reduces to

$$T(20 + \log_{10} t) = K$$

If master curves could be drawn for the short term rupture stresses at elevated temperatures against the constants K, it would then be possible to determine the long term rupture stresses at room temperature. Goldfein[63] suggests that the equation could be applied to the tensile and compressive rupture of FRP composites.

The author, working with several of his students both in England and in the States, has developed an approach that is capable of taking into account the varying stress levels on the structure when determining the reduction in ultimate strength with time. This is particularly realistic for a practical structure, where the stress levels would constantly vary depending on the snow load, wind load, and the like. For the maximum stress levels on the structure, a knowledge of the nature of the loads and their seasonal variation would be necessary, as well as a structural analysis of the structure itself. If there were available a record of the variation of the stress levels that the structure was likely to undergo over a period of one year, the pattern could then be identically repeated by the computer for as many years as necessary, up to the total life of the structure.

It has been seen that the experimental results on static fatigue, or reduction in ultimate strength with time, are represented in the form of a curve as shown typically in Fig. 7. The author has found, however, that the reduction in ultimate strength with time for FRP laminates is not related to time by the type of curve shown in Fig. 7 at any particular stress level. Instead of reducing constantly over time, as Fig. 7 implies, the ultimate strength remains fairly constant until a certain percentage of the life of the laminate has been exhausted, after which time the laminate suffers rapid mechanical failure. Depending on the stress level, therefore, there are a family of curves that more accurately depict this behavior, as shown in Fig. 8. If the behavior is considered at any one stress level σ_0, results show that the ultimate strength of the material after t is given by

$$\sigma_{ult(T)} = \sigma_{ult(I)} - K(\sigma_{ult(I)} - \sigma_0) \tag{34}$$

where

$K = 1 - \sqrt{1 - (T/TL)^2}$,
T = time, in hours, the composite was held at stress σ_0,

TL = total life, in hours, of the composite at stress σ_0,
$\sigma_{ult(I)}$ = initial ultimate strength of the composite before application of stress σ_0, and
$\sigma_{ult(T)}$ = ultimate strength of the composite after T hours at stress σ_0.

If the composite has therefore been stressed for a particular time T hours at a particular stress level, the loss in strength together with the fractional loss of life can be computed. For this, however, the static fatigue curve of the material, as given in Fig. 7, has to be imputed into the computer as data; as has been seen, such a curve is easily available. With the next stress level, the computer then considers the appropriate curve transferring to the point on the curve that corresponds to the same loss in ultimate strength, but with a different fractional loss of life, and the further reduction in ultimate strength at that new stress level, over that new period of time, is then computed. The process can be easily repeated for any variations in stress levels over the entire life of the structure. If at any time the fractional loss of life is calculated by the computer to be 1, the computer indicates that the life of the laminate is over and that rupture has occurred.

Associated with the long term behavior of FRP is also the creep of the material. It is essential to be able to determine this creep in order to estimate the increase in deflections of the structure with time. The creep of FRP is dependent on the creep characteristics of the constituent materials. The resin, which serves to distribute the forces into the reinforcement, is subject to creep. The creep in the fibers and the resin therefore influence each other, leading to a redistribution of stresses and a transfer of load to other fibers, which in turn is accompanied by further creep. This leads to a gradual rupture of the bond between resin and fibers, with attendant slip, leading to eventual failure of the composite.

The increase in tensile strain with time for composites at 18°C (64°F) and 60°C (140°F) is shown in Figs. 31 and 32, respectively. In both cases, the load was applied in the warp direction. Kabelka's interpretation[49] of his experimental results is shown by the full lines. The author's interpretation of the same experimental results is shown by the dotted lines, and these fit the experimental results more closely. The dotted lines clearly show that the curves have a tendency to flatten out, and are almost horizontal at about 10^5 hours, indicating little further increase in strain with time. Figure 33 shows the variation of the apparent tensile modulus E_{xt} with time and temperature. This modulus is the creep modulus discussed in Section 1.5.

The increase in compressive strains with time at temperatures of 30°C (86°F), 45°C (113°F), and 60°C (140°F) are shown respectively, in Figs. 34, 35, and 36. Once again, the full lines represent Kabelka's interpretation[49]

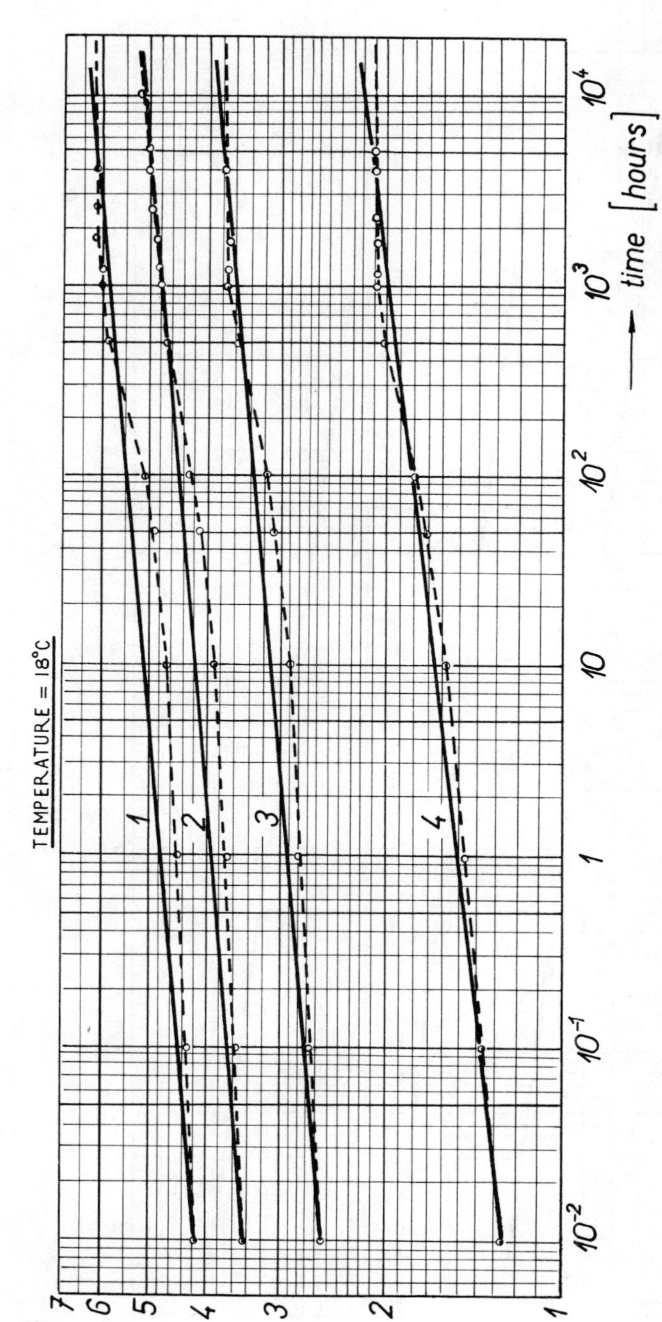

TEMPERATURE = 18°C
$10^3 \cdot \varepsilon$
[1]
1
2
3
4
5
6
7
10^{-2}
10^{-1}
1
10
10^2
10^3
10^4
time [hours]

Fig. 31

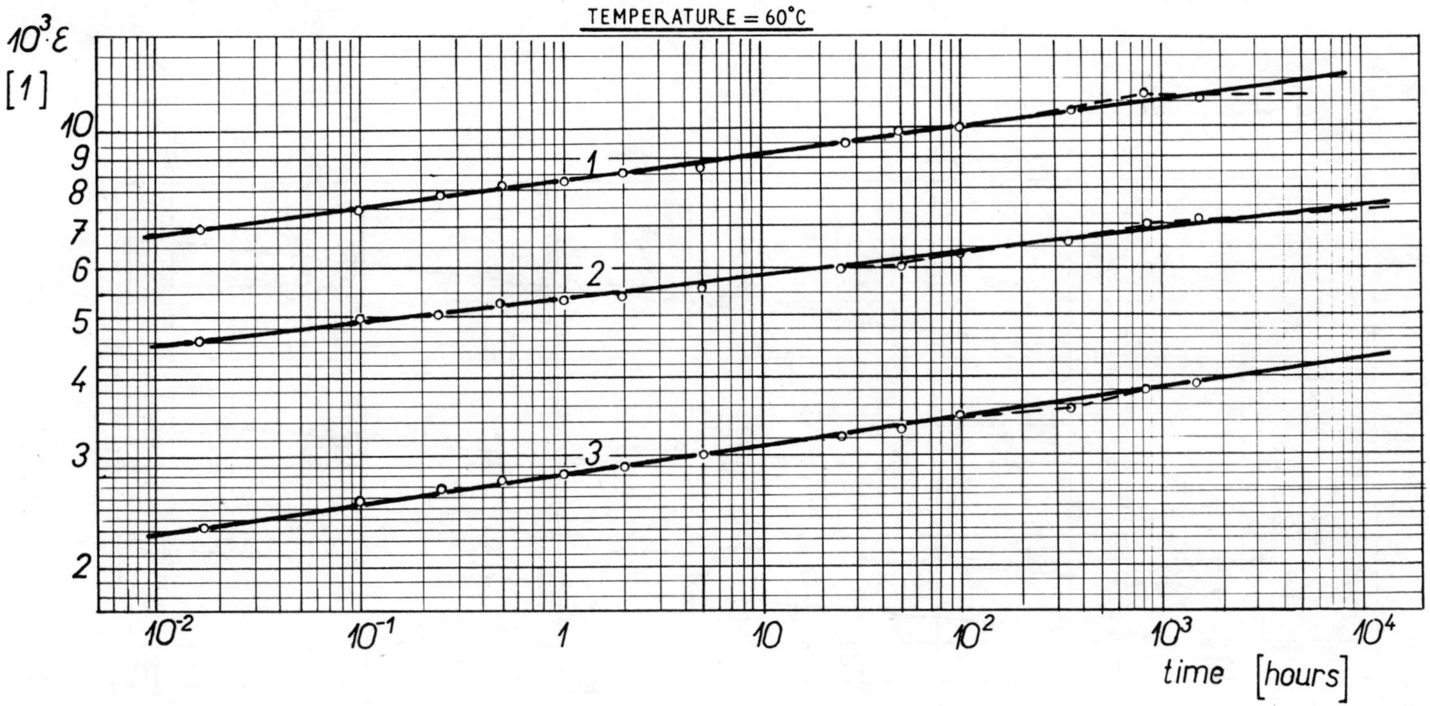
TEMPERATURE = 60°C
10³·ε
[1]
10
9
8
7
6
5
4
3
2
1
2
3
10⁻²
10⁻¹
1
10
10²
10³
10⁴
time [hours]

Fig. 32

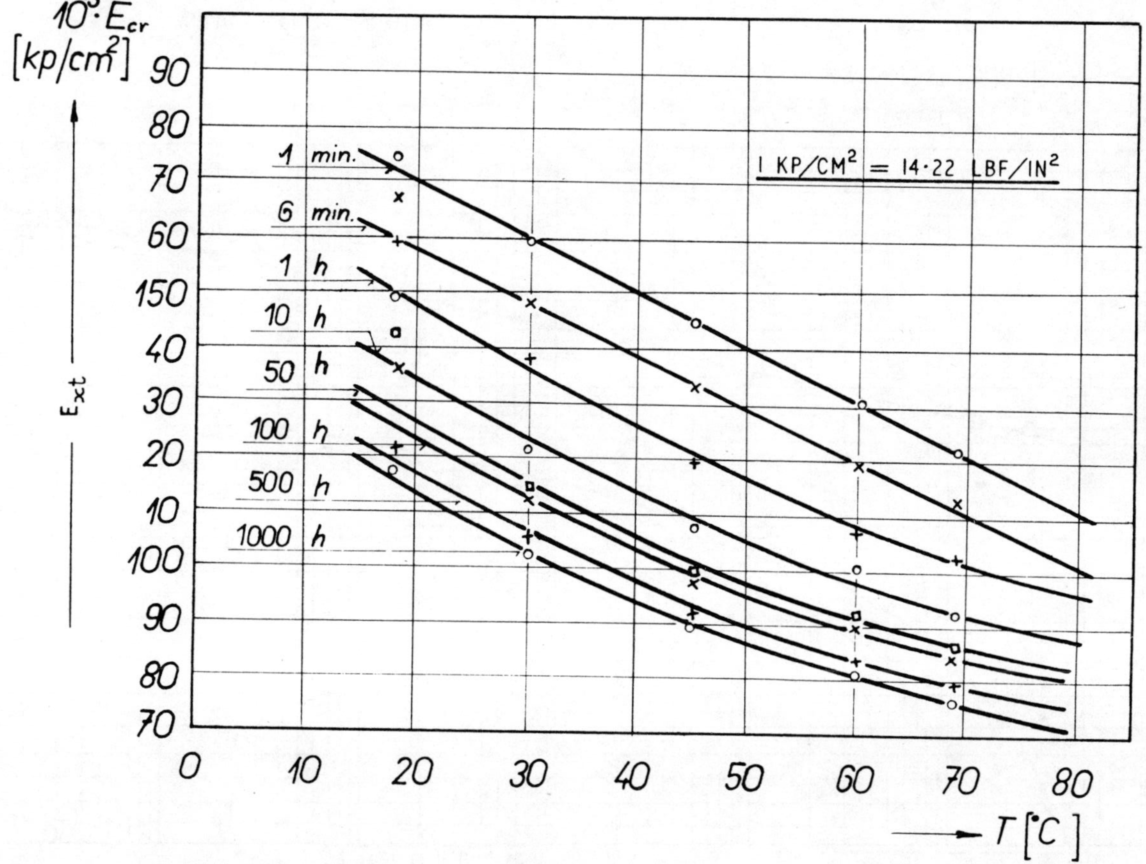

$10^{-3} \cdot E_{cr}$ [kp/cm²]
E_{xt}
90
80
70
60
150
40
30
20
10
100
90
80
70
1 min.
6 min.
1 h
10 h
50 h
100 h
500 h
1000 h
1 KP/CM² = 14·22 LBF/IN²
0
10
20
30
40
50
60
70
80
T [°C]

Fig. 33

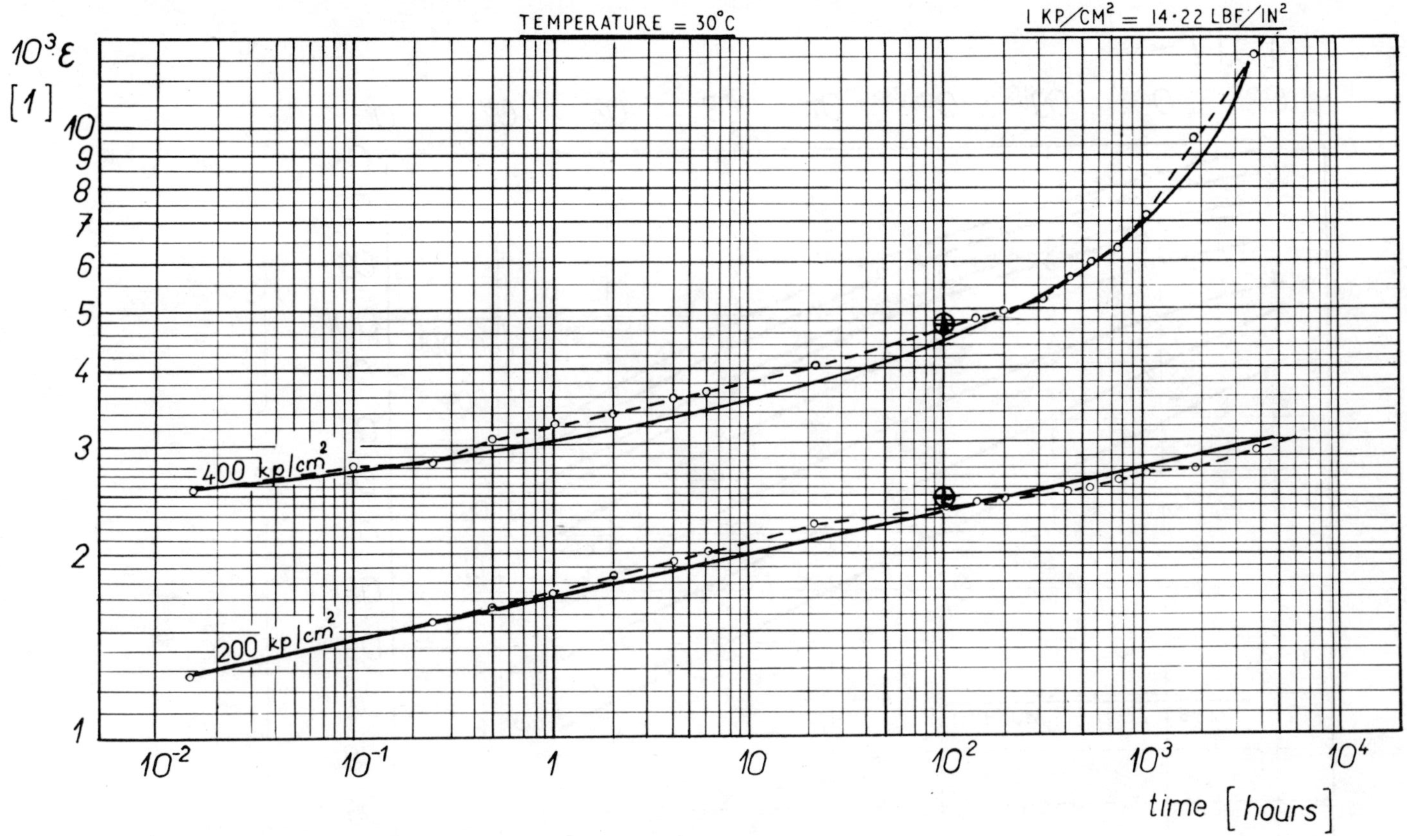

⊕ VALUES FROM EQUATION 35

Fig. 34

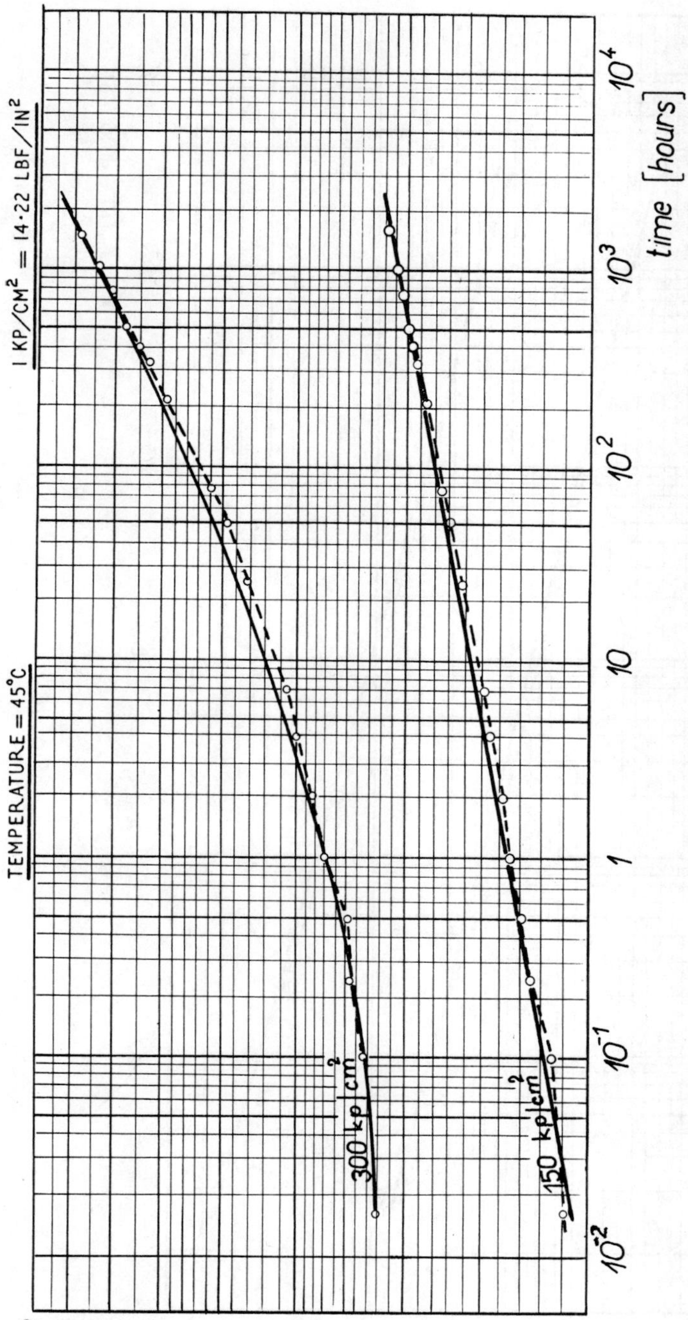
TEMPERATURE = 45°C
1 KP/CM² = 14·22 LBF/IN²
$10^3 \cdot \varepsilon$ [1]
8
7
6
5
4
3
2
1
300 kp/cm²
150 kp/cm²
10^{-2}
10^{-1}
1
10
10^2
10^3
10^4
time [hours]

Fig. 35

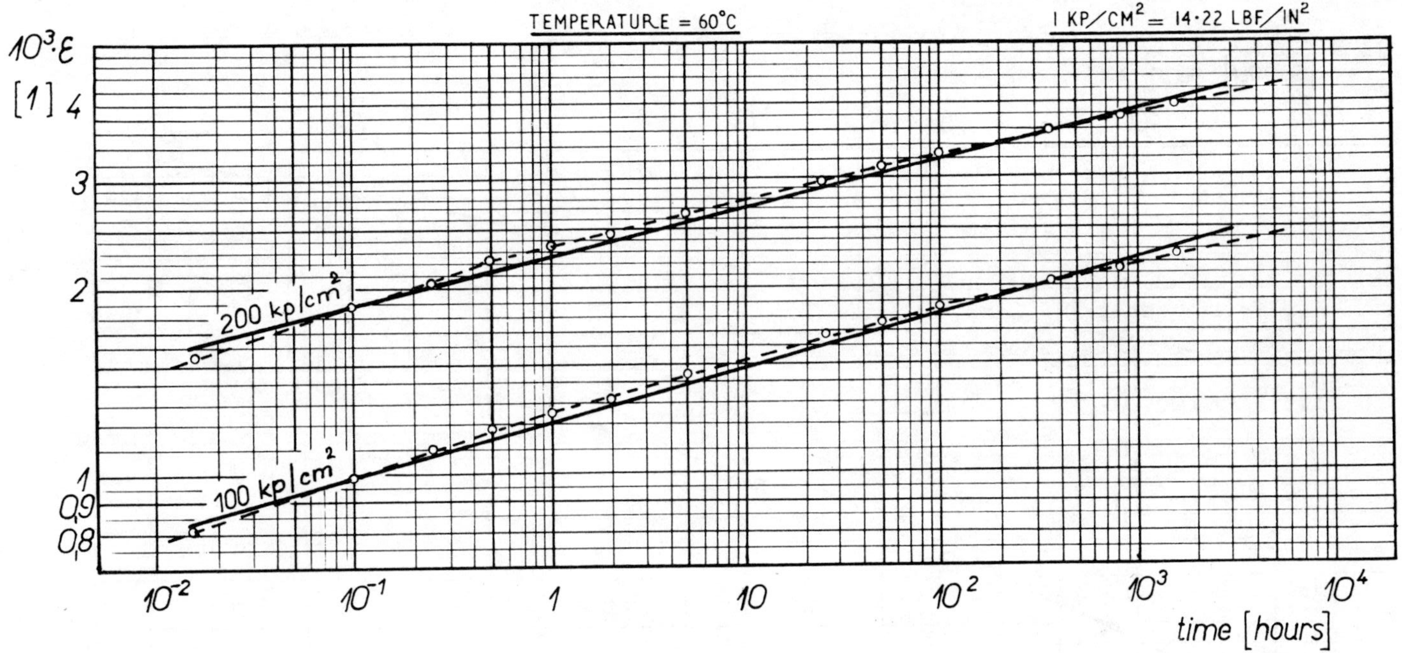

TEMPERATURE = 60°C
1 KP/CM² = 14·22 LBF/IN²
10³·ε
[1]
4
3
2
1
0,9
0,8
200 kp/cm²
100 kp/cm²
10⁻²
10⁻¹
1
10
10²
10³
10⁴
time [hours]

Fig. 36

of his experimental results, while the dotted lines represent the author's interpretation of the same results. The stresses at which the tests were carried out are also indicated on the graphs.

Some work on the theoretical aspects of creep has been carried out by Findley.[64,65] He has shown that the strain ϵ at any time t can be expressed by the relationship,

$$\epsilon = \epsilon_0' \sinh\frac{\sigma}{\sigma_\epsilon} + m'\left(\frac{t}{t_0}\right)^n \sinh\frac{\sigma}{\sigma_m} \tag{35}$$

where

$t_0 = 1$ hour,
$\sigma =$ stress level in the laminate,

and ϵ_0', σ_ϵ, m', and σ_m are constants that depend on the stress level and the temperature. Finally, n is a dimensionless index independent of the stress and is less than 1. All these constants, particularly n, are further influenced by absorbed water. Absorbed water was found to increase the value of n and cause marked changes in the other constants, leading to an increased creep rate. The values of the constants for use in Equation 35, obtained from glass fiber and asbestos reinforced laminates, are given in Table 18.

It is of interest to relate Equation 35 to the results given in Fig. 34. For a FRP composite reinforced with glass cloth at a temperature of 73°F (23°C) it can be shown from Table 18 that,

$$\epsilon = 0.0034 \sinh\frac{\sigma}{15000} + 0.000445\left(\frac{t}{1}\right)^{0.090} \sinh\frac{\sigma}{14000} \tag{36}$$

Figure 34 gives the increase in compressive strain due to creep at a temperature of 30°C (86°F). As an example, therefore, at a stress level $\sigma = 400$ kp/cm² (5690 psi) the strain at time $t = 100$ hours, and with $t_0 = 1$ hour is determined from Equation 36 as

$$\epsilon = 0.001322 + 0.000281 = 0.001603$$

The total strain is equal to strain at 1 hour plus the value of ϵ. Therefore,

$$\text{total strain} = 0.0031 + 0.0016$$
$$= 0.0047$$

Table 18

Material	Temperature °F	Relative Humidity %	Index n	ϵ_o' in. per in.	m' in. per in.	σ_ϵ lbf/in.^{2}	σm lbf/in.^{2}
Melamine resin, glass fabric laminate[a]	77	50	0.0186	0.00575	0.00575	35000	35000
Silicone resin, glass fabric laminate[a]	77	50	0.0283	0.00203	0.00203	7800	7800
Asbestos reinforced laminates	77	50	0.2670	0.00311	0.000127	4000	2400
Polyester resin, glass fabric laminate	73	50	0.0900	0.00340	0.000445	15000	14000
Polyester resin, glass fabric laminate	73	immersed in water	0.2100	0.03300	0.000170	80000	13000
Polyester resin, glass mat laminate	73	50	0.1900	0.00670	0.001100	8500	8500
Epoxy resin, glass fabric laminate	73	50	0.1600	0.00570	0.000500	25000	50000
Epoxy resin, glass fabric laminate	73	immersed in water	0.2200	0.02500	0.000055	80000	11000
Polyester resin, glass woven roving	73	50	0.2000	0.01800	0.001000	40000	22000
Polyester resin, glass woven roving	73	immersed in water	0.2300	0.02060	0.014600	40000	40000

[a] In the analysis of these tests it was assumed that $\epsilon_o' = m'$ and $\sigma_\epsilon = \sigma_m$.

As another example, at a stress level σ = 200 kp/cm^2, the strain ϵ at t = 100 hours and with t_0 = 1 hour, can be determined from Equation 36 as

$$\epsilon = 0.000787$$

Therefore,

$$\begin{aligned}\text{total strain} &= 0.001700 + 0.000787 \\ &= 0.002487\end{aligned}$$

These values have been shown plotted as points on Fig. 34. It can be seen that correlation between the theoretical values determined by the use of Equation 35 and the experimental curves is excellent. This shows quite clearly that the creep of FRP laminates can be readily estimated over long periods of time and at different stress levels if there is available basic information regarding the creep at 1 hour after loading.

Creep has also been studied by other research workers, on a variety of materials including the reinforced thermoplastics, and some of this work is discussed under the relevant material.

Finally, in this discussion on the strength properties of FRP, consideration must be given to the shear strength of FRP composites. Kabelka[49] found that the short term shearing strength is a function of the rate of loading of the specimen. The variation in shearing strength with rate of loading for laminates tested by Kabelka[49] at 20°C (68°F) is shown in Fig. 37. In structural engineering practice, structures are loaded at differing rates over their life, and the short term shearing strength may, therefore, be assumed to be about 520 kp/cm².

Torsion tests were also run by Kabelka[49] to determine the creep of the material in shear. Thin walled cylindrical test pieces were used, and the warp of the reinforcing fabric was parallel to the axis of the test piece. The tests were all conducted at 20°C (68°F). The increase in the relative displacement γ with time for various shear stress levels has been shown in Fig. 38.

It would be appropriate to close this article on FRP by stating that fiberglass reinforced polyester is a versatile material with a large variety of uses and

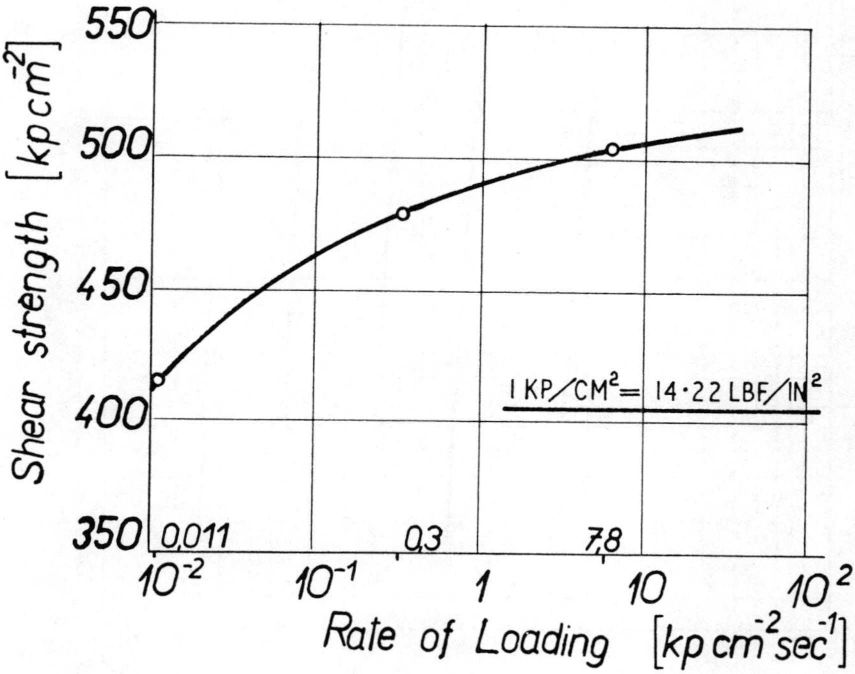

Fig. 37

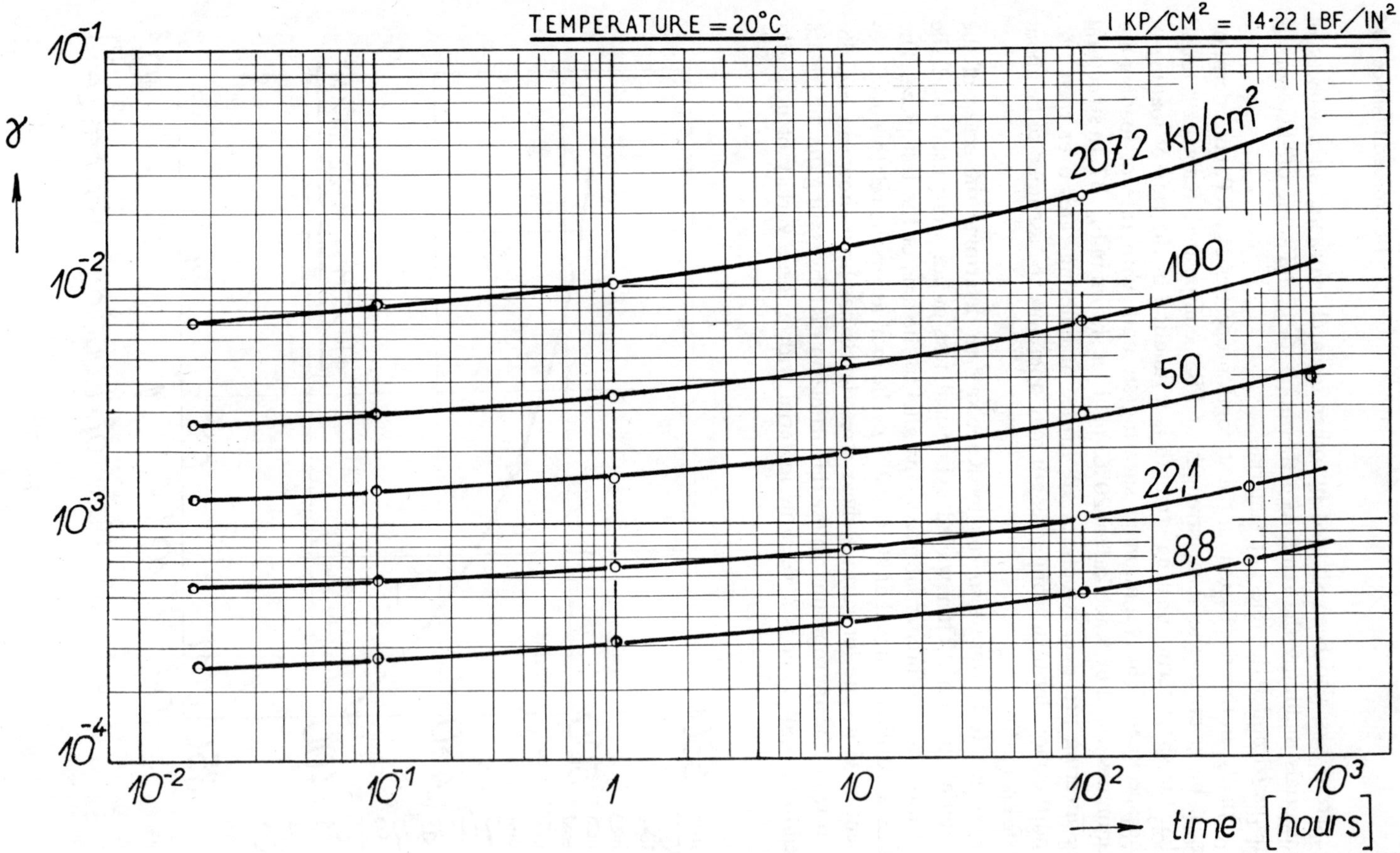
TEMPERATURE = 20°C
1 KP/CM² = 14·22 LBF/IN²
γ
10⁻¹
10⁻²
10⁻³
10⁻⁴
207,2 kp/cm²
100
50
22,1
8,8
10⁻²
10⁻¹
1
10
10²
10³
time [hours]

Fig. 38

end applications. The material can also be used in direct combination with other materials and reinforcements to improve even further specific properties of the composites. FRP can be used with polyester fibers as reinforcement to improve energy absorption and abrasion resistance. McCorsley[66] has reported on polyester composites reinforced with woven roving thermoplastic polyester reinforcement and has compared the properties of the composites with polyester composites using fiber glass woven roving and fiber glass chopped strand mat. Results are given in Table 19. The chemical resistance of the laminates after 60 day immersion in various reagents at 23°C (73°F) is shown in Table 20.

FRP can also be used in combination with thermoplastic sheet such as PVC or acrylic. The filament winding of FRP on PVC pipe considerably

Table 19

Property	Woven Roving Thermoplastic Polyester (PET)	Fiberglass Woven Roving (F.G.W.R)	Fiberglass Mat (F.G.Mat)
Thickness per layer, in.	0.055	0.030	0.052
Fiber content, % weight	40	50	25
Weight per layer, lb/yd²	3.74	3.00	3.38
Flexural strength, psi	15,700	48,000	18,500
Flexural modulus, psi × 10⁶	0.650	2.175	0.980
Tensile strength, psi	16,000	52,000	12,000
Tensile modulus, psi × 10⁶	0.704	2.20	1.015
Compressive strength, psi	14,000	32,000	11,600
Compressive modulus, psi × 10⁶	0.750	3.40	1.20
Shear strength, psi	12,100	18,400	13,800

Table 20

Test Reagent at 23°C	PET Laminate Flexural Strength psi	% of Original	F.G.W.R. Laminate Flexural strength psi	% of Original
Untreated	15,700	100	48,000	100
10% Hydrochloric acid	15,700	100	46,500	97
10% Hydrofluoric acid	15,100	96	38,400	80
10% Sulfuric acid	15,100	96	36,500	74
10% Acetic acid	15,400	98	48,000	100
10% Sodium hydroxide	14,700	94	32,200	67
Saturated NaCl solution	15,600	99	47,500	99
Distilled H_2O	15,700	100	46,500	97

improves the strength characteristics of such pipe, yet retains the excellent corrosion resistance of PVC. Haddad[67] gives the long term stress rupture behavior of both the FRP and the PVC liner, and these are shown in Fig. 39.

The use of FRP with acrylic has been studied by Montella.[68] The acrylic sheet is thermoformed to the required shape and then functions as the mold for the FRP. It also serves as the outside surface to provide excellent surface finish. The FRP improves the load carrying capacity of the material. The properties of the unrigidized acrylic and the rigidized acrylic (with FRP) are shown in Table 21.

For sewer and irrigation applications, pipe can also be filament wound

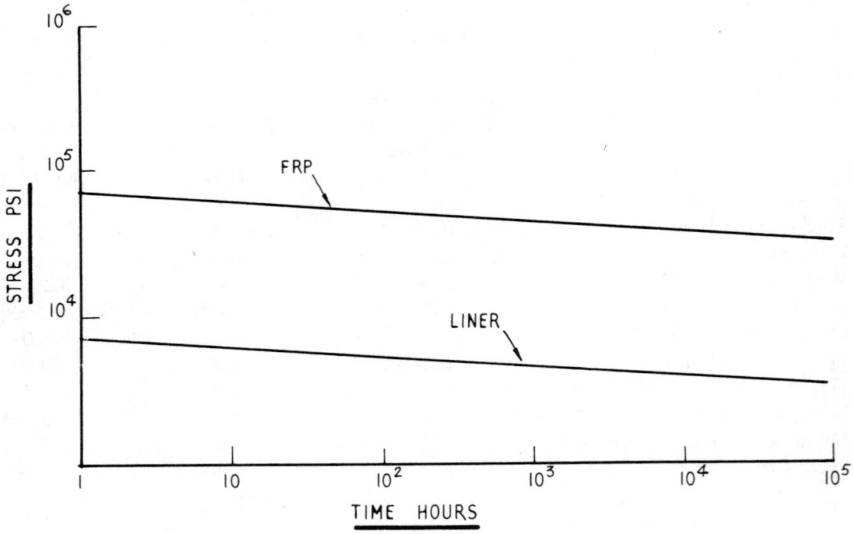

Fig. 39

Table 21

Property	Unreinforced Acrylic Sheet	Rigidized Acrylic, Sheet 0.040 in. + 0.125 in. FRP (25% glass)
Tensile strength, psi	5,400	8,000
Tensile modulus, psi	2.2×10^5	7.4×10^5
Flexural strength, psi	7,000	13,500
Flexural modulus, psi	2.2×10^5	5.4×10^5

with glass, and filled with fine, clean, well graded, high silica sand as a filler to form a reinforced plastic mortar pipe. Rolston and Gilbu[69] tested sand-filled laminates with different sand contents, shown in Table 22. The short term properties of the materials of types I to V are shown in Table 23. Creep tests on RPM pipe were also carried out, and these are shown in Fig. 40. The projected life of such pipe is 100 years.

Finally, the weathering of FRP under the natural environment is important and should always be taken into account. The author[70] studied the weathering of a FRP roof structure, without special ultraviolet additives, to determine the reduction in mechanical properties, and found a reduction of perhaps

Table 22

Construction Type	Sand (% Weight)	Continuous Glass (% Weight)	Chopped Glass (% Weight)	Resin (% Weight)
I	0	11.7	13.2	75.1
II	9.6	11.1	12.6	66.7
III	18.9	10.5	11.6	59.0
IV	25.7	8.8	9.7	55.8
V	34.9	6.2	15.8	43.1

Table 23

Property Hoop Direction (Parallel to continuous strands)	Construction Type I	II	III	IV	V
Tensile strength, psi × 10^3	21.1	16.8	15.7	13.5	12.0
Tensile modulus, psi × 10^6	1.90	1.75	1.81	1.81	2.45
Compressive strength, psi × 10^3	30.6	26.7	31.9	28.4	26.0
Compressive modulus, psi × 10^6	1.28	1.28	1.47	2.11	2.46
Flexural strength, psi × 10^3	21.8	28.1	21.8	24.9	—
Flexural modulus, psi × 10^6	1.26	1.36	1.26	1.84	—
Property Axial Direction (perpendicular to continuous strands)					
Tensile strength, psi × 10^3	9.0	6.1	6.6	6.2	8.4
Tensile modulus, psi × 10^6	1.34	1.32	1.45	1.50	2.08
Compressive strength, psi × 10^3	23.4	22.5	25.9	21.2	—
Compressive modulus, psi × 10^6	0.92	1.04	1.23	1.45	—
Flexural strength, psi × 10^3	9.6	8.0	11.2	12.7	—
Flexural modulus, psi × 10^6	0.87	0.89	1.29	1.33	—

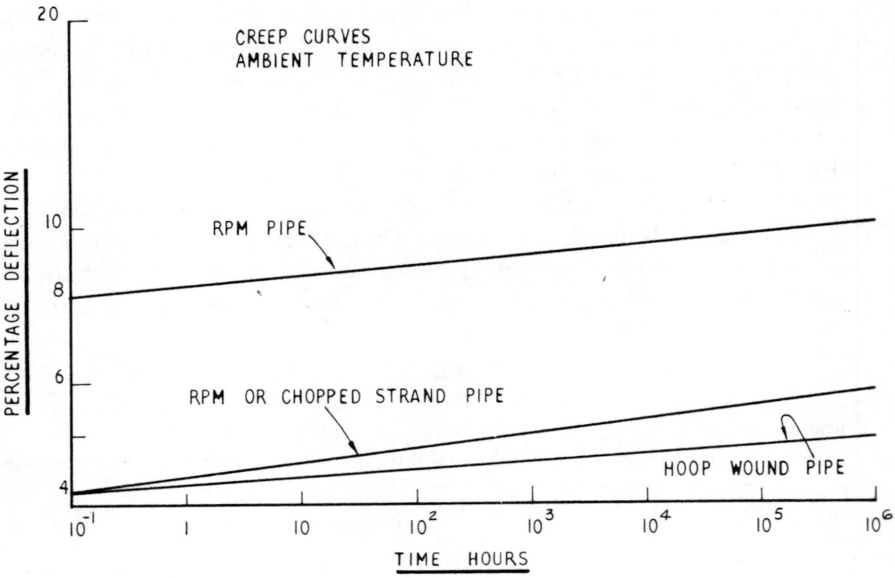

Fig. 40

10% over 3½ years. The structure itself had weathered severely with changes in color and fiber exposure.

2.13 JUTE FIBER REINFORCED POLYESTER

One of the advantages of JRP is its low specific gravity. FRP reinforced with chopped strand mat and with a glass fiber content of about 30% by weight, has a specific gravity of about 1.55. In comparison, JRP reinforced with Hessian and with the same resin content of 70% has a specific gravity of about 1.2. If jute is used in combination with glass fiber in a hybrid composite, the specific gravity will lie between these values. In such cases, the jute reinforcement is placed between layers of glass fiber reinforcement to provide a thick composite that has better strength properties, chemical resistance, and weathering than just a JRP composite. It is essential that the jute be sandwiched between the glass reinforcement, because jute absorbs water readily and has to be well protected. In some applications jute can be used by itself. In other applications it can be used to provide bulk to the composite at reduced cost. For this application, heavy jute twill is the most suitable reinforcement.

Some work has also been performed on the use of jute yarn for filament

Table 24

Reinforcement	Resin	Gel Coat	Flexural Strength in Warp Direction (lbf/in.^{2})	Flexural Strength in Weft Direction (lbf/in.^{2})
Hessian, warps in line	Polyester	no	12869	11526
Hessian, warps at right angles in successive layers	Polyester	no	12547	12214
Hessian, warps in line	Polyester	yes (in compression)	11008	11833
Hessian, warps at right angles in successive layers	Polyester	yes (in compression)	12195	12433
Hessian, warps in line	Polyester	yes (in tension)	13078	11199
Unidirectional, warps in line	Polyester	no	18029	3381
Unidirectional, warps at right angles	Polyester	no	14958	12481
Unidirectional, warps in line	Polyester	yes (in compression)	22695	4932
Heavy twill, warps in line	Polyester	no	10298	11165
Heavy twill, warps in line	Polyester	yes (in compression)	9721	11258
Hessian, warps in line with glass mat outer layers	Polyester	no	22120	19469
Hessian, warps at right angles with glass mat outer layers	Polyester	no	19737	19767
Unidirectional, warps in line with glass mat outer layers	Polyester	no	30389	18709
Unidirectional, warps at right angles with glass mat outer layers	Polyester	no	24817	20399
Heavy twill, warps in line with glass mat outer layers	Polyester	no	16804	16981

Table 24. Cont.

Reinforcement	Resin	Gel Coat	Flexural Strength in Warp Direction (lbf/in.^{2})	Flexural Strength in Weft Direction (lbf/in.^{2})
Hessian, warps in line with glass woven roving outer layers	Polyester	no	26521	20564
Heavy twill, warps in line with glass woven roving outer layers	Polyester	no	17315	19348

winding. Here again only the core would be wound with jute, to provide bulk, the inner and outer layers being filament wound with glass roving.

In JRP, the resin/jute cloth ratio varies with the type of reinforcement. In a hand lay-up process using Hessian or unidirectional cloth, the resin/jute cloth ratio varies between 1.25:1 and 1.85:1. Heavy twill is harder to impregnate than other forms of reinforcement. However, twill is itself much heavier than Hessian or unidirectional cloth, and the resin/jute cloth ratio generally works out to about 1:1. The properties of JRP have been studied by Phillip[71] in a paper in the EM&D journal. The strength of the composites depends on the type of jute cloth reinforcement, and the choice of the reinforcement fixes the resin/jute cloth ratio. The flexural strengths for various types of reinforcements are shown in Table 24. The tensile strengths of three of the important types of composites listed in Table 24 are given in Table 25. As might be expected, unidirectional cloth shows considerably reduced strengths in the weft direction, as can be seen in both Tables 24 and 25. The flexural modulus of JRP composites reinforced with Hessian is about 1×10^6 psi (6.89 GPa).

Finally, the effects of environment on the behavior of JRP composites has been briefly studied,[71] particularly the effects of water immersion and weathering. In the case of water immersion, tests show that degradation takes place because of the absorption of water through the edges of the composite. The provision of a suitable gel-coat and the sealing of the edges of the laminates go a long way towards reducing degradation in laminates immersed in water. This is very clearly shown in Table 26. This shows the reduction in flexural strength after 6½ weeks and 13 weeks of immersion in water at 20°C (68°F). The flexural strength of the laminate has been

Table 25

Reinforcement	Resin	Gel Coat	Tensile Strength in Warp Direction (lbf/in.^{2})	Tensile Strength in Weft Direction (lbf/in.^{2})
Hessian, warps in line	Polyester	no	10295	6505
Unidirectional, warps in line	Polyester	no	17716	1512
Hessian, warps in line with glass mat outer layers	Polyester	no	11436	8693

Table 26

				Flexural Strength as a Percentage of the Original Flexural Strength of the Laminate	
Reinforcement	Resin	Gel Coat	Type of Edge	After 6½ Weeks Immersion in Water	After 13 Weeks Immersion in Water
Glass, woven roving	Polyester	no	unsealed	97.48	—
Glass, woven roving	Polyester	yes	sealed	98.94	96.64
Hessian	Polyester	no	unsealed	83.78	81.81
Hessian	Polyester	yes	sealed	93.56	87.12

expressed as a percentage of the original flexural strength as determined from nonimmersed specimens.

Natural weathering tests on samples of JRP composites tested to a period of about 10^4 hours show a reduction in flexural strength, due to weathering, and this is shown in Table 27. The flexural strength has been expressed as a percentage of the original flexural strength of the composites determined from samples retained in a conditioned atmosphere. It can be seen that in samples with all jute reinforcement, there is a loss of about 18% in the flexural strength. The loss is apparently due the degradation of the composite by light. Pigmenting the gel-coat resin reduced the loss in strength to about 10%. Wells et al.[29] have given the mechanical properties of laminates reinforced with square weave fabric and felt with varying fiber volume content, and these are shown in Table 28.

Table 27

Reinforcement	Resin	Gel Coat	Flexural Strength at the End of 64 Weeks of Weathering Expressed as a Percentage of the Original Flexural Strength of the Laminate
Jute cloth (various types)	Polyester	yes	82.8
Jute cloth (various types) with glass mat outer layers	Polyester	yes	100.1

Table 28

Fabric	Fiber Volume (Content)	Flexural Strength (MN/m^2)	Flexural Modulus (GN/m^2)	Density $10^3 kg/m^3$
Square weave, jute hessian	25%	65 ± 5	6.0 ± 3	1.2
Square weave, jute hessian	50%	90 ± 4	8.0 ± 1	1.2
Jute felt	60%	100 ± 5	10.0 ± 0.5	1.2

2.14 ASBESTOS REINFORCED POLYESTER

Asbestos reinforcement can be used with polyester resin in almost every technique applicable to glass fiber chopped strand mat reinforcement. For contact molding, it is advisable to use the asbestos mat in combination with glass reinforcement. For press molding, this is not necessary.

Because of the high surface areas of the fibers, it is necessary to use low viscosity polyester resins with asbestos reinforcement. The resin/asbestos fiber ratio is also kept quite high to ensure thorough impregnation. For contact molding a resin/asbestos fiber ratio of 3:1 or 4:1 is normal. Thixotropic additives are not necessary, even on vertical surfaces. The typical short term properties of asbestos reinforced polyester (ARP) with crocidolite fiber reinforcement will now be briefly discussed. They were determined by Rassell[31] on composites using mat reinforcement.

The thickness of the composite is a function of the resin/reinforcement ratio or the reinforcement content and is shown in Fig. 41. The flexural strength, flexural modulus, tensile strength, and tensile modulus, once again expressed as functions of the reinforcement content, are shown in Figs. 42–

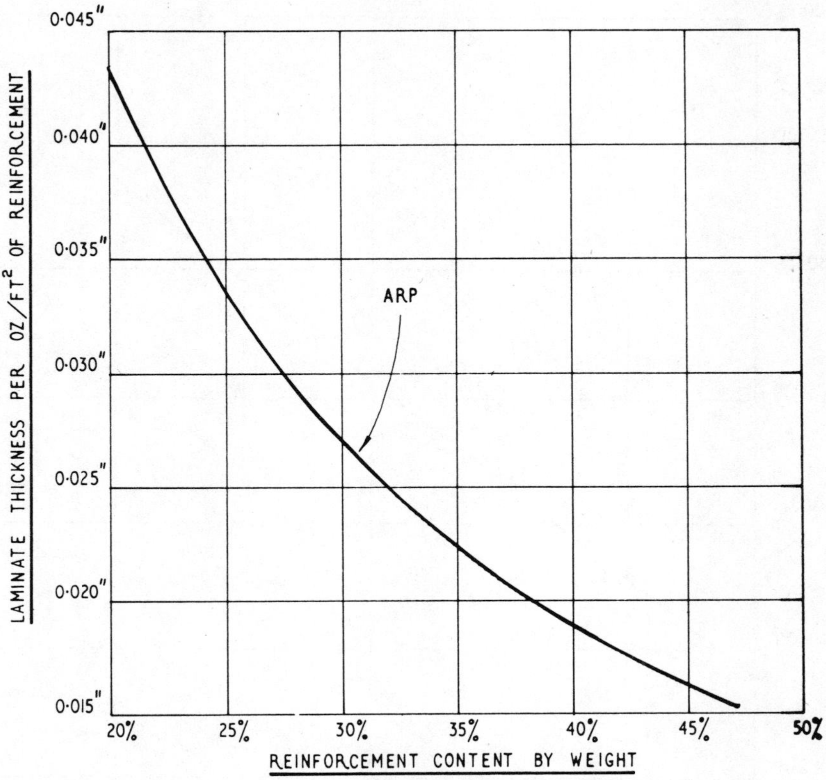

Fig. 41

45. In Fig. 45 the tensile modulus of FRP, using chopped strand mat reinforcement, expressed as a function of the reinforcement content has also been shown. The superior properties of ARP can be easily seen.

One of the advantages of using ARP composites is their good chemical resistance, which is superior to that of FRP. The percentage retention of flexural strength with time, for both crocidolite and amosite mat composites, has been shown in Figs. 46–52. The polyester resin used was Atlac Chemical Atlac 382. The composites were continuously boiled in the various chemical solutions such as 20% H_2SO_4, 5% H_2SO_4, 20% HCl, and so on, and the flexural strength after periods of immersion was compared with the flexural strength of nonimmersed composites.

The values of the constants to be used in the creep equations, for ARP, are given in Table 18.

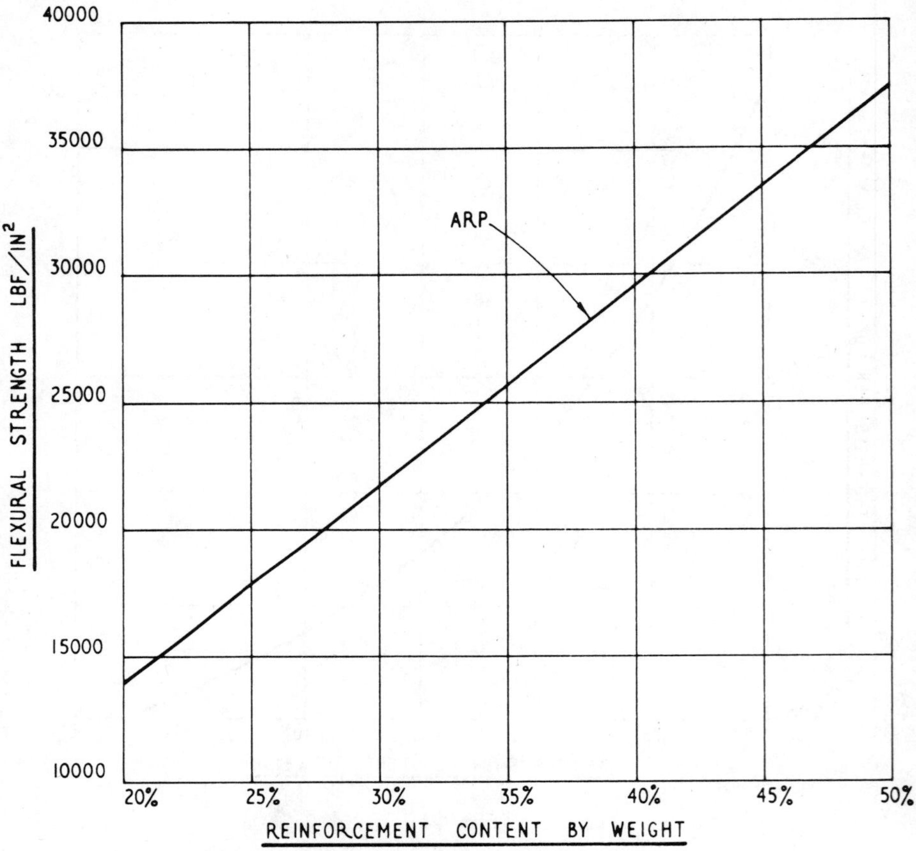

Fig. 42

2.15 GLASS FIBER REINFORCED EPOXY COMPOSITES

The epoxy resins are costlier than the polyesters and, hence, are less frequently used for building applications, such as roof sheeting. However, they have excellent mechanical properties, particularly at elevated temperatures, and find good application for high-performance composites, particularly with filament winding processes for the manufacture of pressure vessels, rocket motor casings, and so on. The high-performance composites are considered separately in Section 2.18.

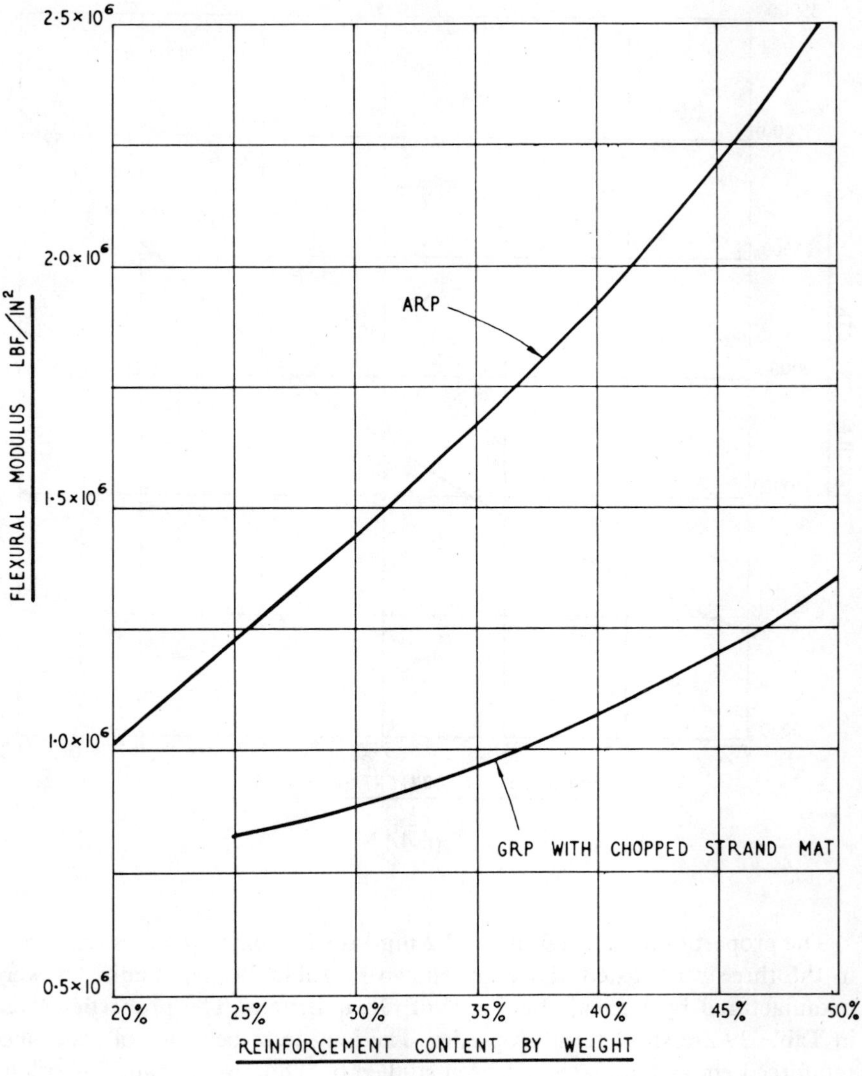

Fig. 43

Glass fiber reinforcement for epoxies can be of three main types:

1. All the roving reinforcement in one direction.
2. Equal roving reinforcement in two directions at right angles to one another.
3. Fine weave cloth reinforcement.

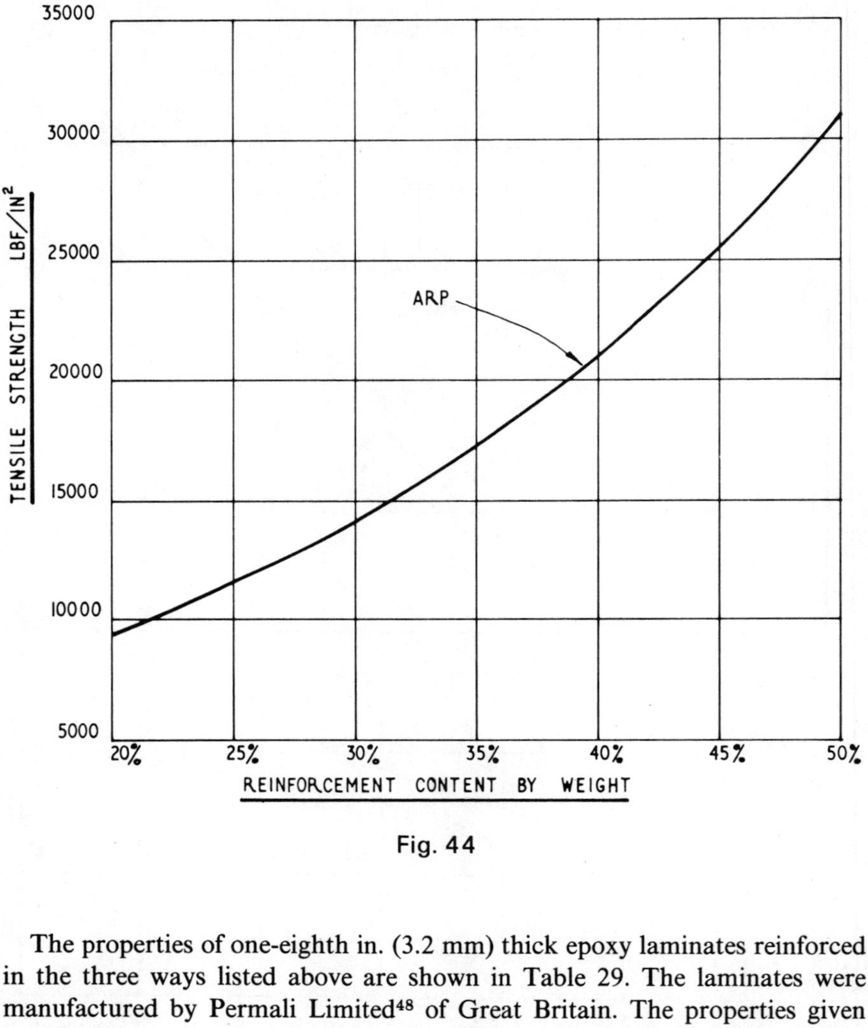

Fig. 44

The properties of one-eighth in. (3.2 mm) thick epoxy laminates reinforced in the three ways listed above are shown in Table 29. The laminates were manufactured by Permali Limited[48] of Great Britain. The properties given in Table 29 are short term properties. The long term behavior of glass fiber reinforced epoxy laminates has been studied by Thompson,[59] and his results are shown in Fig. 53. Figure 53 shows the reduction in ultimate tensile strength with time for two laminates called Epoxide "A" and Epoxide "E" laminates. The Epoxide "A" laminate was made with a hand lay-up process. It had a short term ultimate tensile strength of 42500 psi (293 MPa) and a resin content of 51%. Further, the epoxide resin was a highly plasticized one. This led to very poor strength retention characteristics, as can be seen in Fig. 53. The Epoxide "E" laminate, on the other hand, was preimpregnated and press molded, and had an ultimate tensile strength of 46300 psi (319

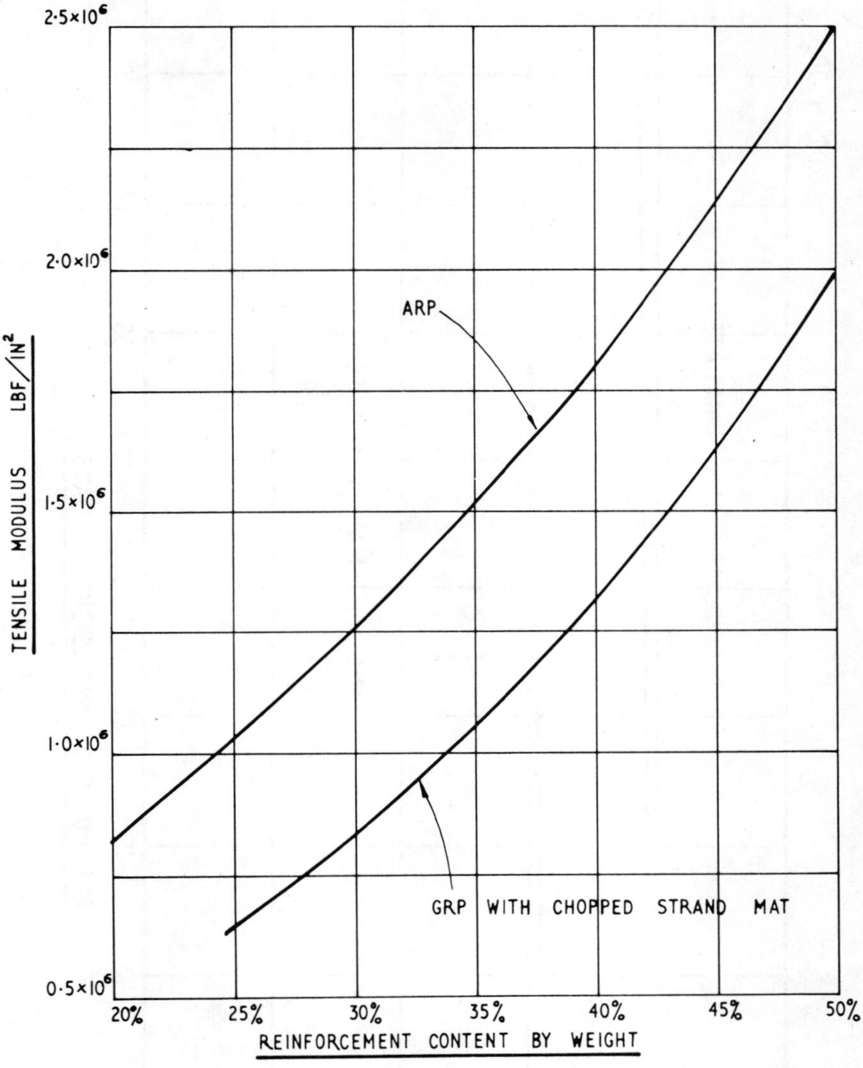

Fig. 45

MPa). It had a resin content of only 29%. The Epoxide "E" laminate had very good strength retention characteristics, and Thompson[59] concludes that, designing for 100,000 hours, the ultimate tensile strength can be assumed to be about 65% of the short term ultimate tensile strength, provided the resin has not crazed. If the resin crazes, moisture attack can considerably reduce the strength, as was seen in Section 2.12. If the curve in Fig. 53 is

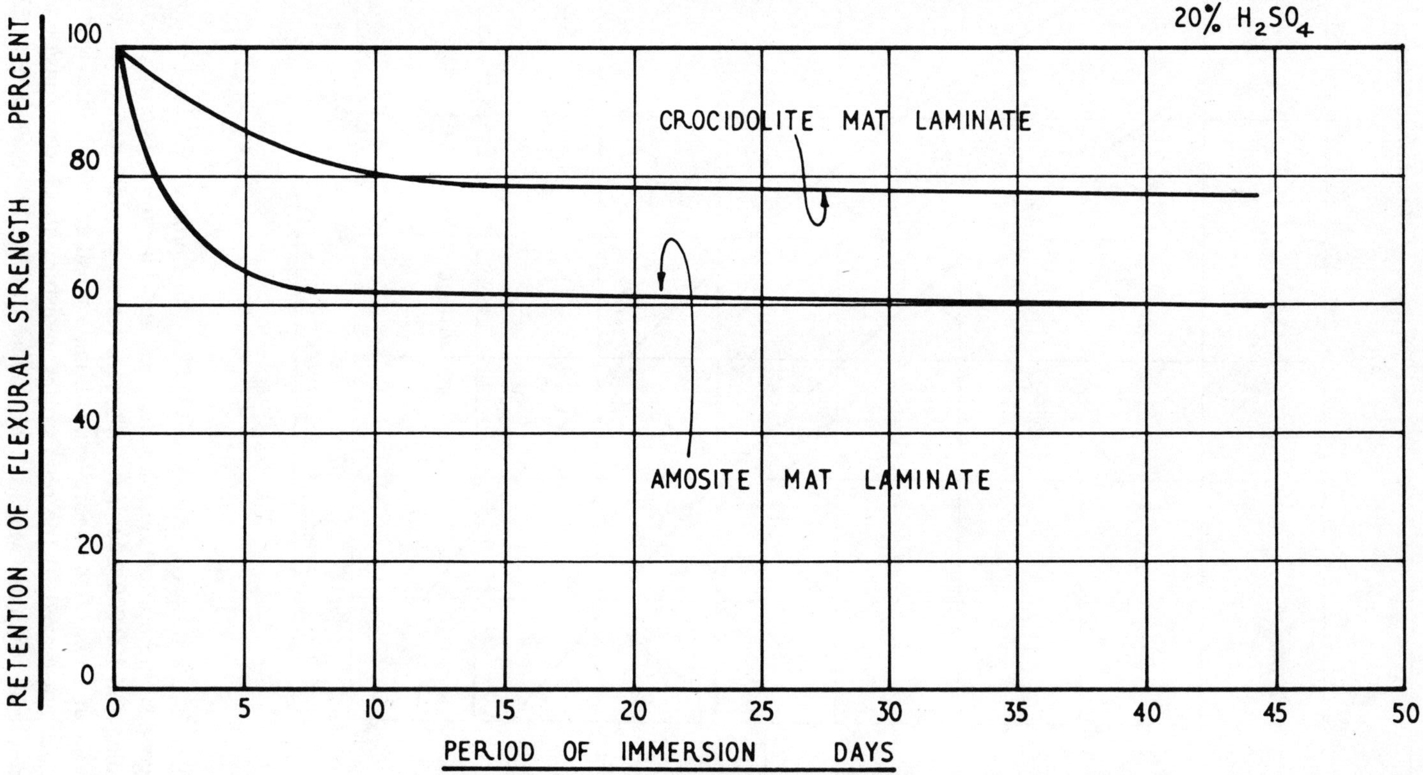
20% H_2SO_4
RETENTION OF FLEXURAL STRENGTH PERCENT
100
80
60
40
20
0
CROCIDOLITE MAT LAMINATE
AMOSITE MAT LAMINATE
0
5
10
15
20
25
30
35
40
45
50
PERIOD OF IMMERSION DAYS

Fig. 46

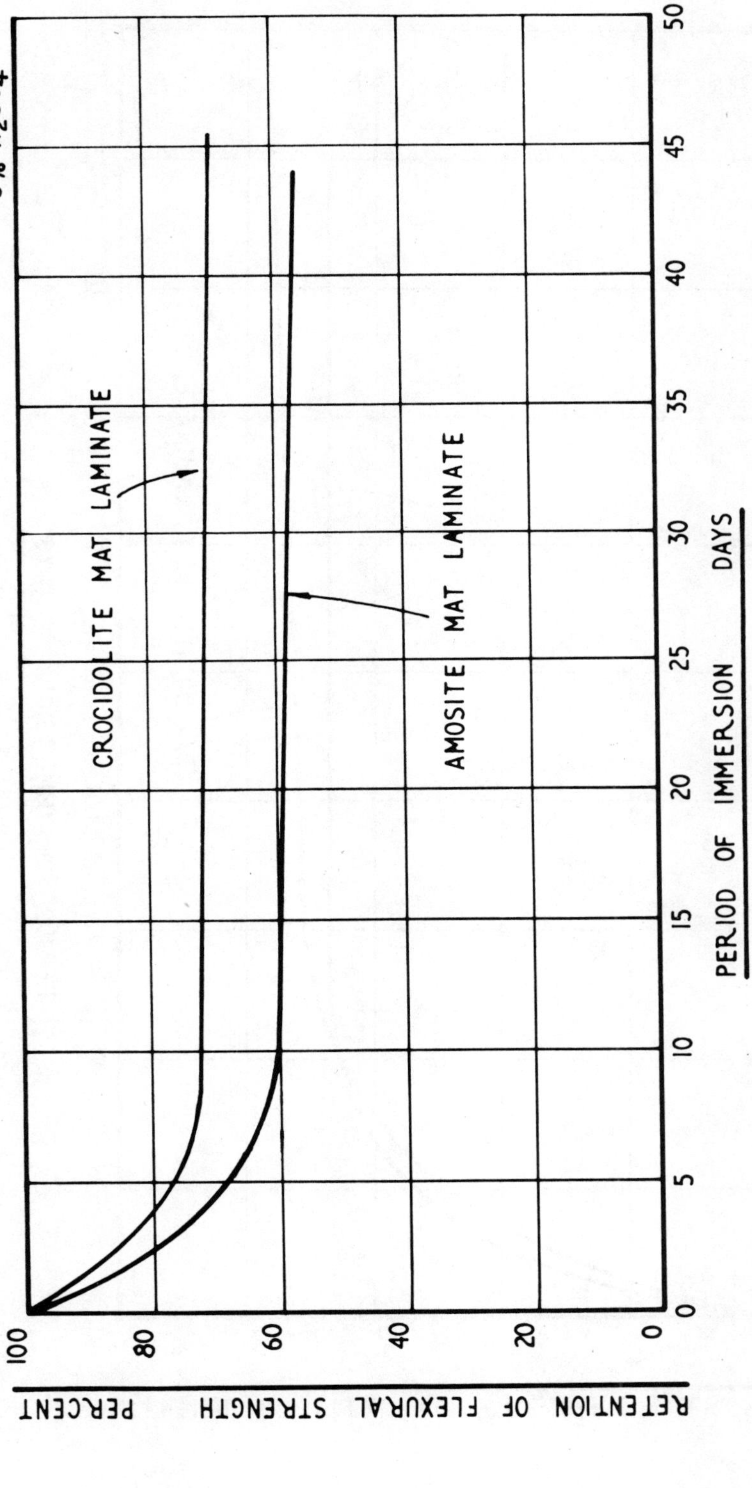
5% H2SO4
CROCIDOLITE MAT LAMINATE
AMOSITE MAT LAMINATE
PERIOD OF IMMERSION DAYS
0
5
10
15
20
25
30
35
40
45
50
PERCENT RETENTION OF FLEXURAL STRENGTH
100
80
60
40
20
0

Fig. 47

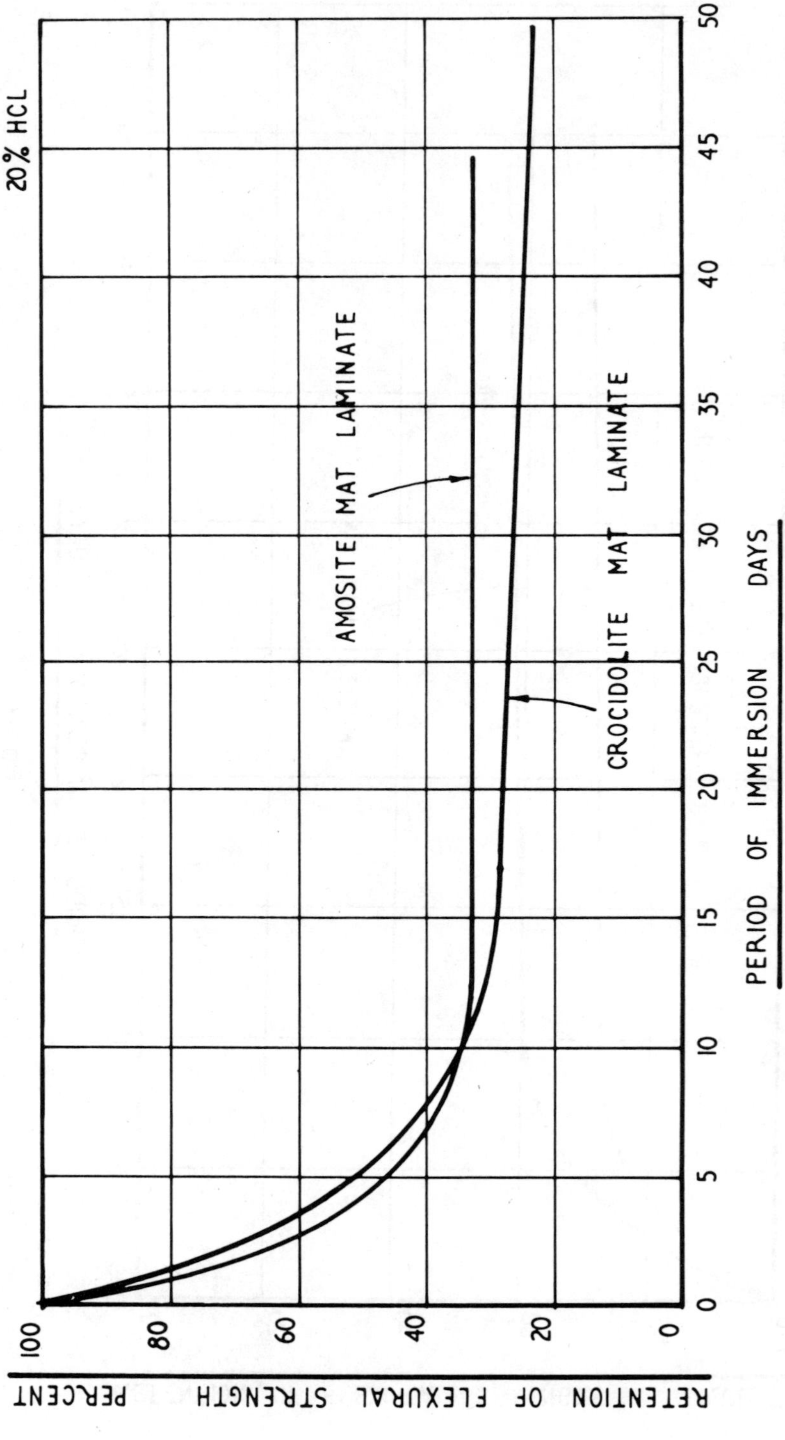
20% HCL
AMOSITE MAT LAMINATE
CROCIDOLITE MAT LAMINATE
RETENTION OF FLEXURAL STRENGTH PERCENT
100
80
60
40
20
0
0
5
10
15
20
25
30
35
40
45
50
PERIOD OF IMMERSION DAYS

Fig. 48

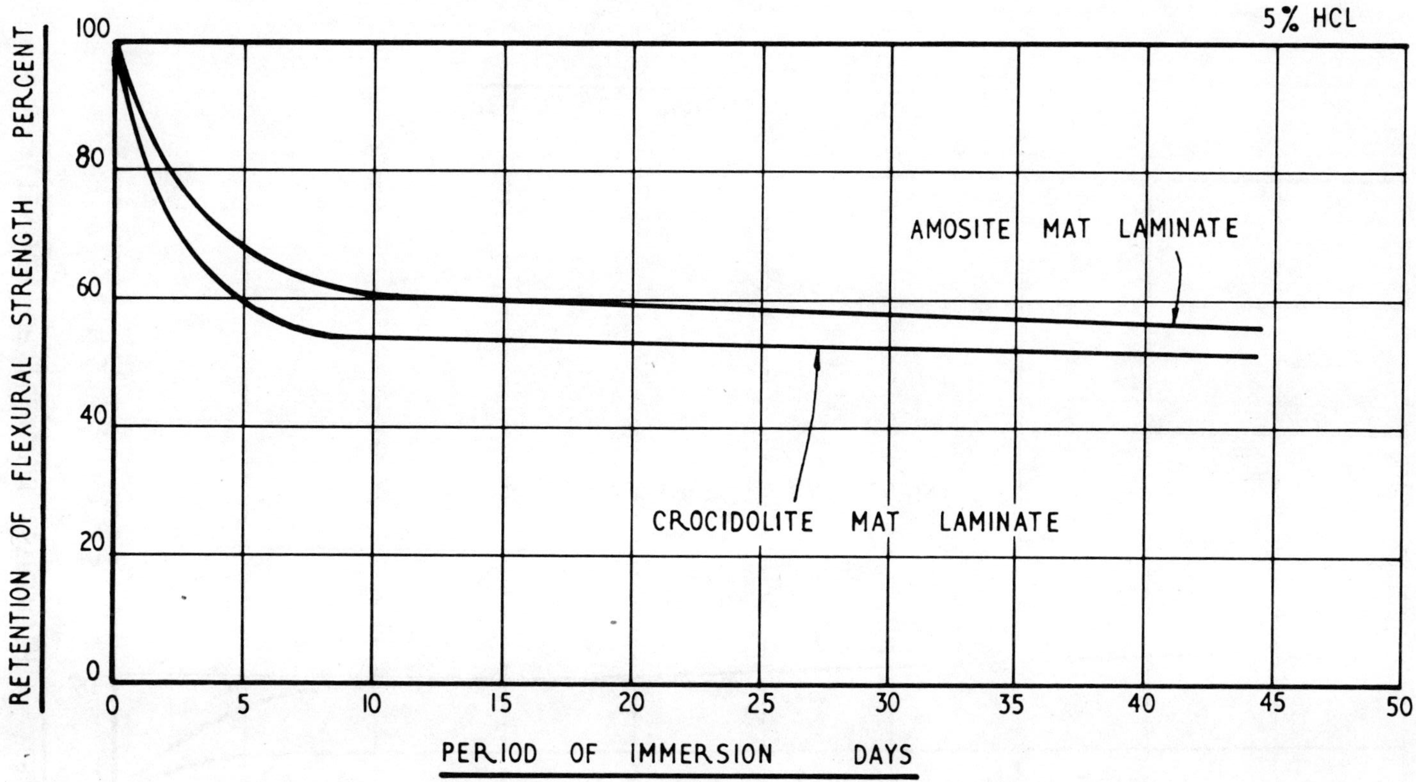

5% HCL
AMOSITE MAT LAMINATE
CROCIDOLITE MAT LAMINATE
RETENTION OF FLEXURAL STRENGTH PERCENT
100
80
60
40
20
0
0
5
10
15
20
25
30
35
40
45
50
PERIOD OF IMMERSION DAYS

Fig. 49

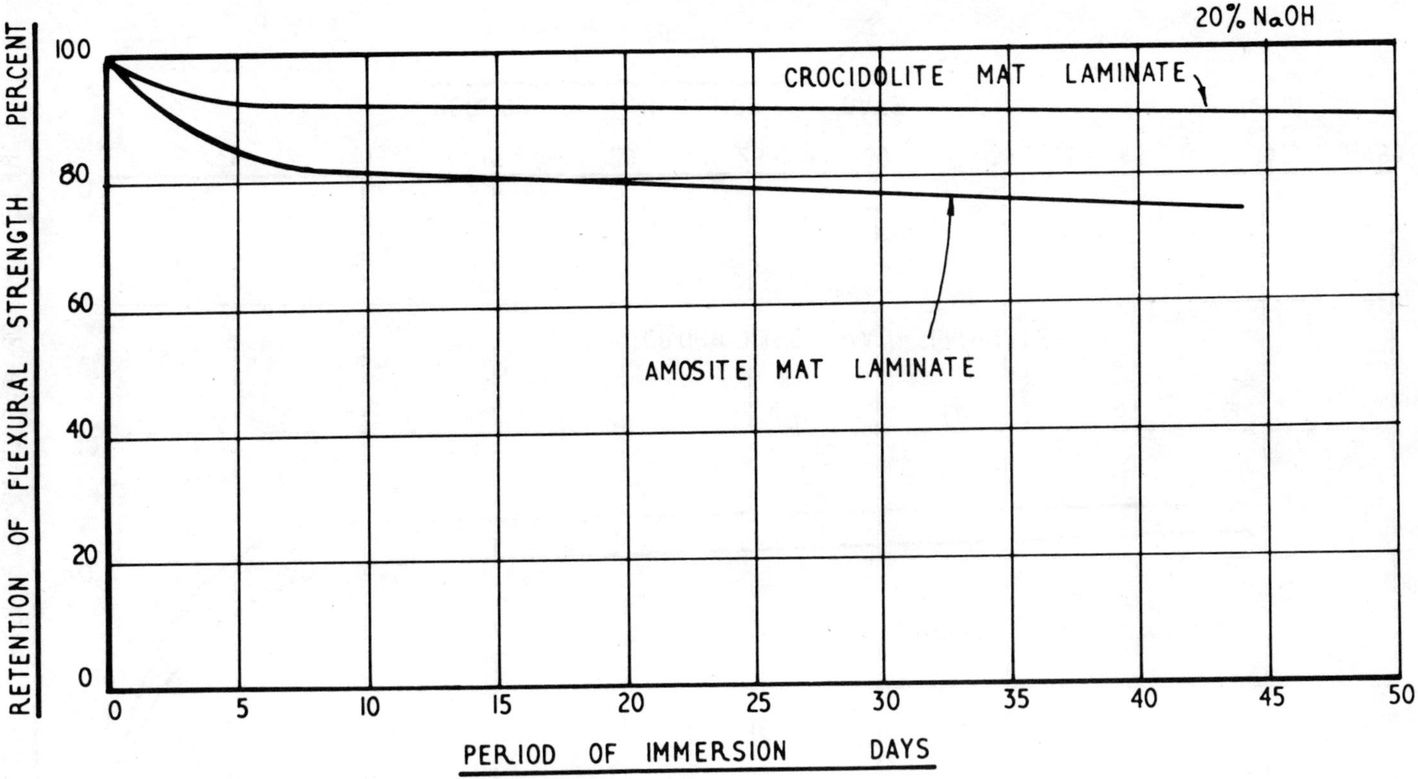
20% NaOH
CROCIDOLITE MAT LAMINATE
AMOSITE MAT LAMINATE
RETENTION OF FLEXURAL STRENGTH PERCENT
100
80
60
40
20
0
0
5
10
15
20
25
30
35
40
45
50
PERIOD OF IMMERSION DAYS

Fig. 50

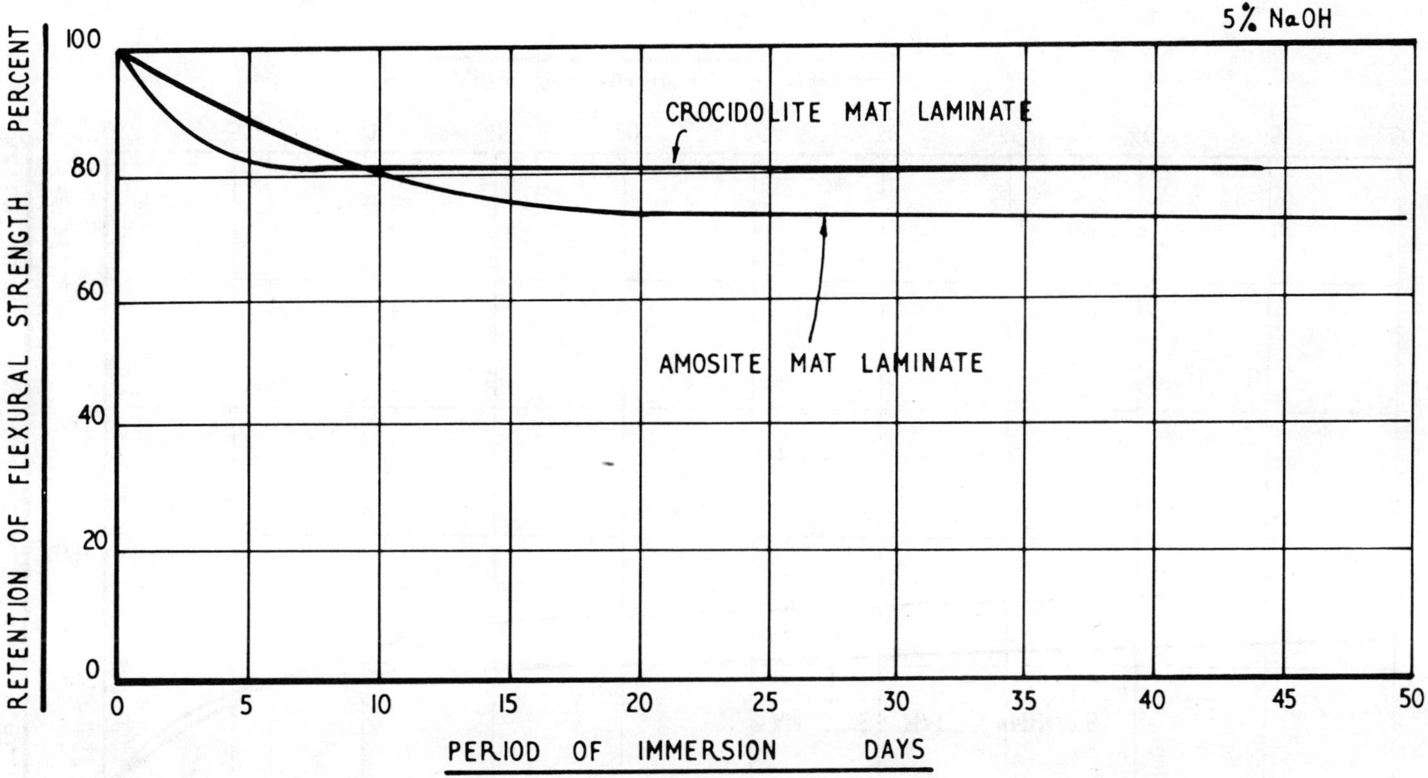
5% NaOH
CROCIDOLITE MAT LAMINATE
AMOSITE MAT LAMINATE
RETENTION OF FLEXURAL STRENGTH PERCENT
100
80
60
40
20
0
0
5
10
15
20
25
30
35
40
45
50
PERIOD OF IMMERSION DAYS

Fig. 51

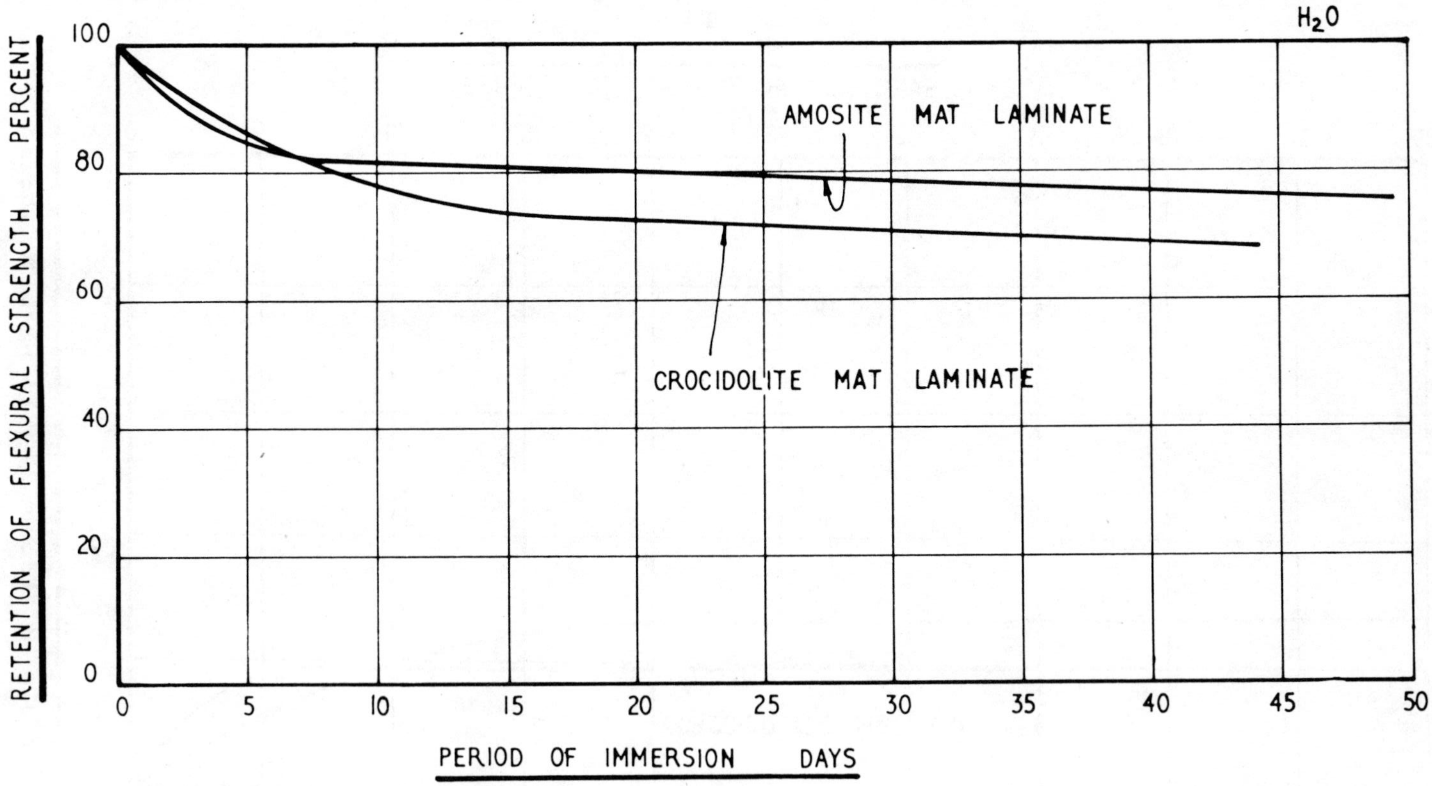
H$_2$O
AMOSITE MAT LAMINATE
CROCIDOLITE MAT LAMINATE
RETENTION OF FLEXURAL STRENGTH PERCENT
100
80
60
40
20
0
0
5
10
15
20
25
30
35
40
45
50
PERIOD OF IMMERSION DAYS

Fig. 52

Table 29

	Type 1	Type 2	Type 3
Specific gravity	2.10	2.10	1.76
Flexural strength, lbf/in.^{2}			
longitudinally	185000	110000	60000
transversely	8000	82000	60000
Flexural modulus, lbf/in.^{2}			
longitudinally	7×10^6	4.5×10^6	2.5×10^6
Tensile strength, lbf/in.^{2}			
longitudinally	100000	70000	35000
Compressive strength, lbf/in.^{2}			
flatwise	17000	90000	59000
Shear strength, lbf/in.^{2}			
longitudinally	30000	30000	19000
transversely	4000	30000	19000
Impact strength, ft. lbf/½ in. notch, parallel to the laminations	40	25	5

Note: The actual properties of the laminate will depend on the resin used and the exact glass content.

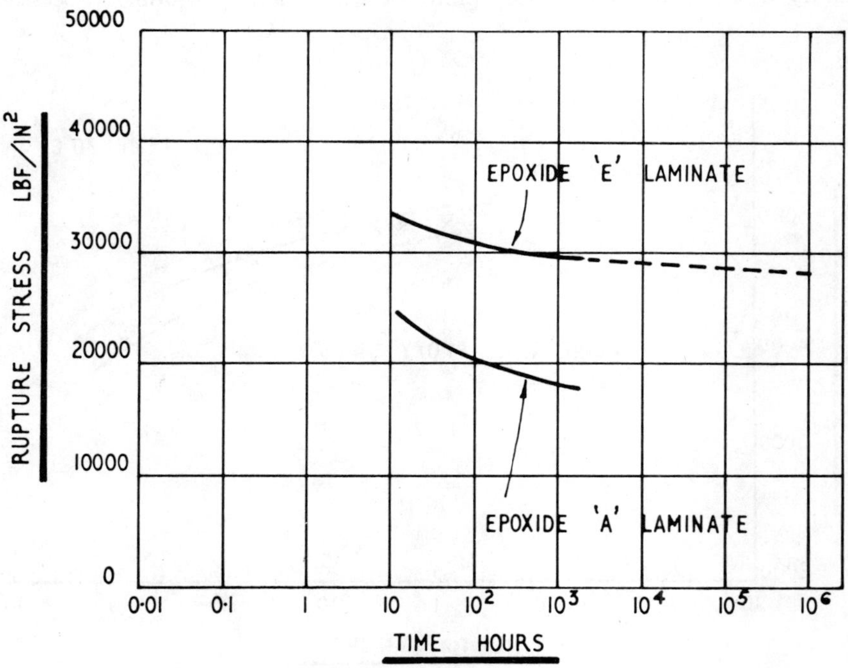

Fig. 53

extrapolated to 10^6 hours (114 years), it can be seen that the ultimate tensile strength may still be assumed to be about 28,000 psi (193 MPa), which is about 60% of the short term ultimate tensile strength of the material. Bax[62] has considered the rupture of glass fiber reinforced epoxy pipes, and his results are shown in Fig. 54.

The creep of glass fiber reinforced epoxy composites has also been studied by Findley,[64] and the values of the constants to be used in the creep Equation 35 are given in Table 18.

It has been mentioned that epoxy composites give a better performance at elevated temperatures than polyester composites. Lant[72] has studied the effects of temperature on the flexural strength and the flexural modulus of epoxy composites reinforced with glass cloth and in the dry condition. His results are shown in Fig. 55. Lant[72] concluded that neither curve indicates a definite softening point, and that the flexural strength results suggest that, provided the stress level is moderate, the system could be used to 200°C (392°F) or more. The author feels, however, that, for structural applications, a temperature of about 120°C (248°F) should not be exceeded for the system shown.

Epoxy resins have great suitability in high-performance applications, particularly when reinforced with graphite, or with filament winding processes,

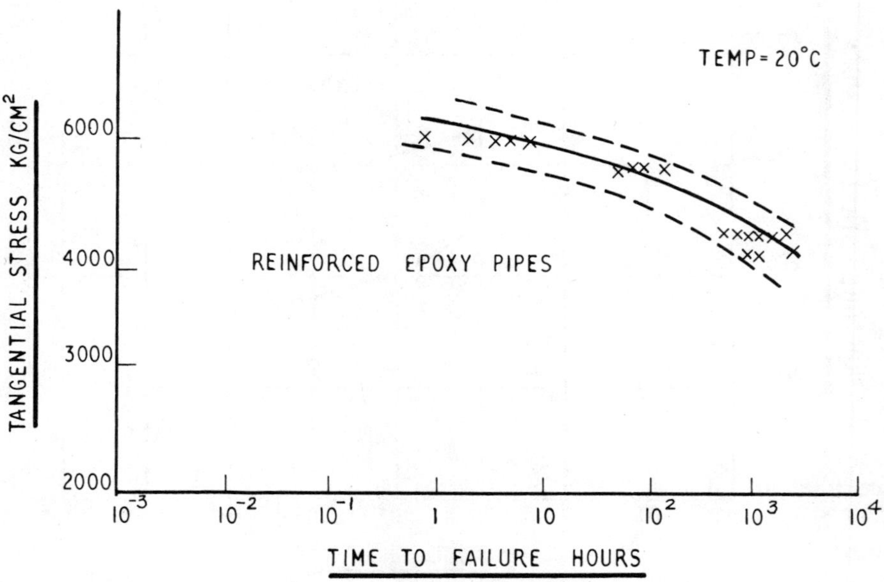

Fig. 54

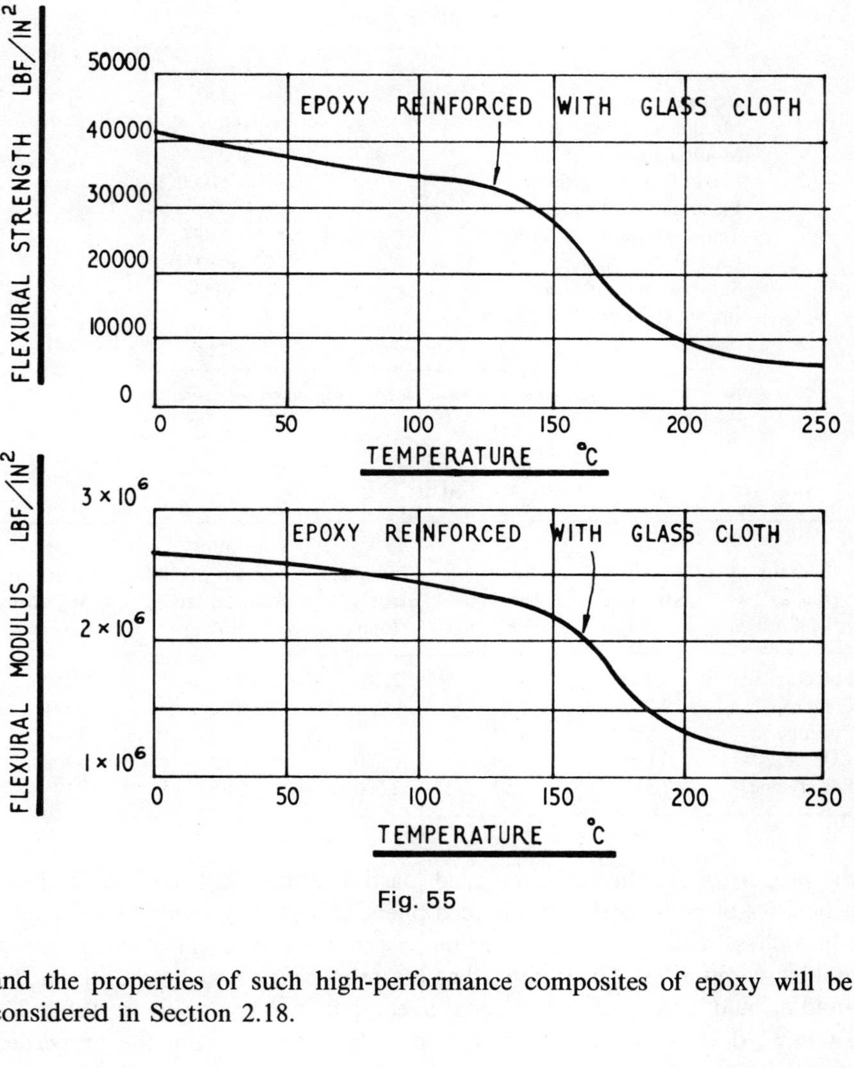

Fig. 55

and the properties of such high-performance composites of epoxy will be considered in Section 2.18.

2.16 PHENOLIC, MELAMINE, AND SILICONE RESIN COMPOSITES

Phenolic resin composites,[73] particularly when reinforced with asbestos, are very suitable for high-temperature work and can be remarkably resistant to chemical attack. The typical short term properties of phenolic resin composites reinforced with mat asbestos reinforcement[31] are shown in Table 30. The properties of the composite vary with the type of phenolic resin and the method of manufacture. Reis[34] gives the properties of phenolic resin reinforced with chrysotile asbestos, and these are shown in Table 31. Also shown are

Table 30

Specific gravity	1.8–2.1
Resin content	35%–45%
Moulding temperature, °C	145–150
Moulding pressure, lbf/in.^{2}	500–800
Flexural strength, lbf/in.^{2}	38000–62000
Flexural modulus, lbf/in.^{2}	$2.5 \times 10^6 - 3.5 \times 10^6$
Tensile strength, lbf/in.^{2}	26000–32000
Compressive strength, lbf/in.^{2}	35000–64000
Shear strength, lbf/in.^{2}	20500–25000
Impact strength, ft. lbf/½ in. notch	6–7

Note: The actual properties of the laminate will depend on the resin used and the exact glass content.

Table 31

Resin (Asbestos Filled)	Flexural Strength (psi)	Flexural Modulus (psi $\times 10^5$)	Tensile Strength (psi)	Impact Strength Notched Izod (ft lb/in.)	Heat Distortion Temperature (°F)
Polyethylene	5,840	4.21	3,380	0.6	157
Polypropylene	9,700	5.8	—	0.7	221
Phenolic	13,500	14.4	—	0.20	—
ABS	8,290	13.9	6,340	0.58	208
Polystyrene	7,410	5.97	2,500	0.56	194

the properties of other asbestos filled plastics. Price and Lucy[74] studied the properties of both asbestos reinforced phenolic and glass reinforced phenolic composites, with particular attention to the molding cycle of compression molded composites. They found that the properties vary, depending on the molding and cure cycles, but some average values are given in Table 32. Table 32 does allow some comparison to be made between the properties of these composites.

Phenolic resin can also be reinforced with mica. Xanthos, Hawley, and Antonacci[41] investigated the use of suzorite mica as reinforcement for plastics. Some properties of compression molded phenolic composites reinforced with suzorite mica and asbestos are compared in Table 33.

The long term behavior of phenolic resin composites reinforced with glass cloth has been studied by Thompson,[59] and his results are shown in Fig. 56. The composite was preimpregnated and press molded, and had an ultimate short term tensile strength of 59,000 psi (407 MPa) and a resin content of

Table 32

Composite	Flexural Strength (MN/m^2)	Flexural Modulus (GN/m^2)	Hardness, (Rockwell F)
Phenolic/asbestos	132	15.3	76.3
Phenolic/glass	181	17.6	51.3

Note: These are average values for a particular molding cycle, as-molded. To convert to psi, divide N/m^2 by 6895.

Table 33

Properties of Compression Molded Phenolic Composites	Reinforcement		
	Suzorite Mica 60-H	Suzorite Mica 325-H	Calidria Asbestos
% Reinforcement, by weight	50	50	50
Flexural strength, psi $\times 10^3$	18.0	13.9	16.7
Flexural modulus, psi $\times 10^6$	4.24	3.11	1.89

28.5%. The composite has very poor strength retention characteristics, and any extrapolation of the curve to 10^5 or 10^6 hours would be ill-advised.

Melamine and silicone resin composites are nonflammable and are excellent for use at high temperatures. However, high cost and processing requirements limit them to aerospace applications. The typical short term properties of melamine and silicone resin composites reinforced with fine weave glass cloth reinforcement[48] have been shown in Table 34. Also given are the typical short term properties of phenolic resin composites reinforced with fine weave glass cloth.

Finally, the creep behavior of melamine and silicone resin composites reinforced with glass fabric has also been studied by Findley,[64] and the constants to be used in the creep Equation 35 are given in Table 18.

2.17 REINFORCED VINYL ESTER COMPOSITES

Vinyl ester resins have become widely accepted because of their superior chemical resistance and increased toughness and good resiliency. The increase in physical properties and toughness, as compared to the polyesters, can be attributed to the epoxy resin backbone of the vinyl ester resin. Vinyl esters are used for sheet molding composites or compounds (SMC), and this is

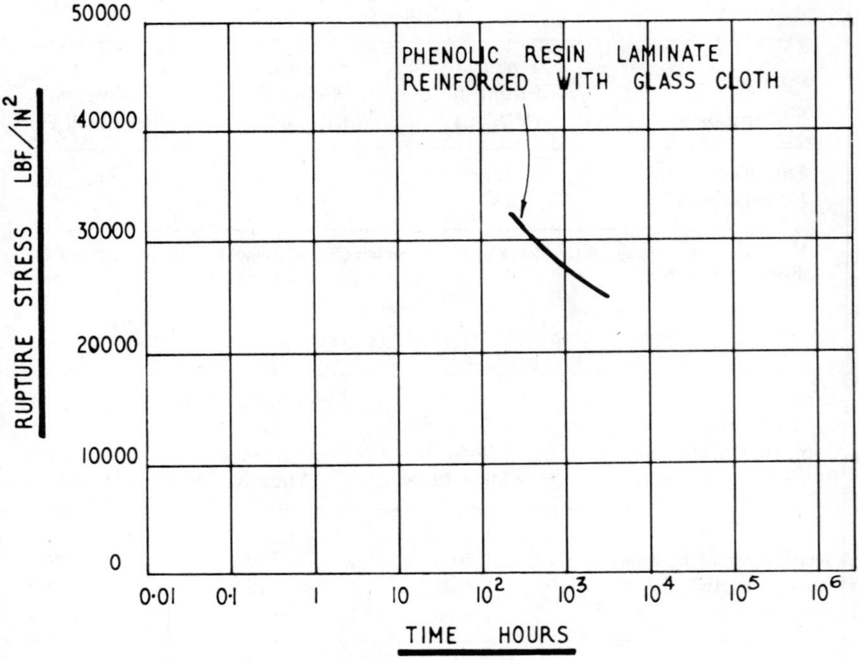

Fig. 56

Table 34

	Phenolic Resin Laminate	Melamine Resin Laminate	Silicone Resin Laminate
Reinforcement	Fine Weave Glass Cloth	Fine Weave Glass Cloth	Fine Weave Glass Cloth
Specific gravity	1.86	1.88	1.78
Flexural strength, lbf/in.^{2}	45000	45000	20000
Flexural modulus, lbf/in.^{2}	2.2×10^6	2.0×10^6	—
Tensile strength, lbf/in.^{2}	27000	25000	22000
Compressive strength, lbf/in.^{2}	60000	45000	38000
Shear strength, lbf/in.^{2}	20000	13000	10000
Impact strength, ft lbf/½ in. notch	5	4	6

Note: The actual properties of the laminate will depend on the resin used and the exact glass content.

discussed in Section 2.19. However, the properties of vinyl ester composites were studied by Thomas and Enos[75] for very high strength SMC in automotive structures: their results are given in Table 35. The resin, XD-9013, was reinforced with 50% glass, and a 25% calcium carbonate filler was used. Three types of glass reinforcement were used. Random 1 in. (25 mm) and 2 in. (51 mm) glass gave lesser values than a combination of 30% continuous glass with 20% random 1 in. (25 mm) glass. The increase of tensile and flexural strength and moduli with percentage glass reinforcement is shown, respectively, in Figs. 57 and 58. The behavior of the composites at elevated temperatures, with and without 25% calcium carbonate filler, is shown, in Figs. 59 and 60. The behavior at elevated temperatures of vinyl ester composites has also been studied by Breitigam and Brass[76] on Shell Chemical Co. Epocryl resins reinforced with mat and woven roving. The laminates were one-quarter in. (6 mm) thick. The reduction in strengths and moduli with temperature are shown in Figs. 61–66, while the creep behavior is shown in Figs. 67 and 68.

The chemical resistance of vinyl ester composites with new resins was considered by Szymanski and Vaccarella,[77] and the percentage reduction in flexural strength and modulus after 1 year of exposure to various chemical reagents is shown in Fig. 69 and 70. The resins used were Hetron 901, 902, and 903 and the reinforcement was glass. Jaspers and Wynes[78] studied Diacryl resins, and some results after 2 years of testing in various chemicals, are shown in Figs. 71 and 72.

2.18 HIGH-PERFORMANCE COMPOSITES

It has been seen in this chapter that the use of certain types of reinforcements can lead to very high strength composites. Graphite, boron, or aramid fibers

Table 35

	Type of Glass		
Property	Random 1 in. Length	Random 2 in. Length	30% Continuous Glass with 20% Random 1 in. Glass
Tensile strength, psi	25,000	29,000	35,000
Tensile modulus, psi	2.5×10^6	2.5×10^6	2.5×10^6
% Elongation	1.75	1.85	2.1
Flexural strength, psi	49,000	54,000	71,000
Flexural modulus, psi	2.5×10^6	2.6×10^6	2.5×10^6

Note: Resin XD-9013 25%, with Calcium carbonate 25%, and with 50% Glass reinforcement.

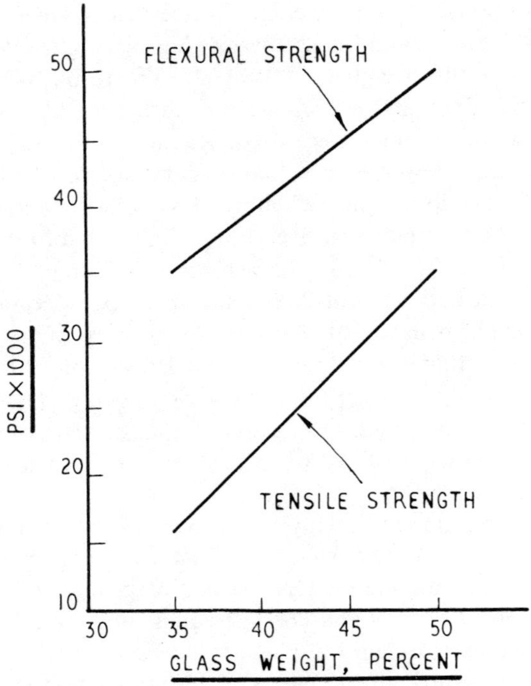

Fig. 57

used as reinforcing materials for epoxy resin all provide high-performance materials. In addition, composites may be required to function at elevated temperatures, and the use of these reinforcements with other resins such as polyimides also lead to high-performance composites. It would not be appropriate to use such high-performance materials in applications where the polyesters and glass reinforcement would be adequate, or with manufacturing processes, such as hand lay-up, where quality control would be difficult. Press molding and filament winding are very suitable for such high-performance composites.

Penn and Chiao[79] considered the use of fiber composites for energy storage flywheels, and a comparison of the specific tensile strengths and modulus of some of these high-performance composites[80] is provided in Fig. 73. It can be seen that the use of aramid fibers (Kevlar 29 or Kevlar 49) leads to very high specific strength composites, whereas the use of graphite can lead to very high specific modulus composites. The strength of graphite composites can of course be improved by using HTS graphite.

The construction of high-performance composites often involves the use

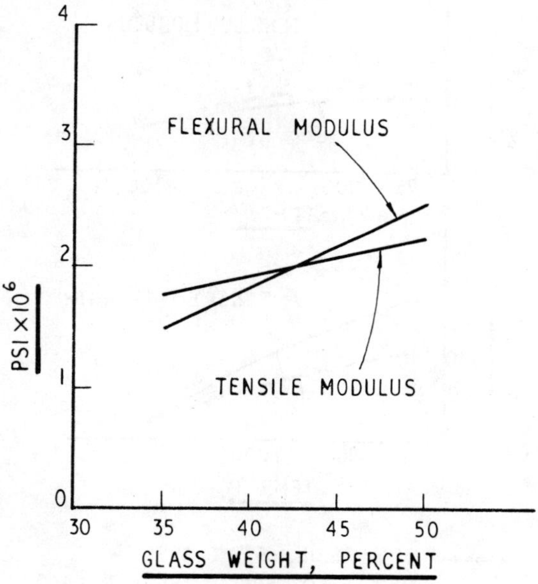

4
3
2
1
0
PSI × 10^6
FLEXURAL MODULUS
TENSILE MODULUS
30
35
40
45
50
GLASS WEIGHT, PERCENT

Fig. 58

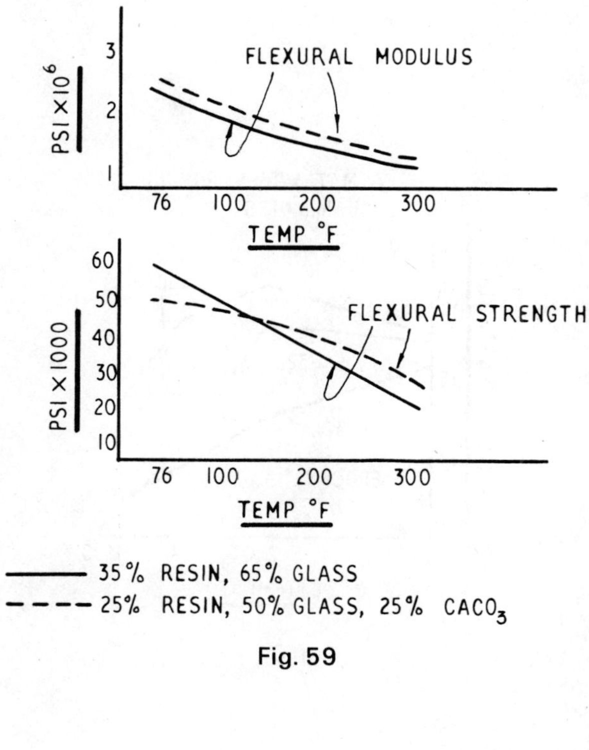

3
2
1
PSI × 10^6
FLEXURAL MODULUS
76
100
200
300
TEMP °F
60
50
40
30
20
10
PSI × 1000
FLEXURAL STRENGTH
76
100
200
300
TEMP °F
35% RESIN, 65% GLASS
25% RESIN, 50% GLASS, 25% CACO3

Fig. 59

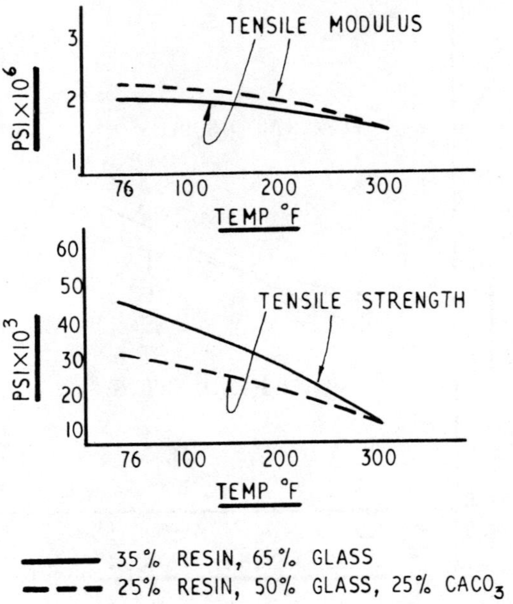
TENSILE MODULUS
PSI×10^6
3
2
1
76
100
200
300
TEMP °F
TENSILE STRENGTH
PSI×10^3
60
50
40
30
20
10
76
100
200
300
TEMP °F
35% RESIN, 65% GLASS
25% RESIN, 50% GLASS, 25% CACO3

Fig. 60

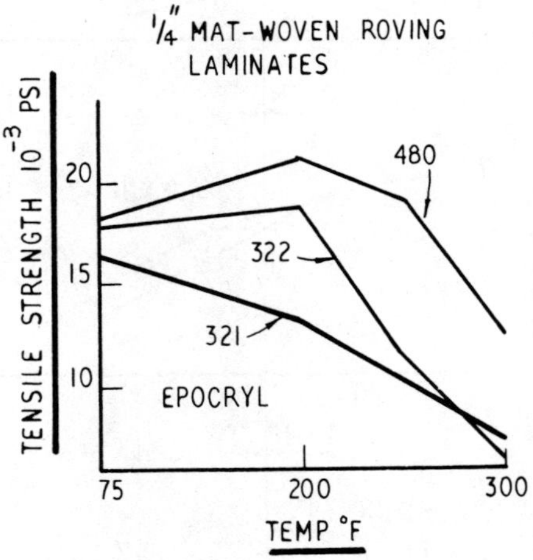
1/4" MAT-WOVEN ROVING
LAMINATES
TENSILE STRENGTH 10^-3 PSI
20
15
10
480
322
321
EPOCRYL
75
200
300
TEMP °F

Fig. 61

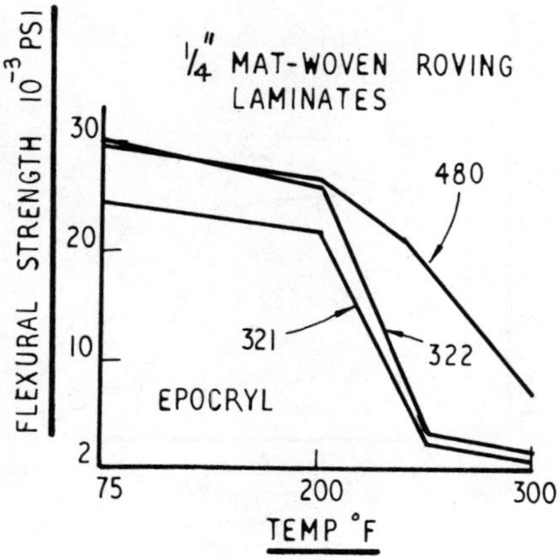

Fig. 62

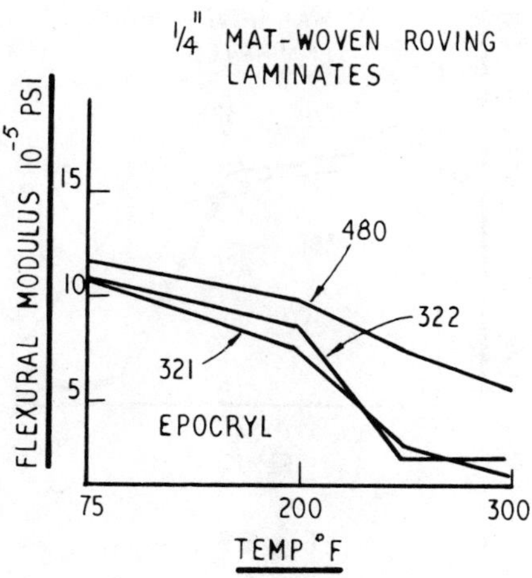

Fig. 63

of angle plies to get specific strengths in required directions. The plies can be made of preimpregnated tape that is then oriented in specific directions. For example, 13(+40, −40, 9(0), −40, +40) defines 13 plies, with 9 plies at 0° orientation and 4 plies at ±40° orientation. The ply orientation is shown in Fig. 74. Hanson and Chamis[81] studied, both theoretically and experimentally, HT-S graphite/polyimide composites with various ply orientations, and the properties along the 0° ply direction are shown in Table 36. It can be seen that 6(0) is unidirectional and the tensile and flexural strengths are

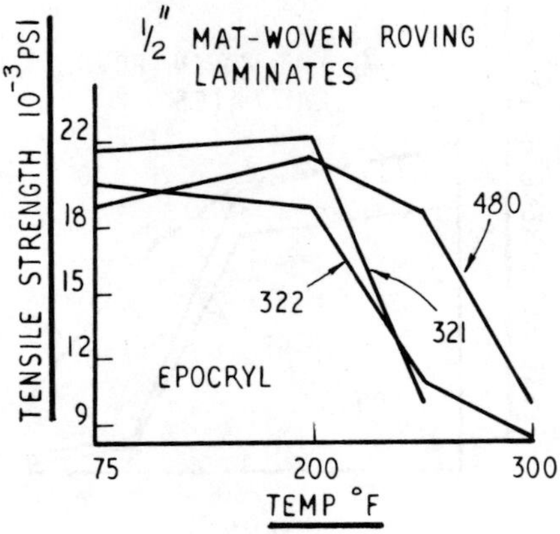

Fig. 64

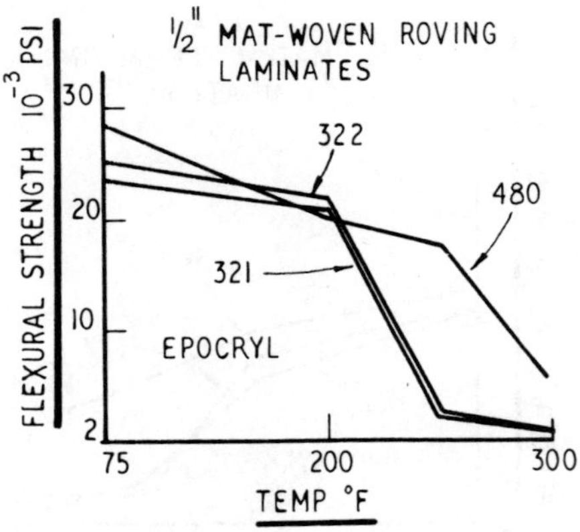

Fig. 65

very high along the 0° ply direction. However, 6(90), with the plies oriented at 90°, gives extremely low values of the properties along the 0° direction.

It is also interesting to note that Hanson and Chamis[81] found that the stress/strain diagrams for the HT-S composites, obtained by tests on the angle plied laminates, were straight lines.

The properties of some of the high-performance composites will now be considered. Epoxy resin can be reinforced with graphite, aramid, or boron fibers, and several researchers have studied all these types of composites. Liber and Daniel[82] also tested eight-ply angleplied laminates, with ply orientations at 0° and ± 45° $[\pm 45/0_2]_S$ and also found the stress/strain diagrams

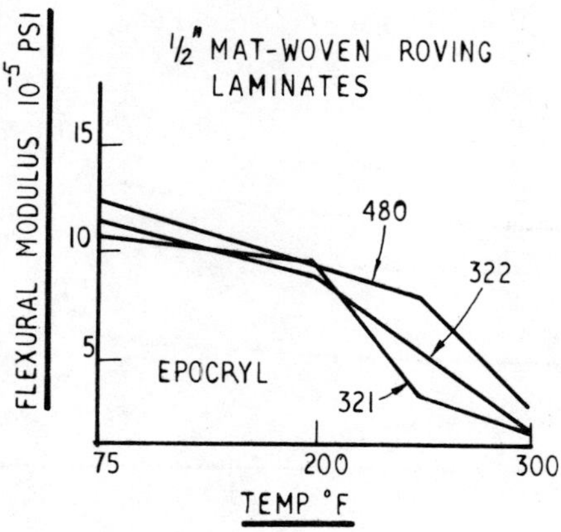

Fig. 66

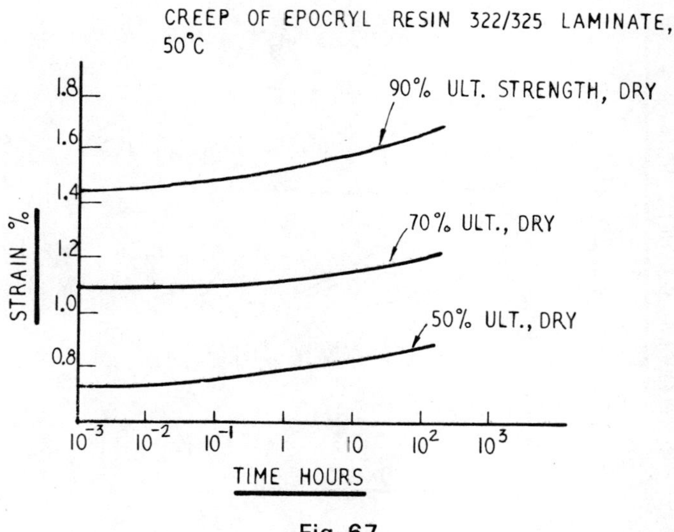

Fig. 67

to be linear. Liber, Daniel, and Chamis[83,84] extensively tested high-performance composites of graphite/epoxy, boron/epoxy, as well as graphite/polyimide, and the initial properties of the angle ply laminates are shown in Table 37. Also included are properties of S-glass/epoxy and boron/polyimide composites. All the composites were of eight ply construction.

High-performance composites of hybrid construction using two or more types of high-strength reinforcement in a single resin matrix can also be

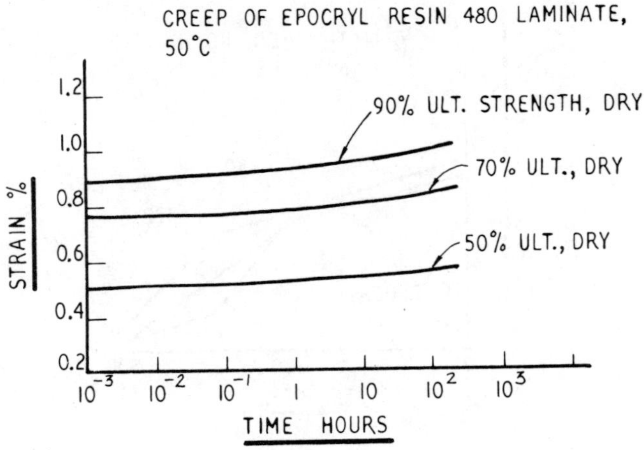

Fig. 68

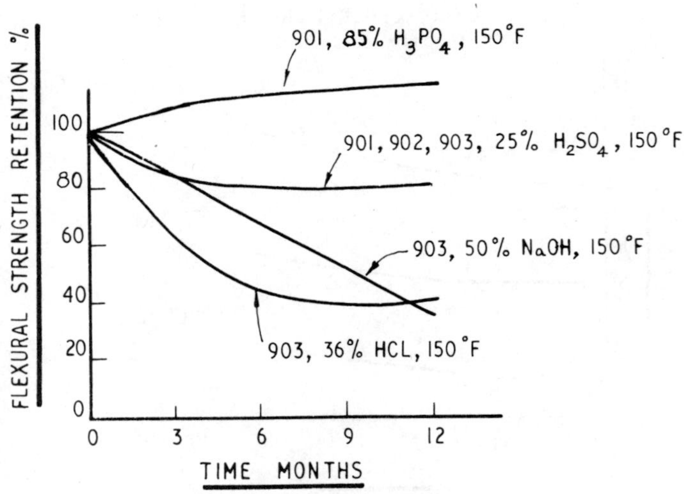

Fig. 69

used. The materials can be intimately mixed, alternately layered, or otherwise placed. The great advantage of hybrids is that they can be tailored to meet specific requirements. Liber, Daniel, and Chamis[83] also tested hybrid laminates with Kelvar 49 (aramid fiber), graphite (carbon fiber), and glass reinforcements with epoxy resin in eight ply construction, and the properties of the various configurations are shown in Table 38. Riewald and Zweben[38] tested Kevlar 29 and Kevlar 49, aramid fiber reinforcement, in combination with graphite

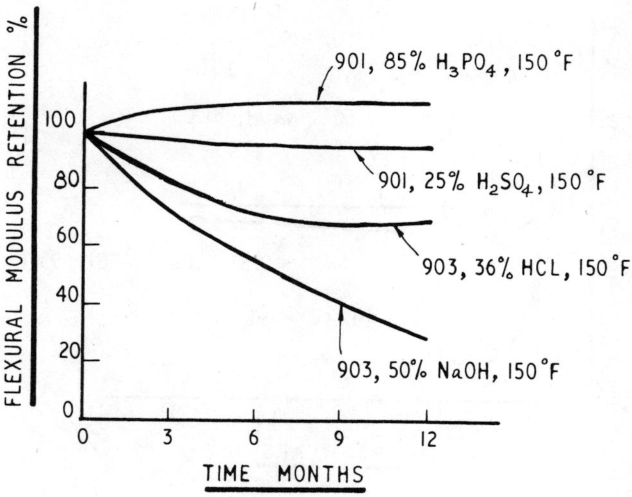

Fig. 70

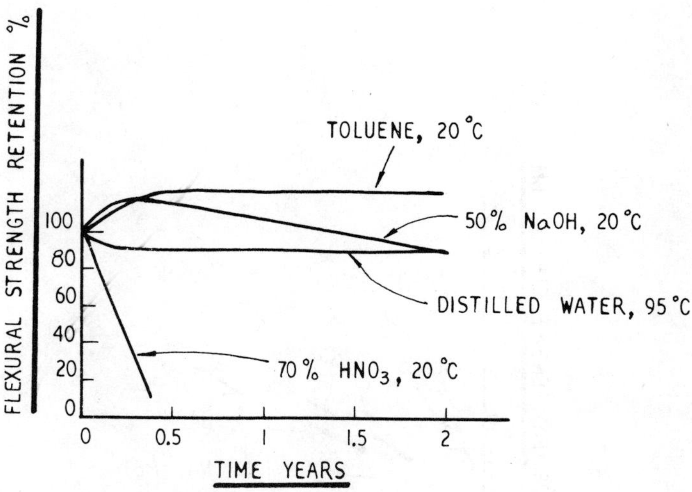

Fig. 71

reinforcement as hybrid composites with epoxy resin. The typical properties of unidirectional composites with hybrid reinforcement are shown in Table 39. The data have been normalized to 60 v/o total fiber content. They found that the behavior of the hybrids showed improved impact resistance and greater fracture toughness than all-graphite composites. The tensile strengths

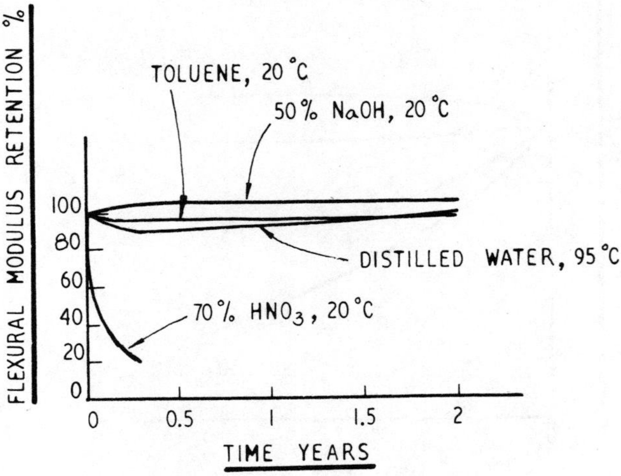
FLEXURAL MODULUS RETENTION %
TOLUENE, 20 °C
50% NaOH, 20 °C
DISTILLED WATER, 95 °C
70% HNO3, 20 °C
100
80
60
40
20
0
0
0.5
1
1.5
2
TIME YEARS

Fig. 72

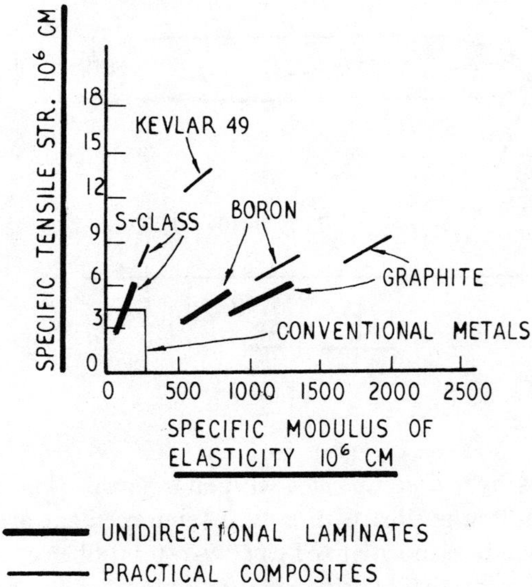
SPECIFIC TENSILE STR. 10⁶ CM
18
15
12
9
6
3
0
KEVLAR 49
S-GLASS
BORON
GRAPHITE
CONVENTIONAL METALS
0
500
1000
1500
2000
2500
SPECIFIC MODULUS OF ELASTICITY 10⁶ CM
UNIDIRECTIONAL LAMINATES
PRACTICAL COMPOSITES

Fig. 73

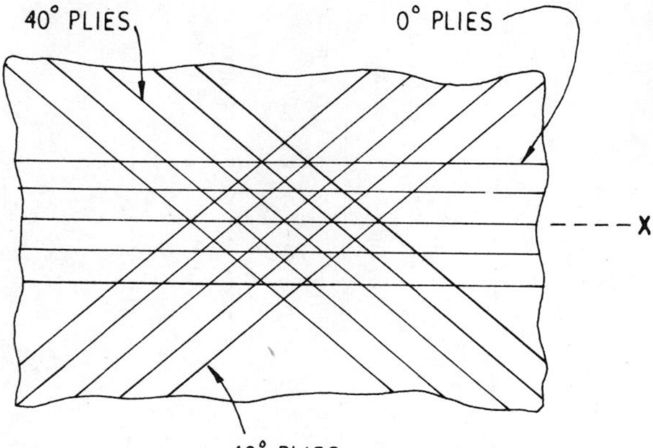

Fig. 74

of the hybrid composites was lower than the unidirectional Kevlar composites whose properties are shown in Table 40. All these properties vary with the variation in the graphite percentage content (expressed as a percentage of the total fiber content) and this is shown for the compressive yield stress (0.02% offset) and the tensile modulus in Figs. 75 and 76. The T-300 stands for Thornel 300 graphite tape.

The family of polyimide resins, of the thermosetting type, are particularly useful for high-temperature work. They have high dimensional stability and low creep. Polyimide resins can be used with glass, but graphite, boron, and aramid fibers are more suitable for high-performance applications. The properties of some boron/polyimide and graphite/polyimide composites have already been given in Table 37. The unfilled resins have a tensile strength at room temperature between 5000 and 17000 psi (34.5–117.2 MPa). The resins modified for easier processing have generally the lower properties. It is the behavior of these composites at elevated temperatures that is of importance, and several researchers have studied this behavior. Wolkowitz and Poveromo[85] tested graphite reinforced polyimide composites at 450°F (232°C) and some results are plotted in Fig. 77. The specimens were tested for both flexural properties and interlaminar shear for periods up to 2000 hours. Serafini and Vannucci[86] also tested graphite reinforced polyimide composites, and some of the properties over a period of 600 hours are shown in Fig. 78. The retained percent tensile strength of polyimide at 500°F (260°C) over 2000 hours given by Hensel,[87] is shown in Fig. 79.

The creep and static fatigue of the high-performance composites is also of some interest. The polyimides show lower creep than the phenolics at

Table 36

Property	Ply Angles, Degrees				
	6(0)	6(90)	4(±10∓10)	8(0, +10, 0, −10, −10, 0, +10, 0)	13[±40, 9(0), ∓40]
Tensile strength, psi	180,000	9,740	150,000	191,000	124,000
Tensile modulus, psi	21.7×10^6	1.15×10^6	21.1×10^6	19.0×10^6	14.0×10^6
Compressive strength, psi	135,000	34,000	—	—	—
Flexural strength, psi	206,000	16,350	230,000	197,000	145,000
Flexural modulus, psi	17.6×10^6	1.07×10^6	16.8×10^6	19.0×10^6	9.3×10^6
Short-beam interlaminar shear, psi	16,000	—	—	—	—

Note: 55% fiber volume. All specimens tested along the 0° ply direction.

Table 37

Material	Tensile Strength S_{xx} psi × 10^3	Tensile Modulus E_{xx} psi × 10^6	Poisson's Ratio ν_{xy}
Boron/epoxy	105	16.7	0.70
Graphite high modulus/epoxy	70	14.3	0.72
S-Glass/epoxy	108	5.3	0.52
Boron/polyimide	82	17.0	0.72
Graphite/polyimide	50	16.5	0.71

Note: All laminates $(0_2 \pm 45)_s$.

Table 38

Laminate	Tensile Modulus E_{xx} psi × 10^6	Tensile Strength S_{xx} psi × 10^3	Poisson's Ratio
$(0^k/\pm45^c/0^c)_s$	13.9	56	0.78
	13.6	45	0.72
$(\pm45^c/0^k/0^c)_s$	13.0	54	0.73
	13.3	53	0.87
$(0^k/\pm45^c/0^k)_s$	7.3	112	0.78
	6.7	120	0.85
$(\pm45^c/0_2{}^k)_s$	6.7	105	0.82
	6.6	109	0.86
$(0^g/\pm45^c/0^c)_s$	13.5	50	0.74
	12.4	56	0.73
$(\pm45^c/0^g/0^c)_s$	12.4	47	0.75
	11.8	50	0.73
$(0^g/\pm45^c/0^g)_s$	5.5	127	0.74
	5.3	118	0.85
$(\pm45^c/0_2{}^g)_s$	5.3	142	0.75
	5.5	100	0.73

Notes:
1. The values given are measured values.
2. The superscripts are:
   c = graphite.
   k = Kevlar.
   g = S-glass.

Table 39

Fiber	v/o Fiber	Flexural Strength (psi × 10^3)	Flexural Modulus (psi × 10^6)	Shear Strength (psi × 10^3)	Compressive Strength 0.02% Offset (psi × 10^3)	Compressive Modulus (psi × 10^6)	Impact Izod (ft lb/in.)
Kevlar 49	12.2	104	22.1	10.2	70	22.4	28.7
HM-S	47.8						
Kevlar 49	23.4	92	18.5	9.7	65	20.2	39
HM-S	36.7						
Kevlar 49	13.1	168	14.3	10.9	76	16.5	52
T-300	46.9						
Kevlar 49	29.3	129	12.4	8.0	54	13.1	68
T-300	30.7						
Kevlar 29	13.1	189	15.4	10.4	83	16.5	56
T-300	46.9						
Kevlar 29	31.2	146	12.1	7.8	74	11.9	75
T-300	28.8						

Notes:
1. Properties are for unidirectional hybrid composites with graphites.
2. For Kevlar 49/HM-S hybrid, the resin is PR-386. For Kevlar 49/T-300 and Kevlar 29/T-300, the resin is BP-907.
3. The data is normalized to 60 v/o total fiber content.

Table 40

Property	Kevlar 29/Epoxy Unidirectional Composites	Kevlar 49/Epoxy Unidirectional Composites
Tensile strength 0°, psi × 10^3	196	200
Tensile modulus, 0°, psi × 10^6	7.25	11.0
Compressive strength 0.02% offset stress 0°, psi × 10^3	21	32
Compressive modulus 0°, psi × 10^6	6.8	10.5
Shear strength 0°, psi × 10^3	6.3	8.7
Flexural strength 0°, psi × 10^3	78	90
Flexural modulus 0°, psi × 10^6	7.0	10.0
Impact strength Izod, 0°, ft lb/in.	75	70
Poisson's ratio	0.34	0.34

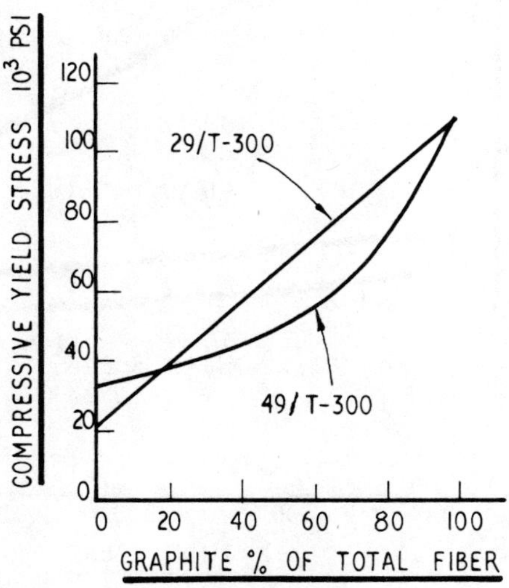

Fig. 75

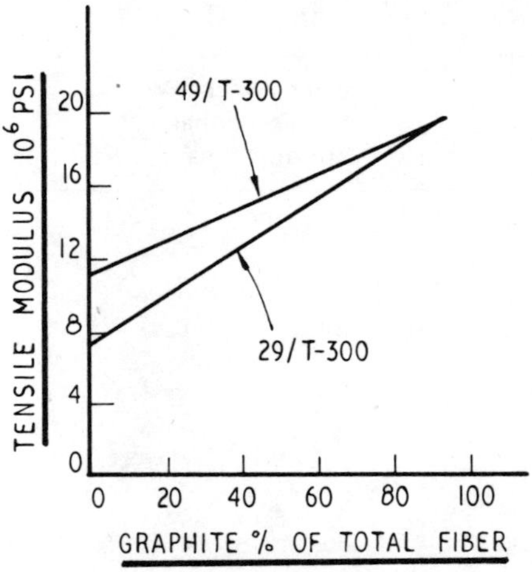
TENSILE MODULUS 10^6 PSI
20
16
12
8
4
0
49/ T-300
29/ T-300
0
20
40
60
80
100
GRAPHITE % OF TOTAL FIBER

Fig. 76

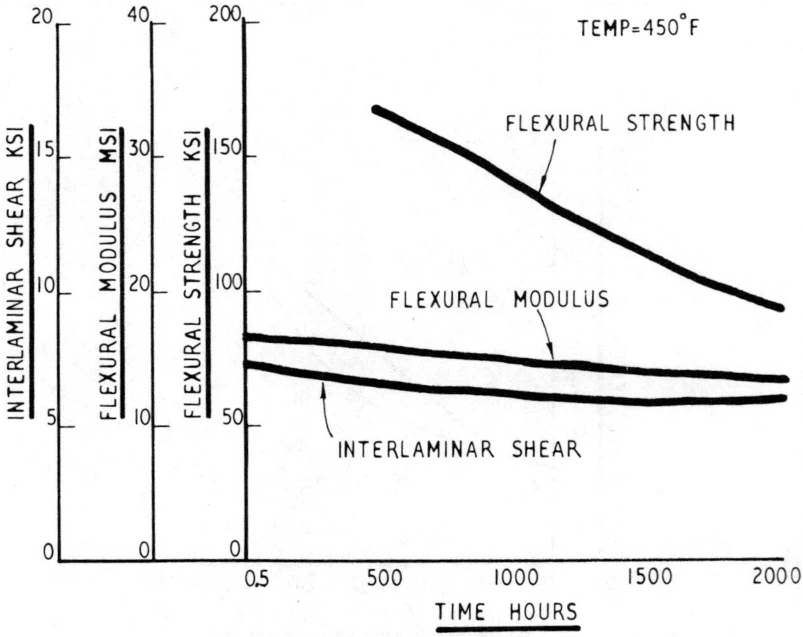
INTERLAMINAR SHEAR KSI
FLEXURAL MODULUS MSI
FLEXURAL STRENGTH KSI
TEMP=450°F
FLEXURAL STRENGTH
FLEXURAL MODULUS
INTERLAMINAR SHEAR
0.5
500
1000
1500
2000
TIME HOURS

Fig. 77

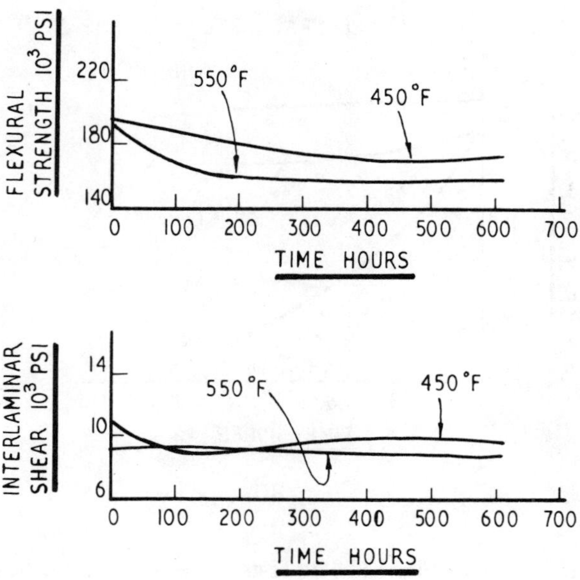

Fig. 78

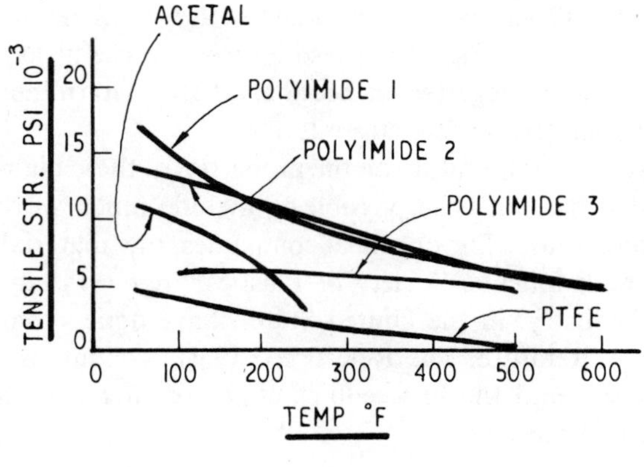

Fig. 79

higher temperatures, and curves given by Alvarez and Darmory[88] are shown in Fig. 80. These curves for polyimide apply to Kinel molding compounds but the results are relevant here as well. Sturgeon[89] did extensive testing on carbon fiber reinforced epoxy laminates. The tests were carried out at 40% and 80% of the ultimate tensile strength of the laminates (which is fairly high) and at temperatures of 21°C (70°F) and 80°C (176°F), and

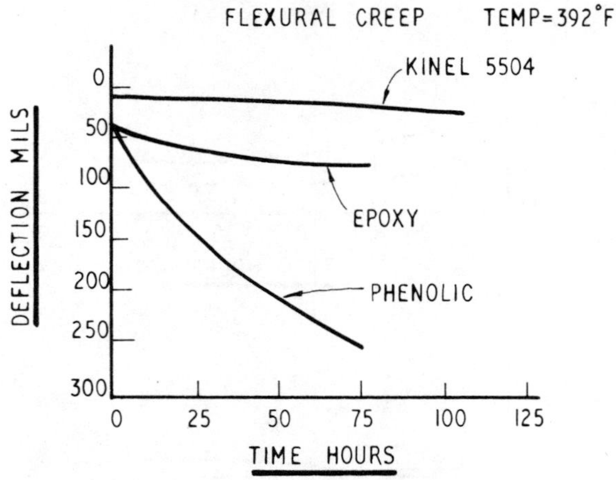

Fig. 80

the results are shown in Fig. 81. The tests were up to 1000 hours, and the creep strains at the end of this period were less than 0.01%.

In their study of high-performance composites for energy storage flywheels, Penn and Chiao[79] also studied the long term rupture stress for epoxy reinforced with Kevlar 49 and epoxy reinforced with S-glass reinforcement, and the behavior is shown in Fig. 82. These curves are useful as raw data for input into the method suggested in Section 2.12 to determine the behavior of the composites under varying stress levels.

There is a vast body of literature on the properties of these high-performance composites, and it has not been possible to include much work that should receive wider recognition. The chemical companies, raw material manufacturers, research establishments, Society of Plastics Engineers, and the Society of the Plastics Industry, in the United States, have done yeoman service in putting forth this literature. The Reinforced Plastics/Composites Institute's Annual Conference, and the Proceedings of that conference published each year, have no equal anywhere.

2.19 SMC/BMC

Much of the information on sheet molding compounds (SMC) and bulk molding compounds (BMC), including aspects such as manufacturing processes, some properties of the materials, and so on, has already been given in earlier articles of this chapter. For instance, compression molding, transfer molding, and the like, are particularly applicable to SMC/BMC. When molded into composites, the properties of these composites, with various

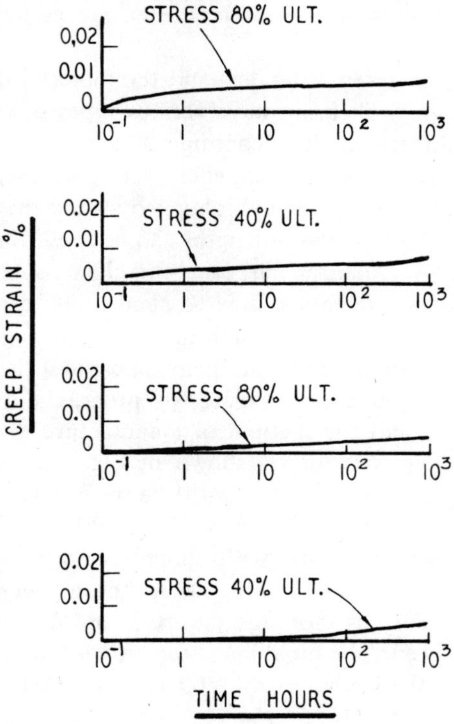
STRESS 80% ULT.
STRESS 40% ULT.
STRESS 80% ULT.
STRESS 40% ULT.
CREEP STRAIN %
0.02
0.01
0
10⁻¹
1
10
10²
10³
TIME HOURS

Fig. 81

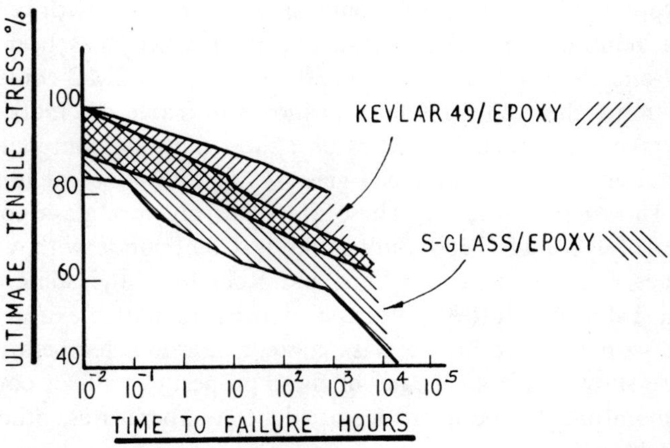
ULTIMATE TENSILE STRESS %
100
80
60
40
KEVLAR 49/ EPOXY
S-GLASS/EPOXY
10⁻²
10⁻¹
1
10
10²
10³
10⁴
10⁵
TIME TO FAILURE HOURS

Fig. 82

resin types and various glass reinforcements, are as has been discussed in this chapter.

It is necessary, however, to understand the material itself. Sheet molding compounds with glass fiber can have various types of reinforcements, such as random fiber, directional fiber, continuous fiber, or a combination of these. Most SMC used is of the random fiber type. The usual reinforcement in this case is chopped roving, though chopped strand mat is also suitable. A single length or a combination of lengths can be used, the usual length being 1 in. (25 mm). For structural purposes, the glass content is 50%–70% by weight, though glass contents as low as about 30% can also be used. The glass used is generally E-glass, though high-strength S-glass and other reinforcements such as graphite, aramid fibers, mica, and the like, are also being used. Jutte[90] has given an excellent materials, process, and performance review of structural SMC, and the method of manufacture of SMC with random fibers of chopped glass roving is shown in Fig. 83. The paste containing the resins, fillers, catalyst, together with the mold release and other agents, are carefully metered onto the carrier films, with the roving fibers fed in, as shown in Fig. 83. For SMC with chopped and continuous rovings, the production process is shown in Fig. 84. At the higher glass contents, it is important that the fiber is thoroughly wetted, and this can be accomplished by a slower line speed. The molding of the part is carried out with molding pressures of 800 psi (5.5 MPa) to 1500 psi (10.3 MPa) and temperatures from 290°F (143°C) to 310°F (154°C).

The choice of the molding process varies with the part. Transfer molding may be considered for small parts with numerous inserts or cavities. Compression molding is more suitable for large parts or for structural purposes with high strength sheet molding compounds.

The properties of SMC/BMC composites or moldings with polyester resin and glass reinforcement, for low shrink applications, has been studied by Conley,[91] and the properties of 20 oz/ft^2 (6.1 kg/m^2) SMC containing 30%, 1 in. (25 mm) chopped glass fiber is shown in Table 41. Gott, Suggs, and Blount[92] give the typical properties of a sheet molding compound based on polyester resin with 30% chopped glass roving and 40% calcium carbonate filler, as shown in Table 42. These are relatively low glass contents. The properties of ROSITE 3550 Bulk molding Compounds with varying glass percentages were obtained by Ward and Robbins,[93] and some of these are shown in Table 43. Jutte[90] gives the increase in both flexural and tensile strengths with increase in the glass content, at room temperature, and his results are shown in Fig. 85. All of these properties will, of course, reduce with temperature; for performance at higher temperatures, other resins are more suitable.

The vinyl ester resin has already been considered in Section 2.17, and

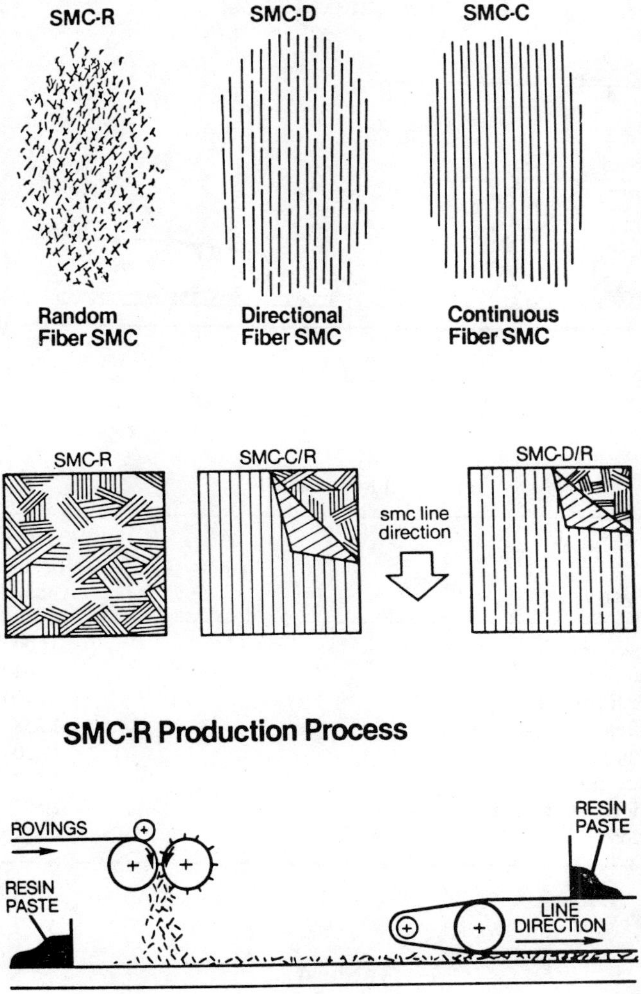

Fig. 83

the properties of high-strength SMC for use in automotive parts are given in Table 35 as well as in Figs. 57–60.

For very high performance applications, the glass percent content is increased to 70%–80%. This leads to high-performance sheet molding compounds that are called HMC,[94] or XMC[95] depending on the trademark. HMC is a trademark of PPG Industries, and in this material the fillers are virtually eliminated, thereby permitting the glass fibers to be wetted more thoroughly. The properties of HMC and SMC, as determined by Maaghul and Potkanow-

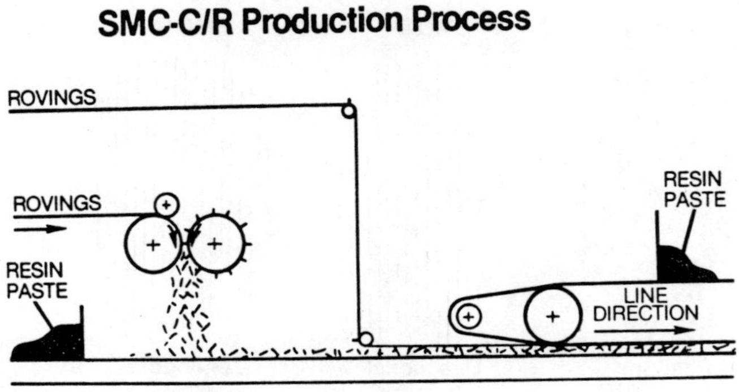

Fig. 84

Table 41

Property	20 oz/ft² SMC with 30%, 1 in. chopped Glass Fiber
Flexural strength, psi	31,000
Flexural modulus, psi	2.663×10^6
Flexural strength (wet), psi	29,700
Flexural modulus (wet), psi	2.258×10^6
Tensile strength, psi	12,200
Tensile modulus, psi	2.587×10^6
Izod impact, notched, ft lb/in.	20.9
Water absoprtion	0.15

Table 42

Property	SMC Based on a Polyester Resin
Flexural strength, psi	20,800
Flexural modulus, psi	1.54×10^6
Tensile strength, psi	12,600
Tensile modulus, psi	1.60×10^6
Izod Impact notched, ft lb/in.	13.2
Heat Distortion temperature, °C	>204
Water Absorption, 24 hr, % wt. gain	0.25

Table 43

Property	3550A	3550C	3550D	3550F
Glass fiber content, %	5	15	20	30
Density, lb/in³	0.075	0.070	0.068	0.065
Specific gravity	2.08	1.94	1.89	1.80
Water absoprtion, 24 hr, %	0.1	0.1	0.1	0.1
Flexural strength, psi × 10³	10	15–20	18–22	25
Compressive strength, psi × 10³	18–20	20–22	21–22	23–25
Tensile strength, psi × 10³	4.5–5.0	5–6	6–7	8–9
Flexural modulus, psi × 10⁵	15	15–17	17–19	25
Impact strength, Izod notched, ft lb/in.	1.5–2.0	3–5	4–6	10–12

icz,[94] are compared in Table 44. XMC is also a high-strength molding compound, normally containing unidirectional continuous glass fibers and polyester resin, with no fillers. With standard XMC, also developed by PPG Industries, the glass content is 70%, but the unidirectional character of the glass reinforcement, which gives it exceptional strength, also makes it difficult to flow into complex dies, Johansen,[95] therefore, used XMC with standard SMC paste to obtain a 50% glass content XMC. The properties of the two materials, again compared to standard SMC, are shown in Table 45.

Glass is not the only reinforcement in SMC/BMC. For low-performance BMC, for instance, it is possible to use fillers such as inexpensive ceramic fillers, silica, clay, asbestos, wollastonite, and the like. Woodhams and Xanthos[40] considered the use of mica flake reinforcement, and this reinforcement was considered in Section 2.9. Polyester resin powders were used to prepare a prepreg suitable for conventional compression molding. The use of mica reinforcement alone leads to poor impact strength values, and the addition of chopped glass fibers improves this property. Table 46 shows the comparative properties of composites molded from mica reinforced polyester compounds with a 60% mica content, and composites with the addition of 50% mica and 10% chopped strands ½ in. (13 mm), to the same resin (40% content).

Kalnin[96] has considered epoxy bulk molding compounds with graphite fiber reinforcement. In all these materials it is also standard practice to specify the reinforcement content by volume percent content. An increase in the fiber length from ¼ in. (6 mm) to 1 in. (25 mm), as also the increase in

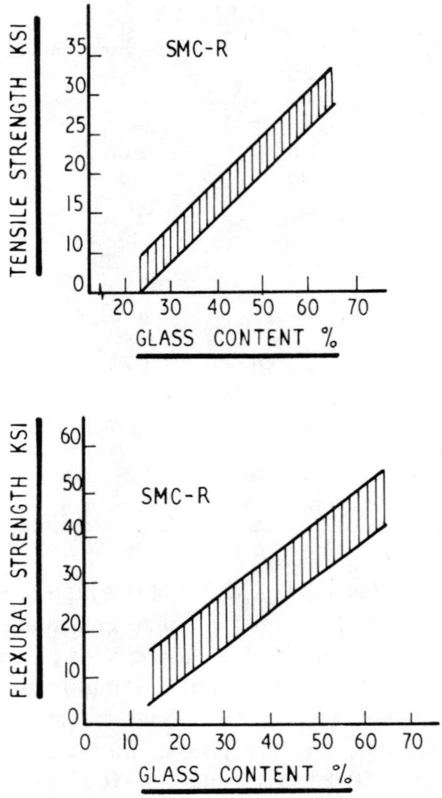

Fig. 85

Table 44

Material	Flexural Strength (psi × 10³)	Flexural Modulus (psi × 10⁶)	Tensile Strength (psi × 10³)	Tensile Modulus (psi × 10⁶)	Impact Izod Unnotched (ft lb/in.)
HMC	55–60	2.1–2.3	30–40	2.6–3.0	25–30
SMC	28–32	1.9–2.1	13–16	1.9–2.1	14–20

percent fiber volume content from 40% to 60% both cause an increase in the flexural strength and modulus as shown in Table 47.

It is interesting to note that Maaghul has found that, in SMC, the sheet density can have an effect on the mechanical properties of the molded panel. This could be due to reduced wetting of the fibers at the higher densities

Table 45

Property	Standard XMC	Filled XMC	Filled SMC
Glass content, %	65	50	30
Specific gravity	1.95	1.90	1.80
Flexural strength, psi	150,000	100,000	30,000
Flexural modulus, psi	4.5×10^6	3×10^6	2×10^6
Tensile strength, psi	60,000	40,000	10,000
Impact strength, Izod notched, ft lb/in.	60	30	15

Table 46

Property	HAR Mica/Resin (Ratio 60/40)	HAR Mica/½ in. Chopped Strand Glass Fibers/Resin (Ratio 50/10/40)
Density, g/cm³	2.01	1.98
Flexural strength, psi × 10³	26.5	22.0
Flexural modulus, psi × 10⁶	7.65	6.75
Impact strength, Izod notched, ft lb/in.	0.6	5.4
Hardness, Barcol	58–62	58–62

Table 47

Nominal Fiber Length	Fiber Content (Vol. %)	Bulk Density (g/cc)	Flexural Strength (psi × 10³)	Flexural Modulus (psi × 10⁶)	Impact Strength Izod Notched (ft lb/in.)
¼ in.	40	1.49	14.4	3.1	0.66
½ in.	40	1.48	18.3	4.1	0.56
1 in.	40	1.48	19.3	4.9	0.50
½ in.	50	1.68	17.3	7.4	
½ in.	60	1.72	30.9	14.9	

or the use of a number of plies with low density mat. Sundstrom, Collister, and Hays[26] also studied SMC/BMC with densities in the range of 75–87.5 lb/ft³ (1201–1402 kg/m³) as opposed to the normal density of 106–118.6 lb/ft³ (1698–1900 kg/m³), which is the typical range of conventional SMC/BMC. The reduction in density was achieved by the use of hollow glass spheres as a filler material. This led to molded parts that were lighter, with a saving in weight of 25%–35%, while retaining almost equivalent physical properties.

Finally, Gruenewald and Walter[97] have studied the weathering of SMC over several years. This weathering could exhibit itself in the form of loss of gloss, color changes including yellowing and fading, glass fiber erosion, and perhaps a reduction in mechanical strength with time. They found that, although after several years the weathering was severe in all of the visual aspects listed above, this was not true of the mechanical properties. The percent retention of flexural strength with time, due to weathering, is shown in Fig. 86. This reduction correlates very well with the reduction in tensile strength with time due to weathering determined by the author[70] on a FRP, folded-plate, conical-dome, roof structure, which was discussed in Section 2.12. Finsel and Younge[98] have reported on a "hybrid BMC," which is a fiberglass reinforced polyester themoset, and the properties of this material compared to other materials are given in Table 48.

2.20 REINFORCED THERMOPLASTICS

The entire discussion in this chapter on reinforced plastics has centered around the thermosetting resins. This is because reinforced thermosets are excellent structural materials. It is quite possible, however, to have reinforced thermoplastics. The advantage of reinforced thermoplastics is that the material is amenable to many of the special processes applicable to thermoplastics, such

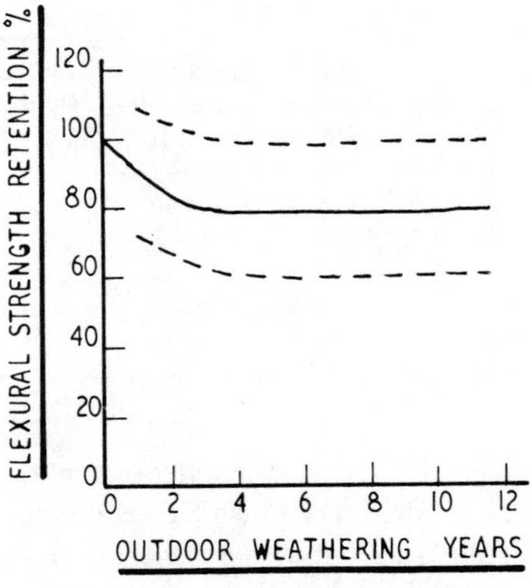

Fig. 86

Table 48

Property	Hybrid BMC	Impact Phenolic	Glass Reinforced Phenolic	Melamine	Urea	Epoxy
Specific gravity	1.80	1.40	1.90	1.73	1.50	1.60
Tensile strength, psi × 10^3	4	5.5	8.0	5.0	7.0	10.5
Compressive strength, psi × 10^3	15	25	38	26	30	24
Flexural strength, psi × 10^3	10	8.5	12	8	12	15.5
Flexural modulus, psi × 10^6	2.0	1.2	2.0	1.9	1.3	1.4
Heat deflection temperature °F, at 264 psi	500	350	350	350	275	320
Impact strength, Izod notched, ft lb/in.	20.0	0.8	0.45	0.45	0.3	0.4
Water absorption, %	0.1	0.8	0.1	0.2	0.6	0.2

as thermoforming, injection molding, and the like. The disadvantage of the materials is that the heat distortion temperature may be quite low.

There are very many types of thermoplastic resins that can be suitably reinforced, such as ABS, acetals, polycarbonate, polyethylene, nylon, polypropylene, PVC, acrylics, and thermoplastics polyester. A wide variety of reinforcements can be used, including glass, asbestos, mica, wollastonite, and graphite. The addition of reinforcement leads to considerable improvement in the mechanical properties of the material and can also improve the heat distortion temperature. Best and Kelly[99] give the percentage improvements for various thermoplastic resins reinforced with 20%–30% fiberglass, as shown in Table 49. Wicker[100] has considered the typical properties of the unreinforced thermoplastics and these are given in Table 50.

The mechanical properties of some reinforced thermoplastics will be considered further. LNP[101] has given the properties of thermoplastics fortified or reinforced with 30% dispersed glass fiber content, and these are shown in Table 51. The heat deflection temperatures are at 264 psi (1820 kPa). There has been considerable research devoted to reinforced nylon. Kjellmark and Johannes[102] studied the properties of nylon 66 reinforced with 13% glass reinforcement, 33% glass reinforcement, and the unreinforced resin, and some of these properties are given in Table 52. Orenski, Berger, and Ranney[103] give the effects of water immersion on 70% wollastonite filled Nylon 6 compared to the unfilled resin, and these are shown in Fig. 87. Maine et al.[104] studied the use of mica flake reinforcement for injection moldable reinforced

Table 49

Reinforced Polymer	Improvement % Flexural Strength	Flexural Modulus	Temperature Resistance (Under Load)
ABS	120	170	8
Acetal	8	210	35
Nylon	300	300	240
Polycarbonate	60	600	7
Polyphenylene oxide (modified)	100	360	19
Polysulfone	170	200	7
PVC	30	120	11
Polystyrene	180	200	10
SAN	45	180	10
Polyethylene	100	50	190
Polypropylene	50	200	100
Polyester (thermoplastic)	85	250	230

Table 50

Material	Tensile Strength (lbf/in.^{2})	Flexural Strength (lbf/in.^{2})	Flexural Modulus (lbf/in.^{2})	Heat Distortion Temperature (°C)	Burning Rate
A B S	5500	8000	0.2×10^6	95	Slow
Polyacetal	9000	13000	0.4×10^6	125	Slow
Polycarbonate	9000	12000	0.3×10^6	140	Self-extinguishing
HD Polyethylene	5000	1000	0.1×10^6	72	Slow
LD Polyethylene	2000	no break	0.03×10^6	50	Slow
Poly methyl methacrylate	10000	15000	0.45×10^6	83	Slow
Nylon	11000	12000	0.40×10^6	175	Self-extinguishing
polystyrene	5000	7000	0.20×10^6	105	Slow
Impact polystyrene	7000	13000	0.50×10^6	97	Slow
PVC	7000	14000	0.50×10^6	66	Self-extinguishing
Phenoxy	9500	14000	0.35×10^6	86	Self-extinguishing

Note: Heat distortion temperature at a stress of 64 lbf/in.^{2}

thermoplastics and found them to be very suitable. When used as reinforcement for ABS, SAN, or Nylon 66 resins, they found that the moduli were generally higher than glass fiber reinforced resins, the strengths were comparable, and the impact strengths were somewhat lower. To improve the impact strengths, it is also possible to produce hybrid reinforced thermoplastics using mica and glass fiber as reinforcement. In order to represent the properties of the composites with mixed ratios of reinforcements, they express the flexural strength and modulus in the form of surfaces as shown in Fig. 88, with the unfilled resin, 60% mica reinforcement, and 60% glass fiber reinforcement forming the apexes of the equilateral triangle. The values of the flexural strength and modulus can then be read off at any hybrid composition.

Cameron, Heron, and Wicker[105] studied the use of asbestos reinforcement with thermoplastics such as nylon 6, polypropylene, and polystyrene. The comparative properties of chrysotile asbestos reinforced nylon 6 and glass reinforced nylon 6, both with 30% reinforcement content are shown in Table 53. Illing[25] has considered the use of epoxy coated glass spheres as reinforcement for nylon 6. He found that powdery or fibrous fillers give increased stiffness and hardness but also a greatly reduce toughness. This disadvantage can be minimized by adding epoxy or phenoxy resins, which improve the impact resistance and permit larger amount of fillers to be incorporated into the resins.

Table 51

Thermoplastic with 30% Dispersed Glass Fiber fortification	Sp.Gr.	Water Absorption 24 Hours (%)	Tensile Strength (psi)	Flexural Modulus (psi × 10^6)	Impact Strength Izod Notched, (ft lb/in.)	Heat Distortion Temp. (°F at 264 psi)
ABS	1.28	0.11	19,000	1.4	1.2	235
SAN	1.31	0.10	17,400	1.6	1.0	215
Polystyrene	1.28	0.05	12,500	1.2	1.0	215
Polycarbonate	1.43	0.07	18,500	1.2	3.7	300
Polyethylene (HDPE)	1.17	0.02	10,000	0.9	1.1	260
Polysulfone	1.45	0.20	18,000	1.2	1.8	365
Acetal	1.63	0.60	13,000	1.3	0.8	325
Polypropylene	1.13	0.03	7,800	0.8	1.3	295
Nylon 6	1.37	1.2	23,000	1.2	2.3	419
Nylon 6/6	1.37	0.80	26,000	1.3	2.0	489
Thermoplastic Polyurethane	1.46	0.75	8,200	0.2	9.5	170
PVC (with 25% glass)	1.53	0.02	16,000	1.25	1.1	160
Polyphenylene oxide (modified)	1.27	0.06	21,000	1.2	1.9	310
Thermoplastic polyester	1.59	0.06	19,500	1.2	1.6	455

Table 52

Property	Nylon 66	Nylon 66 with 13% Glass Reinforcement	Nylon 66 with 33% Glass Reinforcement
Density	1.14	1.22	1.38
Tensile strength at yield, psi	8,500	12,000	18,000
Flexural modulus, psi × 10^6	0.175	0.400	0.900
Impact strength, Izod notched, ft lb/in.	1.0	0.7	2.1
Deflection temperature at 264 psi, °F	200	470	480
Flexural creep, Apparent modulus, psi × 10^6 1000 hr, 1500 psi, 73°F,	0.120	0.300	0.600
1000 hr, 1000 psi, 180°F	0.075	0.225	0.600

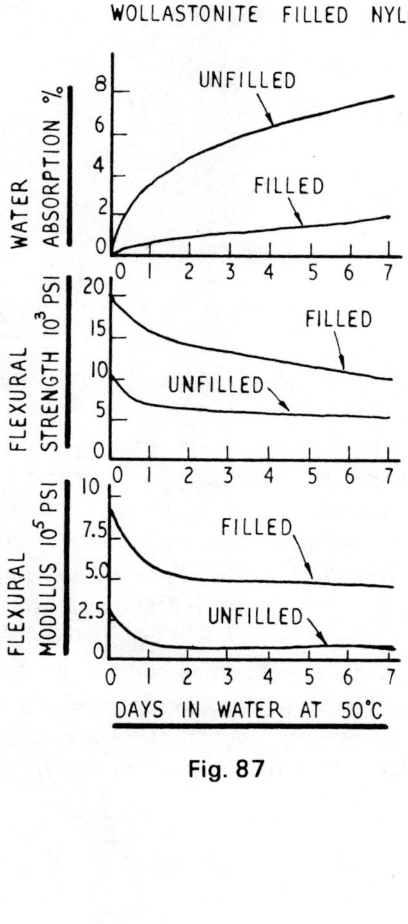

Fig. 87

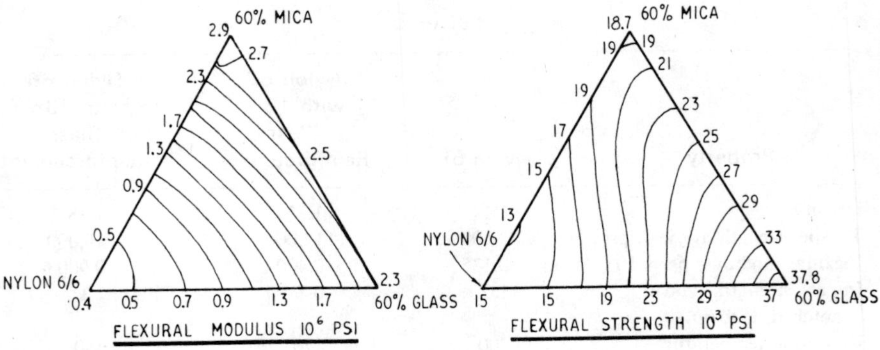

Fig. 88

Table 53

Property	Nylon 6 with 30% Glass Reinforcement	Nylon 6 with 30% Chrysotile Asbestos Reinforcement
Density, g/cc	1.36	1.37
Tensile strength, psi	24,000	19,600
Flexural strength, psi	34,000	29,500
Flexural modulus, psi × 10^6	1.00	1.31
Impact strength Izod notched, ft lb/in.	2.0	1.0
Deflection temperature, at 264 psi, °C	200	215

For high-performance applications, such as at elevated temperatures or higher strength applications, the use of thermoplastic polyimide, thermoplastic polyester and nylon, even reinforced with graphite may be more suitable. Theberge et al.[106] considered the elevated temperature performance of thermoplastic composites reinforced with glass fibers with a glass content up to 50%. They came to the conclusion that the addition of glass fiber reinforcement resulted in improvement for all of the resin systems investigated. The polyimides and nylon had heat distortion temperatures of 530°F (277°C) and 500°F (260°C), respectively, at 264 psi (1820 kPa). Wambach and Kramer[107] have discussed the properties of glass reinforced thermoplastic polyester with a built-in flame retardance, and the reduction of tensile strength and flexural modulus at elevated temperatures are shown in Fig. 89. The use of graphite reinforcement with thermoplastic resins was studied by The-

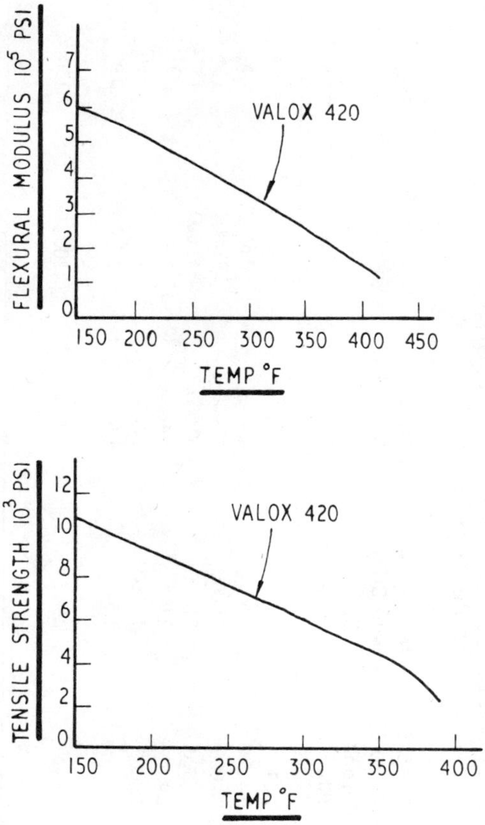

Fig. 89

berge, Arkles, and Robinson,[108] and they came to the conclusion that there was considerable improvement in the flexural modulus, almost double, when compared to glass fiber reinforcement. The tensile strengths were very high. The properties of some thermoplastic composites reinforced with graphite reinforcement are shown in Table 54. However, the impact strength was much lower than with glass reinforcement. This was also found by Menges, Harnier, and Heisse[109] in their work on carbon fiber reinforced polyamides. They recommend hybrid reinforcement with glass fibers and carbon fibers used to reinforce the resin. The variation[108] of flexural strength and modulus with temperature for nylon 66 reinforced with 30% graphite reinforcement is shown in Fig. 90.

The long term creep behavior of reinforced thermoplastics, particularly if available at elevated temperatures, should always be considered in the

Table 54

Property	Nylon 6/6 with 20% Carbon Fiber	Nylon 6/6 with 30% Carbon Fiber	Nylon 6/6 with 40% Carbon Fiber	Nylon 6/6 with 40% Glass Fiber	Nylon 6/6
Specific gravity	1.23	1.28	1.34	1.46	1.14
Water absorption, 24 hr, %	0.6	0.5	0.4	0.6	1.5
Tensile strength, psi	28,000	35,000	40,000	31,000	11,800
Flexural strength, psi	42,000	51,000	60,000	42,000	15,000
Flexural modulus, psi $\times 10^6$	2.4	2.9	3.4	1.6	0.4
Shear strength, psi	12,000	13,000	14,000	12,700	9,600
Impact strength, Izod notched, ft lb/in.	1.1	1.5	1.6	2.6	0.9
Heat distortion temperature at 264 psi, °F	495	495	500	500	150

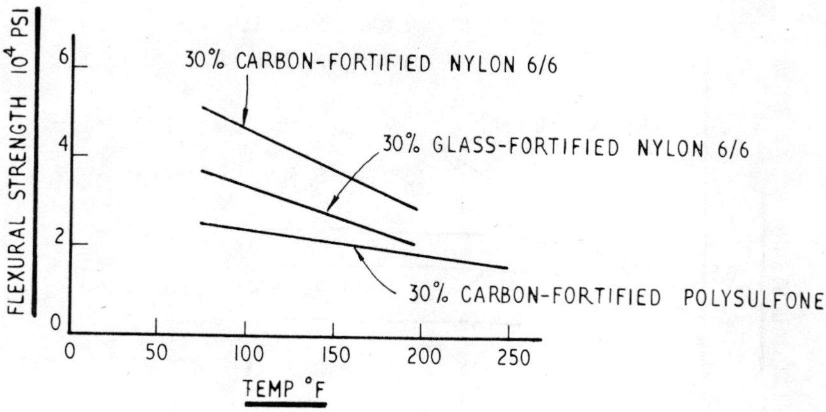

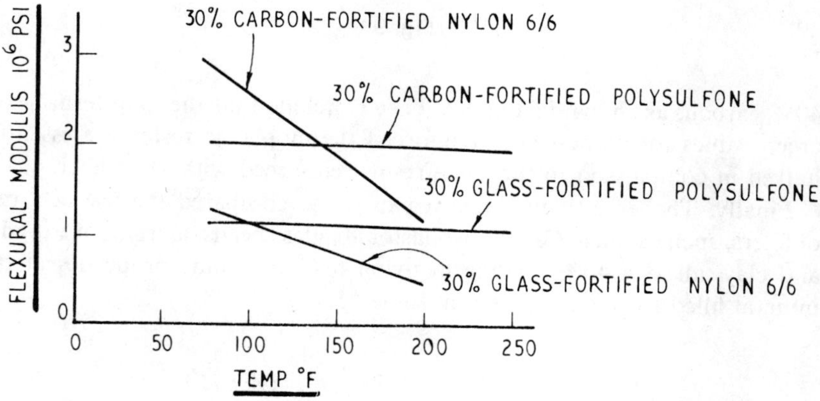

Fig. 90

design. The provision of glass reinforcement reduces the creep of the resin, and the provision of graphite reinforcement reduces the creep even still further. The flexural creep of nylon 6/6 resin with some types of glass reinforcements, in different percentage contents, is shown in Fig. 91. As can be seen, the addition of glass reinforcement reduces the creep considerably. The tests were run at 73°F (23°C), and the results are quoted by Theberge et al.[110] In their work on carbon fiber reinforced polyamides, Menges, Harnier, and Heisse[109] found that the composites exhibited a far better creep behavior than with glass reinforcement. Theberge, Arkles, and Robinson[108] give the comparative creep in flexure of Nylon 6/6 reinforced with 40% glass, and

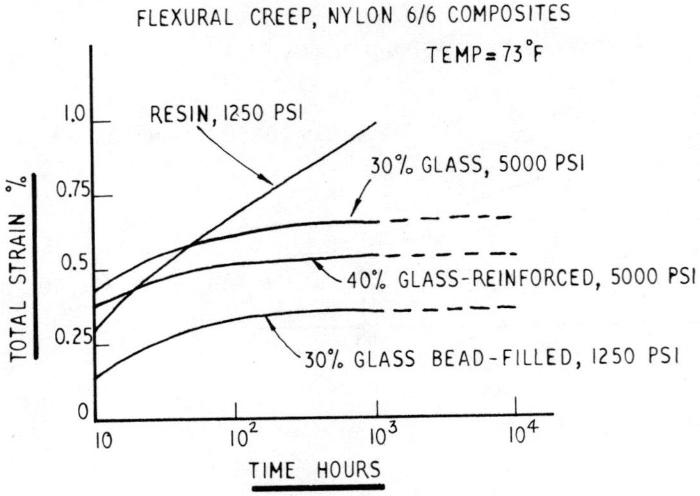

Fig. 91

40% carbon, as shown in Fig. 92. They conclude that the long term flexural creep values for carbon fiber reinforced thermoplastic resins are essentially halved in comparison to the same resins reinforced with glass fiber.

Finally, Theberge,[111] in a recent paper, has compared the use of a range of fillers, such as mica, $CaCO_3$, wollastonite, glass beads, aluminum trihydrate and clay, all at a 40% loading in nylon 6/6, and some properties of these mineral filled nylons are given in Table 55.

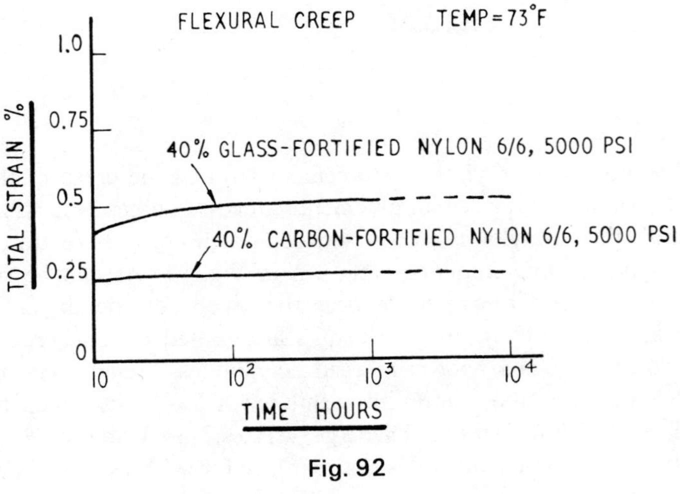

Fig. 92

Table 55

Property	Nylon 6/6 with Mica 60/40	Nylon 6/6 with $CaCO_3$ 60/40	Nylon 6/6 with Wollastonite 60/40	Nylon 6/6 with Glass Beads 60/40	Nylon 6/6 with ATH 60/40	Nylon 6/6 with Clay 60/40
Specific gravity	1.50	1.48	1.51	1.46	1.45	1.47
Water absorption, 24 hr, %	0.6	0.5	0.5	0.6	0.7	0.4
Tensile strength, psi	15,260	10,480	10,480	9,780	9,200	10,850
Flexural strength, psi	26,000	16,500	17,700	15,800	14,700	23,700
Flexural modulus, psi $\times 10^6$	1.540	0.660	0.790	0.615	0.645	1.010
Impact strength, notched, ft lb/in.	0.6	0.5	0.6	0.4	0.5	0.3
Heat distortion temperature at 264 psi °F	460	388	430	407	396	391

CHAPTER 3

LOW DENSITY CORE MATERIALS

3.1 INTRODUCTION

The main types of low-density plastics suitable for use as cores in sandwich construction were briefly mentioned in Section 1.2. To use these materials efficiently, however, it is essential to know their mechanical properties in full, and the complete properties may not be easily available, even from raw material manufacturers. Many researchers have performed considerable work on samples of these materials, and some of the results are discussed in this chapter. It may be noted that the properties given are strictly applicable only to the materials specified, but, for purposes of design, particularly preliminary design, the material properties given can be applied to the same types of materials. For very accurate work, effort should be made to obtain similar properties from the actual supplier of the material to be used.

The main mechanical properties that engineers should know for purposes of design were listed in Section 2.12. When used as cores in sandwich construction, the flexural strength and the flexural modulus of elasticity are of little importance.

The accurate and complete determination of the remainder of these properties raises problems, due to the nature of the material itself. Unlike conventional materials, the properties are affected in varying degrees by a number of factors:

1. Density of the material
2. Method of formation
3. Rate of strain of the material
4. Time after loading
5. Temperature

The properties of these materials will now be considered in detail.

3.2 RIGID POLYURETHANE FOAM

The materials tested by the author[112] were obtained from the Shell Chemical Co. In all cases, a suitable flame retardant was introduced into the formulation. The tests were at ambient temperatures, with values determined immediately after loading.

A typical load/extension curve for the material is shown in Fig. 93. It can be seen that the material yields continuously to fracture. The load *P* at fracture defines the ultimate tensile strength of the material. The tangent to the curve at or near to the origin defines the initial modulus at low strains. A typical load/compression curve is shown in Fig. 94. Rigid polyurethane foam, both against and across the rise of the material, shows a distinct yield line CD. The compressive strength at yield is very clearly defined.

The variations of the tensile and compressive strengths and moduli with density and rate of strain are shown in Figs. 95–102. The material tested is at the lower density ranges, and the properties of the higher densities are also considered later in this article. A look at Figs. 95–98 shows that the increase in strengths and moduli varies almost linearly with the density, except at very low densities, where the curves demonstrate nonlinear behavior. It can also be seen that the direction of foaming and the method of formation are of great importance. The properties of the material vary considerably when measured against the rise and across it. It would seem, therefore, that the size and the shape of the mold would play a part in defining the properties

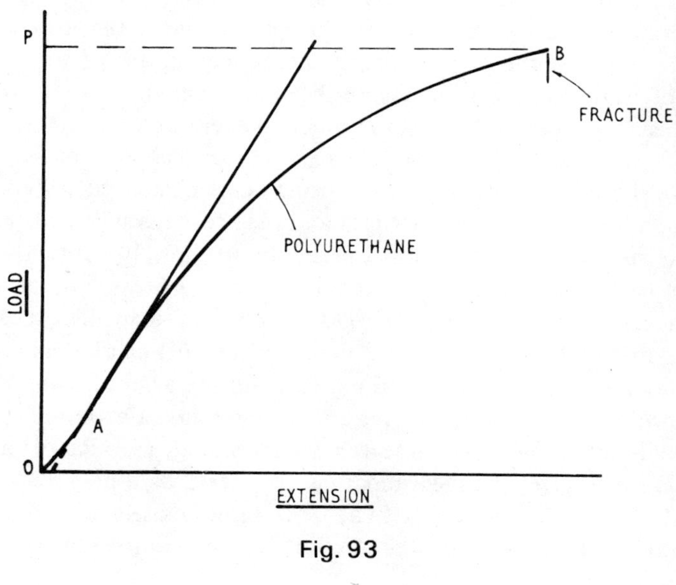

Fig. 93

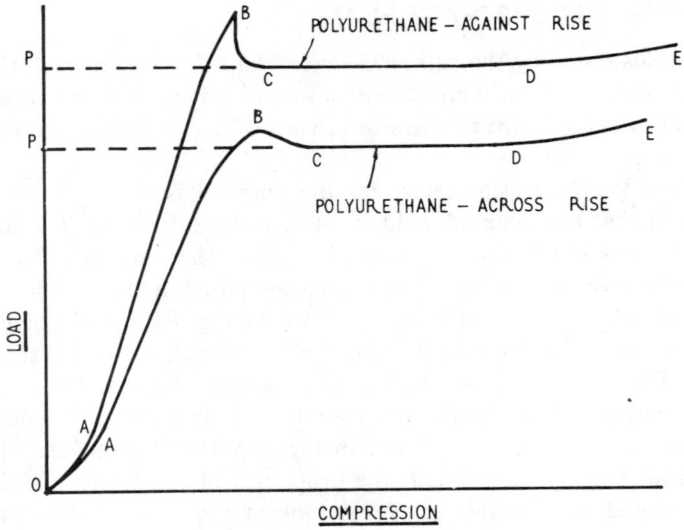

Fig. 94

in various directions. Varying the rate of strain of the material does alter the strengths and the moduli, but the variation is relatively small over a large range of straining rates, as can be seen in Figs. 99–102. The strain rates vary up to 60×10^{-4} per second as shown in Figs. 101 and 102. In civil engineering practice, it is very difficult to define the rate of strain of the material, which is dependent on the rate at which the load is applied to the structure. In the presentation of the remaining properties, therefore, the rate of strain as a variable has not been considered.

The shear strength and the shear modulus were determined by shearing ½ in. (13 mm) thick specimens between two parallel steel plates, the rate of relative slip between the two plates being kept constant at 0.2 cm/minute (0.079 in./minute). The shear strength and the shear modulus, when sheared across the rise, are plotted against the density in Figs. 103 and 104.

It may be mentioned here that the addition of a fire retardant seems to reduce the mechanical properties of rigid polyurethane foam. The shear modulus of the material at a density of 2 lb/ft³ (32 kg/m³) can be obtained from Fig. 104 as about 230 psi (1586 kPa). Tests on the same density foam, but without fire retardant additives, give a shear modulus of about 400 psi (2758 kPa). The Atomic Weapons Research Establishment (AWRE)[113] at Aldermaston, England gives the shear modulus of fire retardant polyurethane foam at 25°C (77°F) as 250 psi (1724 kPa). This agrees closely with the author's value. Other values given by the AWRE for the compressive strength at

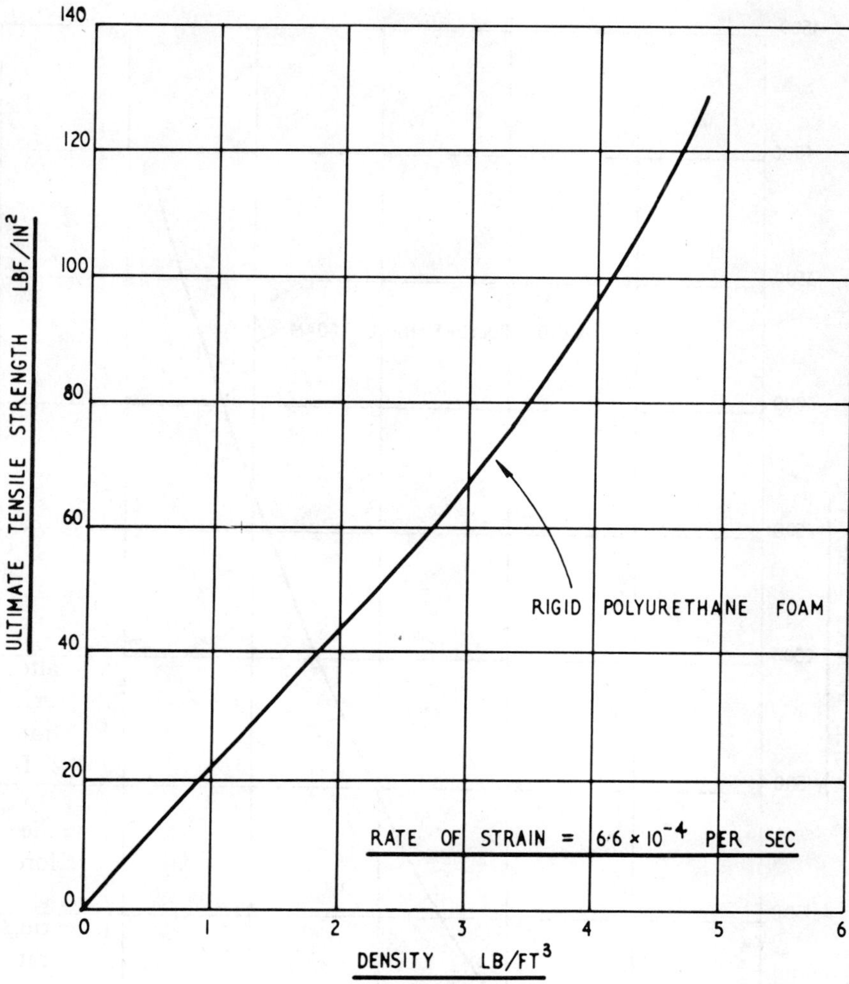

Fig. 95

yield and the compressive modulus also agree fairly well with values obtained from Figs. 97 and 98, the AWRE values being about 15% to 20% higher than the author's values.

For rigid polyurethane foam at these low densities, values are also given by several raw material manufacturers. Celotex[114] gives the compressive strength of self-extinguishing urethane foam core with a density of 1.8 lb/ft³ (29 kg/m³) as 25 psi (172 kPa). Vultafoam[115] gives the compressive strength for a urethane foam system designed for spray application as 30 psi (207 kPa) at a density of about 2 lb/ft³ (32 kg/m³) in place.

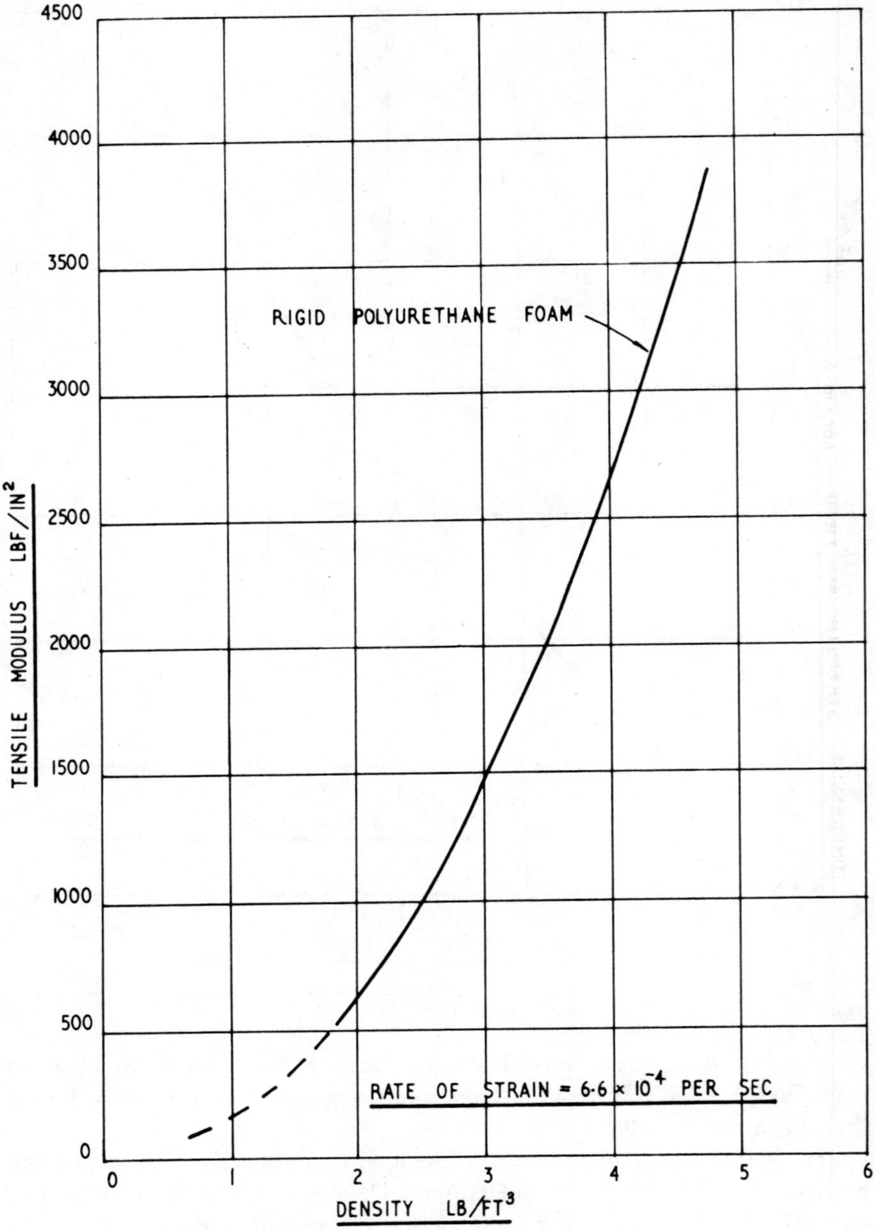
4500
4000
3500
3000
2500
2000
1500
1000
500
0
TENSILE MODULUS LBF/IN²
RIGID POLYURETHANE FOAM
RATE OF STRAIN = 6·6 × 10⁻⁴ PER SEC
0
1
2
3
4
5
6
DENSITY LB/FT³

Fig. 96

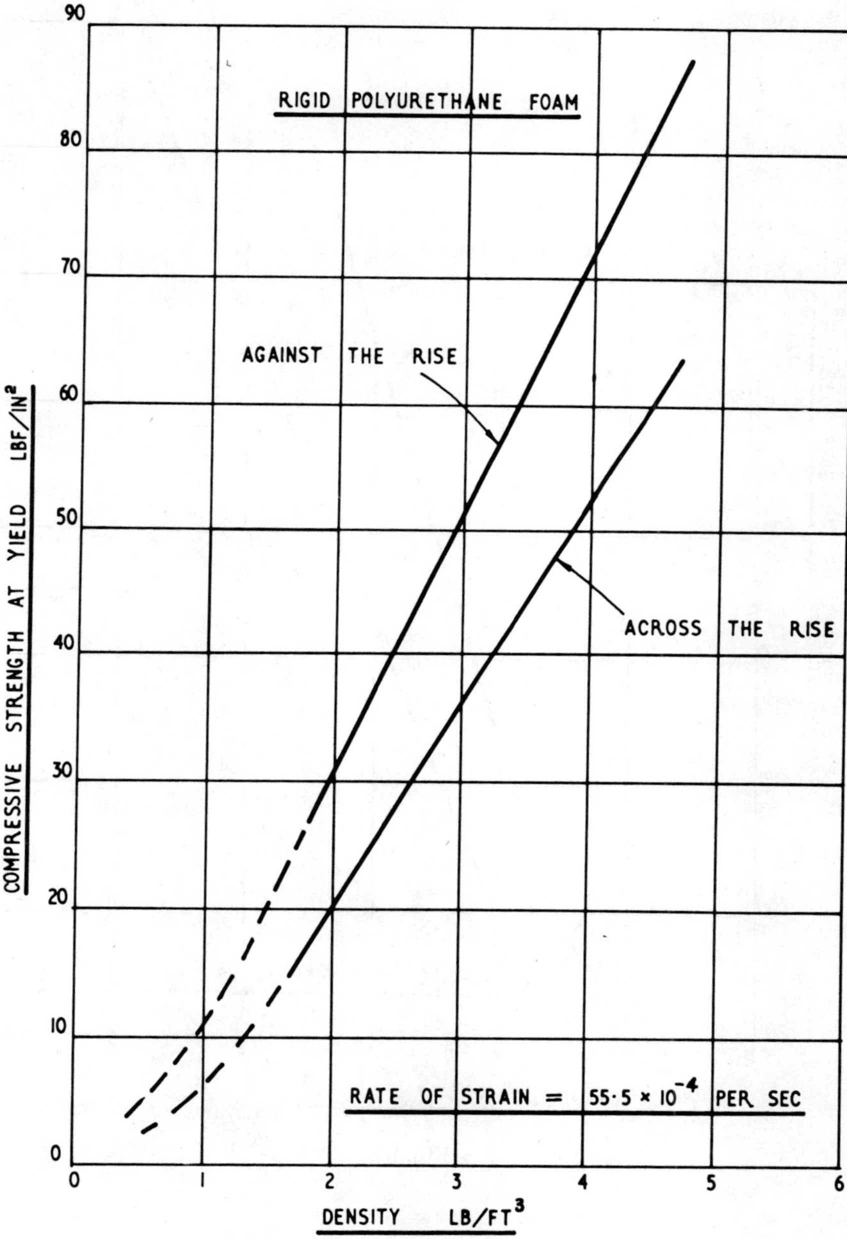
RIGID POLYURETHANE FOAM
AGAINST THE RISE
ACROSS THE RISE
RATE OF STRAIN = 55·5 × 10^-4 PER SEC
COMPRESSIVE STRENGTH AT YIELD LBF/IN²
DENSITY LB/FT³
0
10
20
30
40
50
60
70
80
90
1
2
3
4
5
6

Fig. 97

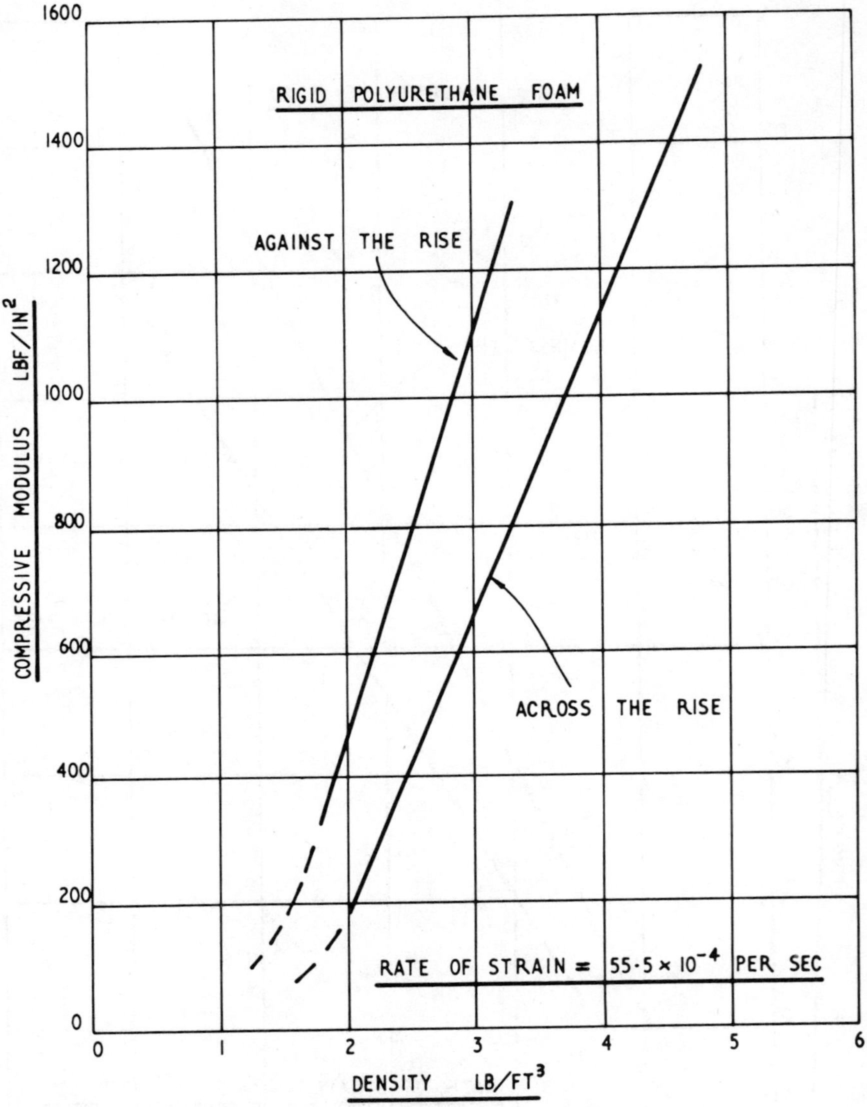
RIGID POLYURETHANE FOAM
AGAINST THE RISE
ACROSS THE RISE
RATE OF STRAIN = 55·5 × 10^-4 PER SEC
COMPRESSIVE MODULUS LBF/IN²
DENSITY LB/FT³
1600
1400
1200
1000
800
600
400
200
0
0
1
2
3
4
5
6

Fig. 98

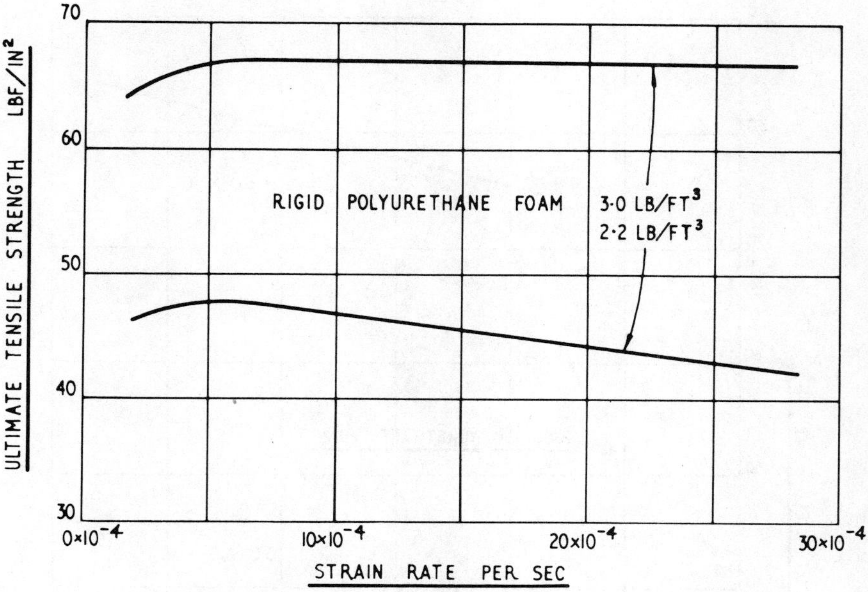

Fig. 99

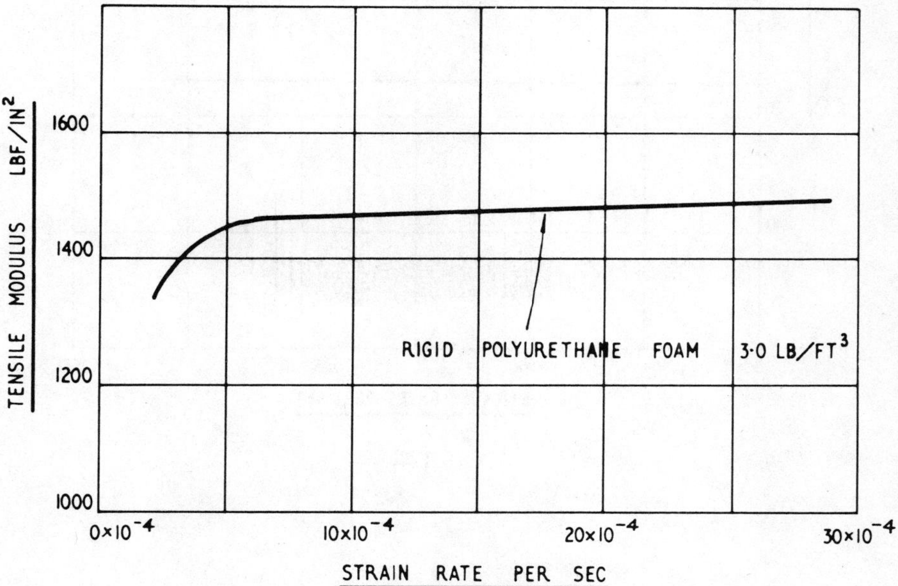

Fig. 100

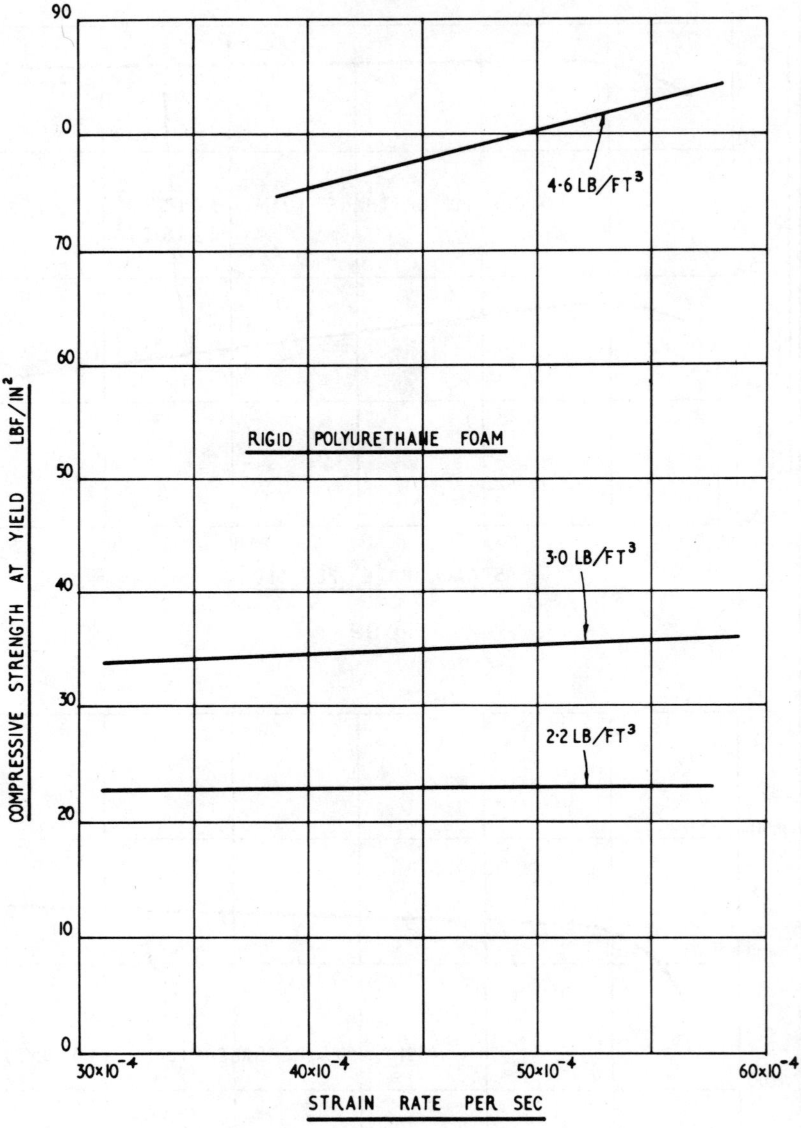

90
0
70
60
50
40
30
20
10
0
COMPRESSIVE STRENGTH AT YIELD LBF/IN²
RIGID POLYURETHANE FOAM
4·6 LB/FT³
3·0 LB/FT³
2·2 LB/FT³
30×10⁻⁴
40×10⁻⁴
50×10⁻⁴
60×10⁻⁴
STRAIN RATE PER SEC

Fig. 101

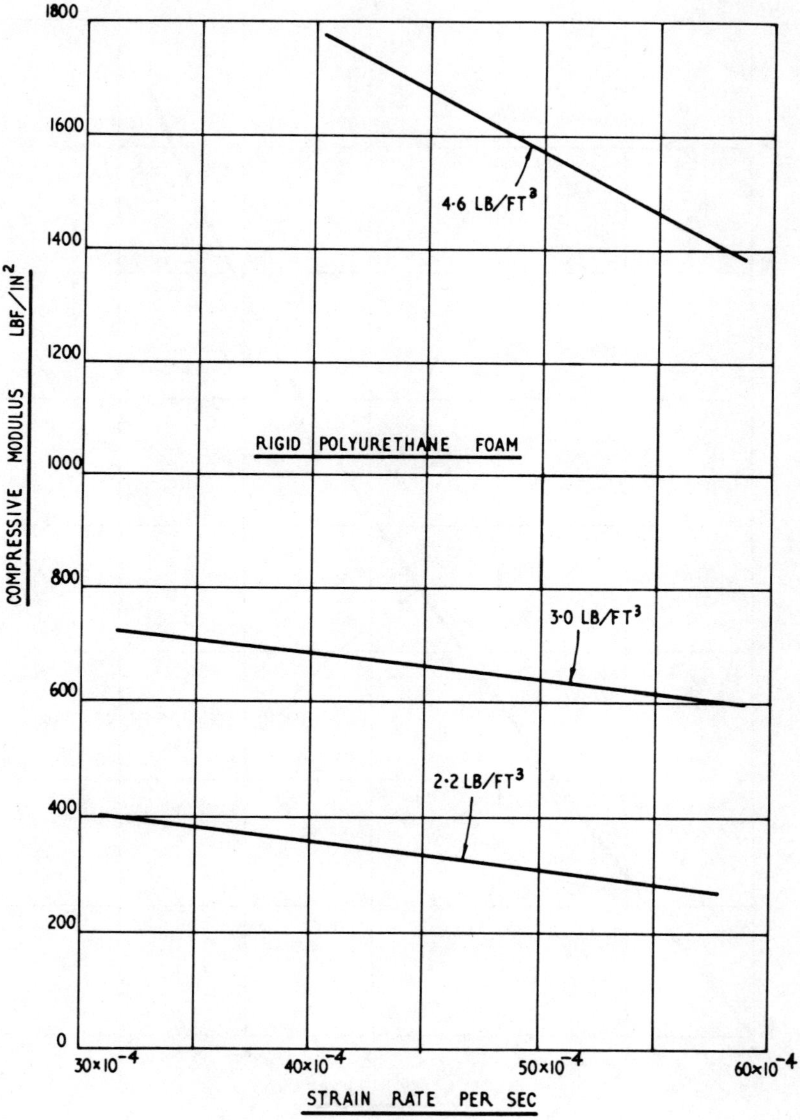
1800
1600
1400
1200
1000
800
600
400
200
0
COMPRESSIVE MODULUS LBF/IN²
4·6 LB/FT³
RIGID POLYURETHANE FOAM
3·0 LB/FT³
2·2 LB/FT³
30×10⁻⁴
40×10⁻⁴
50×10⁻⁴
60×10⁻⁴
STRAIN RATE PER SEC

Fig. 102

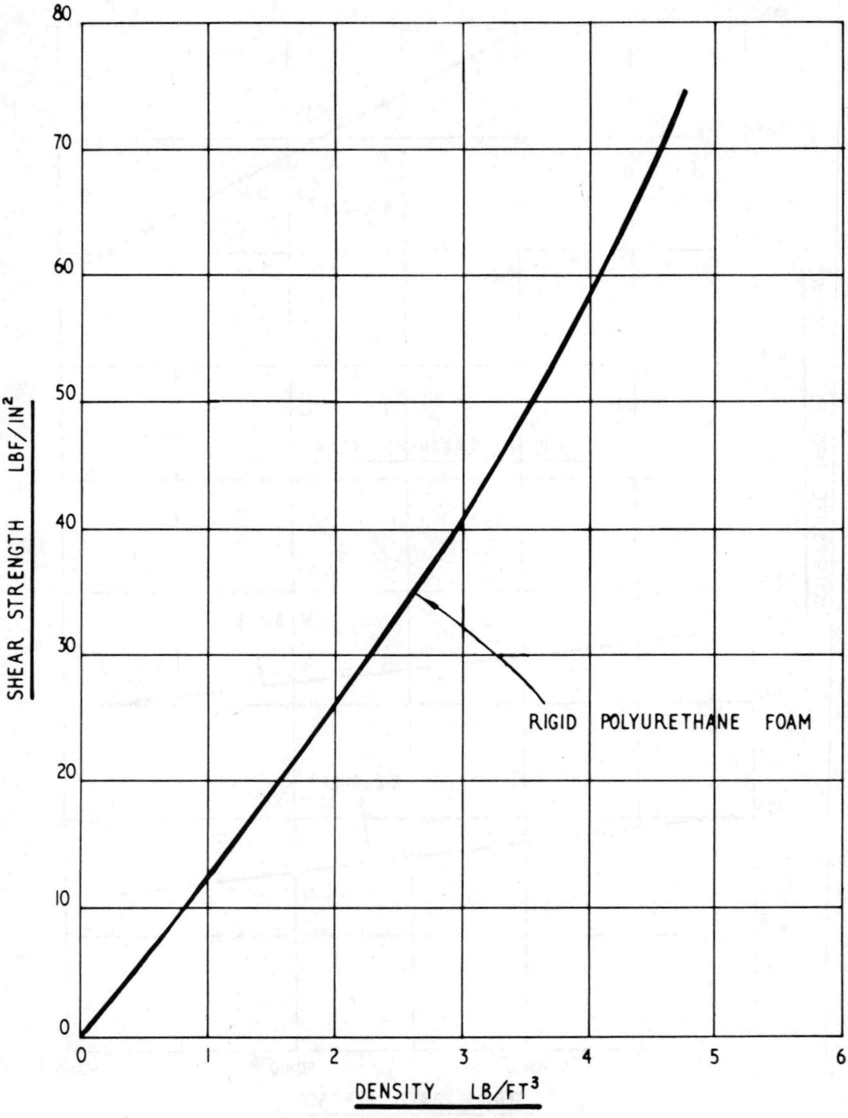
80
70
60
50
40
30
20
10
0
SHEAR STRENGTH LBF/IN²
RIGID POLYURETHANE FOAM
0
1
2
3
4
5
6
DENSITY LB/FT³

Fig. 103

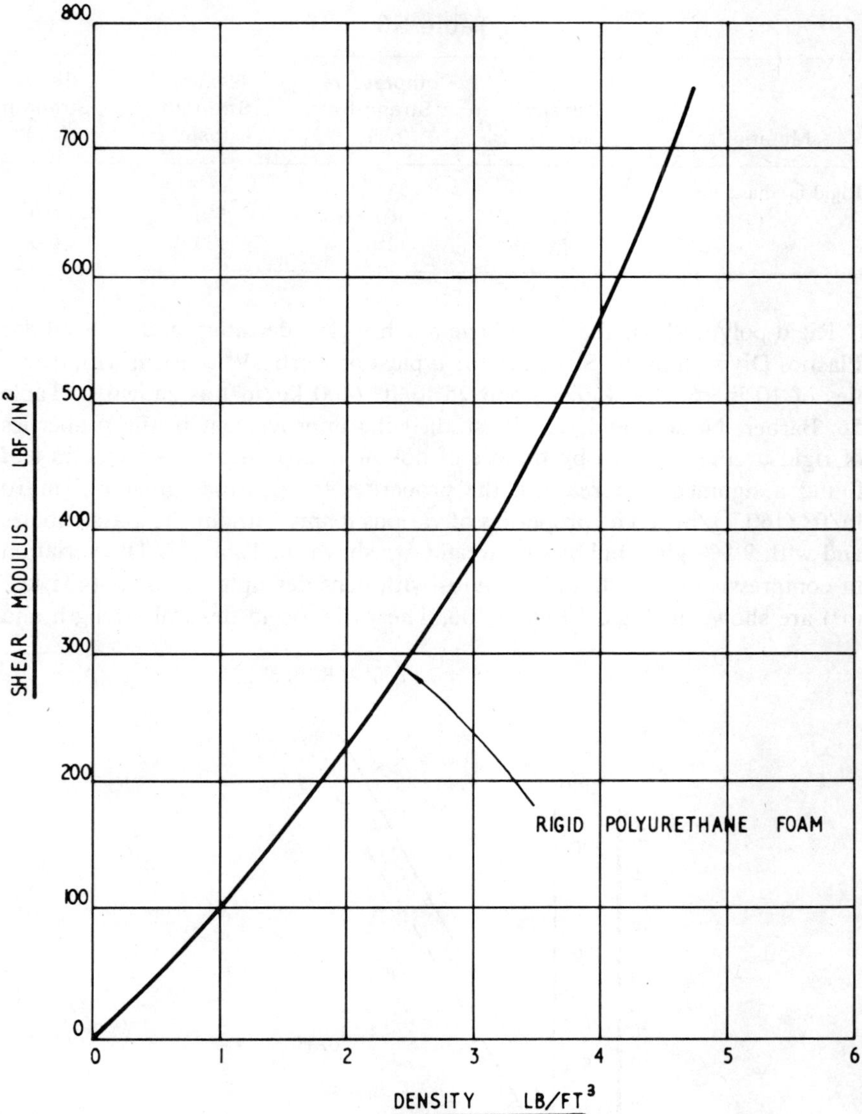
800
700
600
500
400
300
200
100
0
SHEAR MODULUS LBF/IN²
0
1
2
3
4
5
6
DENSITY LB/FT³
RIGID POLYURETHANE FOAM

Fig. 104

Table 56

Material	Density (lb/ft³)	Compressive Strength (psi)	Flexural Strength (psi)	Shear Strength (psi)
Rigid urethane foam	2	35	45	28
	10	360	540	250
	25	1250	1700	450

Rigid polyurethane foam can have much higher densities, and the Cellular Plastics Division of the SPI gives the typical properties[116] of foam with densities of 10 lb/ft³ (160 kg/m³) and 25 lb/ft³ (400 kg/m³) as shown in Table 56. Barber, Nelson, and Beck[117] studied the improvement in the properties of rigid urethane foams by the use of hollow microsphere glass bubbles and found a significant increase in the properties for densities greater than 10 lb/ft³ (160 kg/m³). The properties of various foams without the glass bubbles and with 9.5% glass bubbles by weight are shown in Table 57. The variation in compressive strength and modulus with densities upto 30 lb/ft³ (481 kg/m³) are shown in Figs. 105 and 106. The variation in flexural strength and

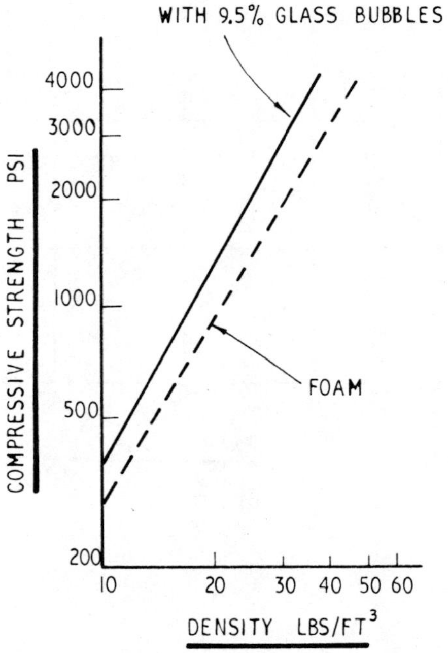

Fig. 105

Table 57

Formulation Used	0.0% Glass Bubbles					9.5% Glass Bubbles				
	Density (lb/ft^3)	Compressive Strength		Compressive Modulus		Density (lb/ft^3)	Compressive Strength		Compressive Modulus	
		‖ to rise (psi)	⊥r to rise (psi)	‖ to rise (psi)	⊥r to rise (psi)		‖ to rise (psi)	⊥r to rise (psi)	‖ to rise (psi)	⊥r to rise (psi)
Isonate CPR–6	10.5	316	373	7880	8006	10.3	367	385	11100	10900
Isonate CPR–10	17.2	737	702	19100	18200	17.8	990	992	29700	27600
Isonate CPR–16	28.1	2036	1950	44600	43000	28.3	2956	3010	63700	61000
Stepanfoam HW 1260	19.9	830	865	21600	21700	21.0	1283	1313	36000	36400
Chempol 32–1760/32–1601	20.1	903	860	24000	21700	19.7	1140	1137	32800	31700

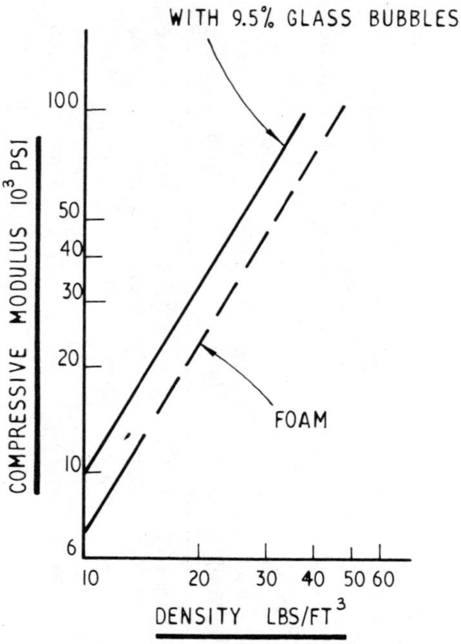

Fig. 106

modulus, with and without the glass bubbles, is shown in Figs. 107 and 108.

The effect of temperature on the mechanical properties of rigid polyurethane foam has been studied by Traeger.[118] He tested foam in the higher densities, ranging from about 4.5 to 26 lb/ft³ (72–417 kg/m³). He has represented his experimental results in the form of gentle curves, and draws the conclusion that the mechanical properties drop rapidly with an increase in temperature. A closer look at his experimental results from P-foam leads the author to conclude that the mechanical properties of this foam are zero at about 500°F (260°C). Furthermore, the variation in mechanical properties with temperature is almost linear between −100°F (−73°C) and 500°F (260°C).

This conclusion by the author is realistic since Traeger's[118] experimental readings show a large scatter. Each of his experimental points is based on an average obtained from five tests. He reports that the properties of the individual test specimens were within ±10% of the mean for flexure and shear tests and ±20% of the mean for the tensile tests. This sort of scatter, however, is to be easily expected in tests on low-density foams, since there are so many intagible factors affecting the property of the foam. The method of formation, the formulations, the cell structure, and the gas phase composition[119] all affect the properties of the material.

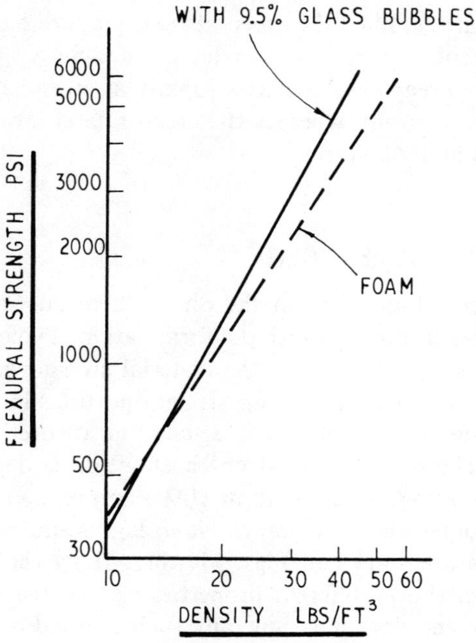

Fig. 107

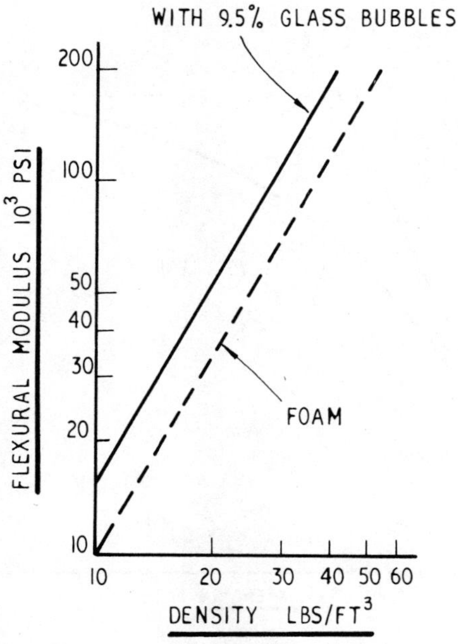

Fig. 108

Traeger[118] has also studied the creep of rigid polyurethane foams at 75°F (24°C). No differences were noted in foams of different formulations. He concludes that the creep rate increases rapidly at stresses above 60%–70% of the stress at 10% strain, whereas the creep rate is low at stresses below 40% of the stress at 10% strain.

3.3 EXPANDED POLYSTYRENE

The materials were obtained from the Shell Chemical Co., and a suitable flame retardant was introduced into the formulation. Typical load/extension and load/compression curves[112] for the material are shown in Figs. 109 and 110. Expanded polystyrene shows two straight portions in its load/compression curve, the intersection of which defines the theoretical yield point Y of the material. The compressive strength at yield, as determined from Y, is less than the compressive strength at 10% compression.

The variations of tensile and compressive strengths and moduli with density and rate of strain are shown in Figs. 111–118. It has been seen that rigid polyurethane foam shows different properties against the rise and across it. Expanded polystyrene does not show any such marked orthotropy. Figures 115–118 show that varying the rate of strain of the material over a large range of straining rates does not have a marked effect on the strengths and

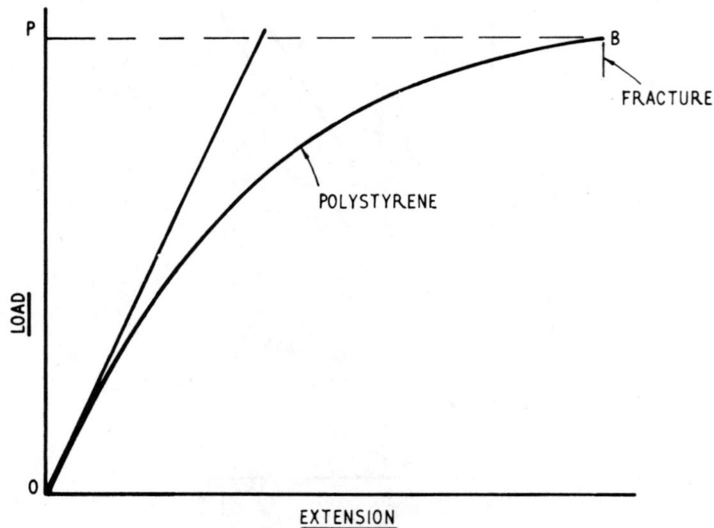

Fig. 109

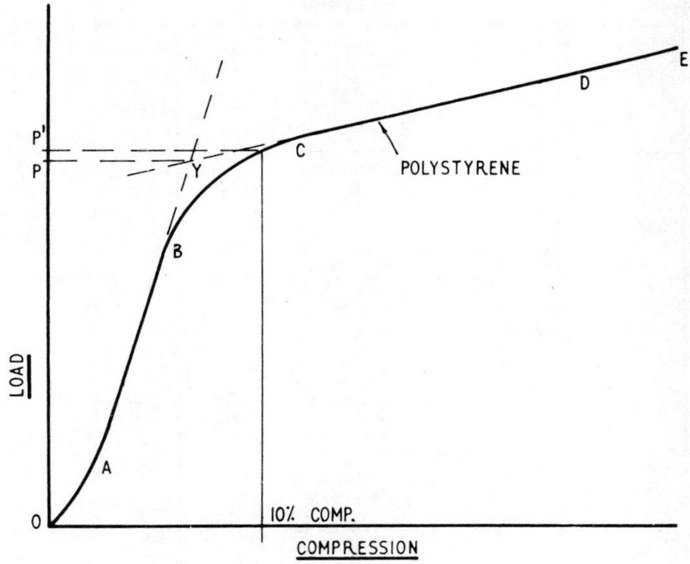

Fig. 110

the moduli of the material. The shear strength and the shear modulus are plotted against the density in Figs. 119 and 120.

Amspec[120] gives the compressive strength of Styrofoam as 40 psi (276 kPa) at a density of 2.1 lb/ft^3 (34 kg/m^3). This is the compressive strength at 10% deformation.

Finally, the long term creep behavior of expanded polystyrene has also been studied in the research laboratories of the Shell Chemical Co.,[121] and the compression creep of the material with time is shown in Fig. 121.

3.4 EXPANDED POLYVINYLCHLORIDE (PVC)

The materials were obtained from BTR Industries Limited of Great Britain, and the tests on this material, and on phenolic foam, were carried out by the author's colleague B. Hughes.[122,123] Typical load/extension and load/compression curves for the material are similar to those for rigid polyurethane foam shown in Figs. 93 and 94. The variation of the elastic modulus in tension with the rate of strain is shown in Fig. 122. It can be seen that the effects are as small as those for the other low-density core materials. The variation with density, however, is of great importance. The variation of

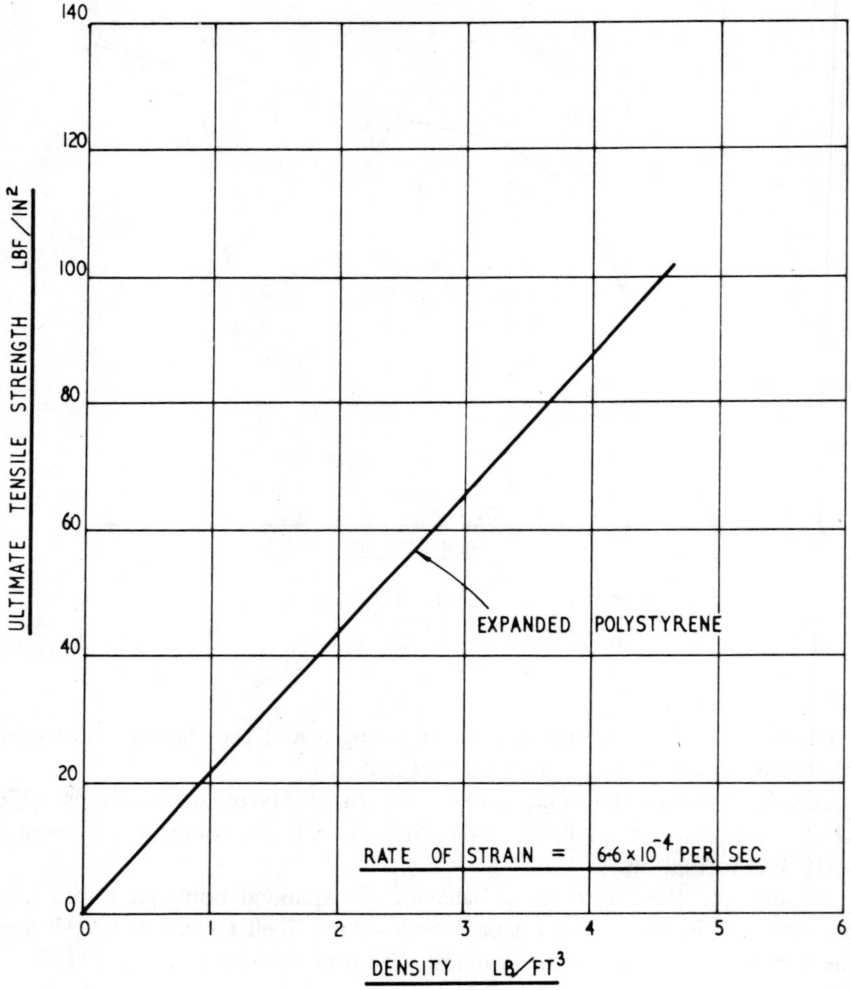

Fig. 111

the tensile, compressive, and shear strengths and moduli with density is shown in Figs. 123–128.

3.5 PHENOLIC FOAM

The material was obtained from Bakelite Plastics Limited of Great Britain. Phenolic foam, like rigid polyurethane foam, is a thermosetting plastics, and

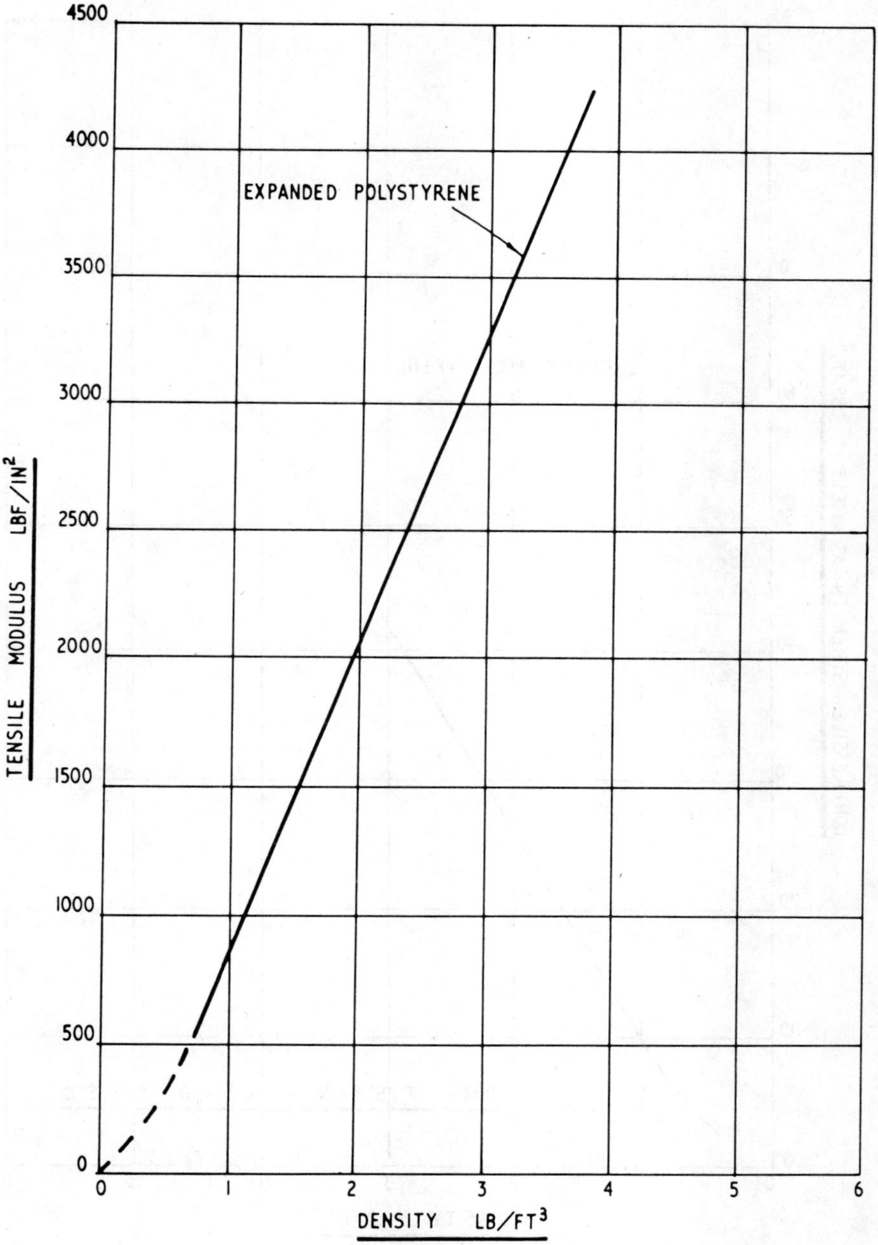
4500
4000
3500
3000
2500
2000
1500
1000
500
0
EXPANDED POLYSTYRENE
TENSILE MODULUS LBF/IN²
0
1
2
3
4
5
6
DENSITY LB/FT³

Fig. 112

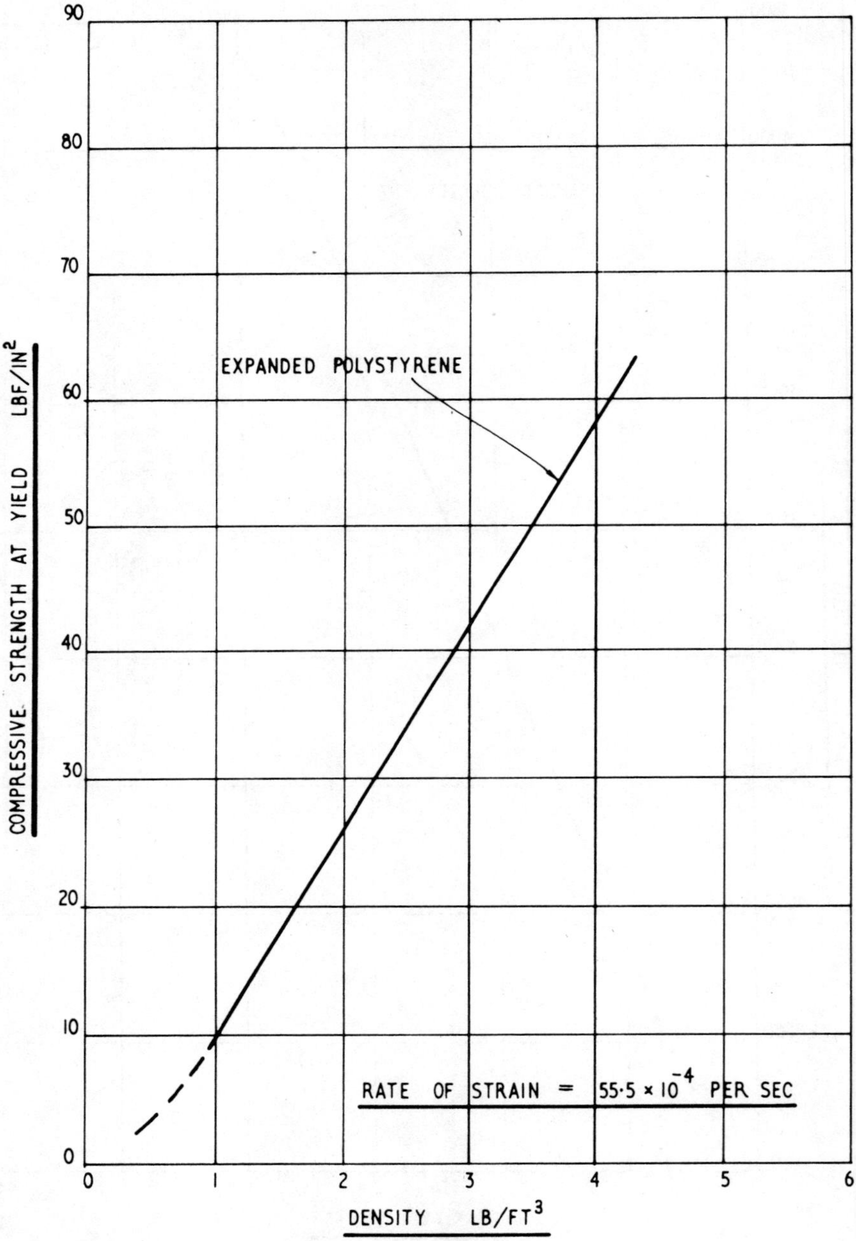

90
80
70
60
50
40
30
20
10
0
COMPRESSIVE STRENGTH AT YIELD LBF/IN²
EXPANDED POLYSTYRENE
RATE OF STRAIN = 55·5 × 10⁻⁴ PER SEC
0
1
2
3
4
5
6
DENSITY LB/FT³

Fig. 113

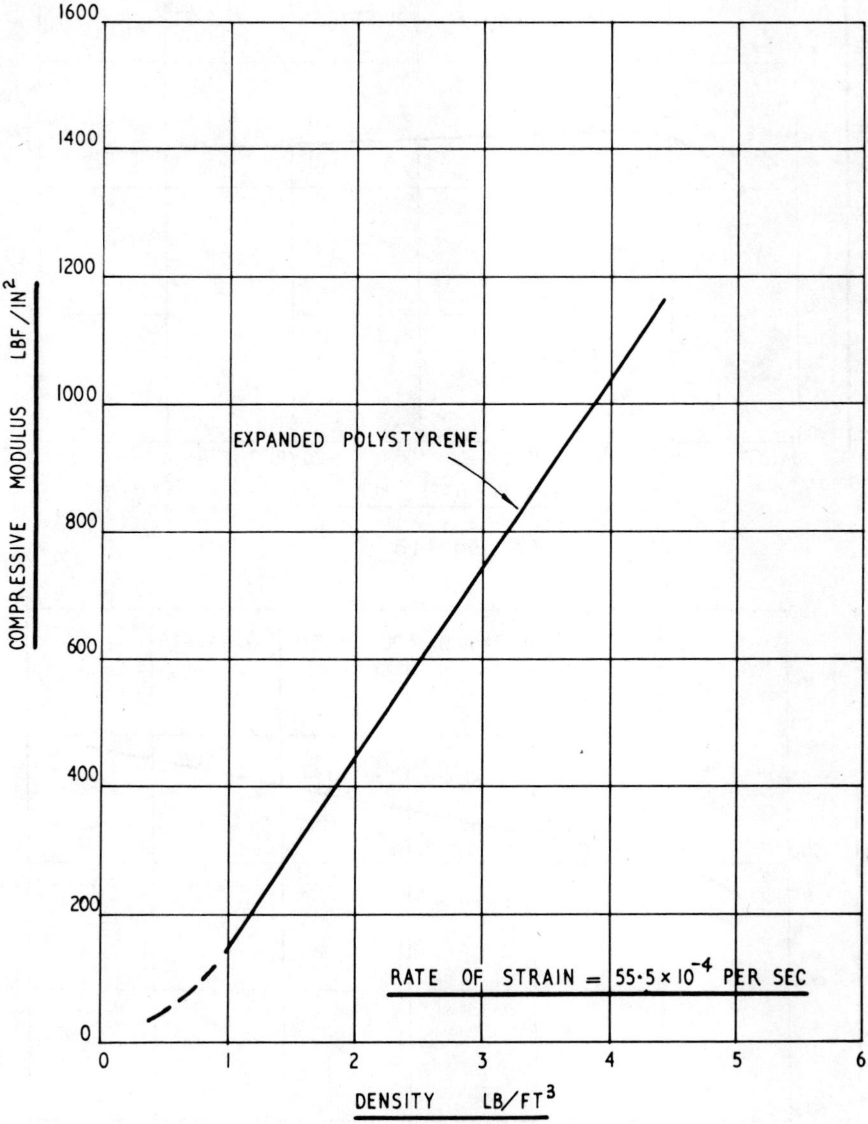
1600
1400
1200
1000
800
600
400
200
0
COMPRESSIVE MODULUS LBF/IN²
EXPANDED POLYSTYRENE
RATE OF STRAIN = 55·5 x 10⁻⁴ PER SEC
0
1
2
3
4
5
6
DENSITY LB/FT³

Fig. 114

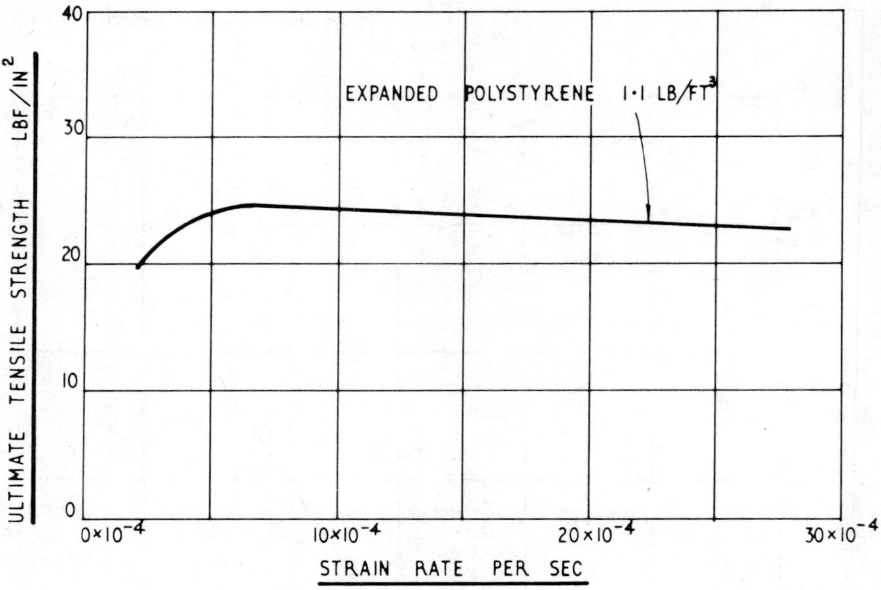

Fig. 115

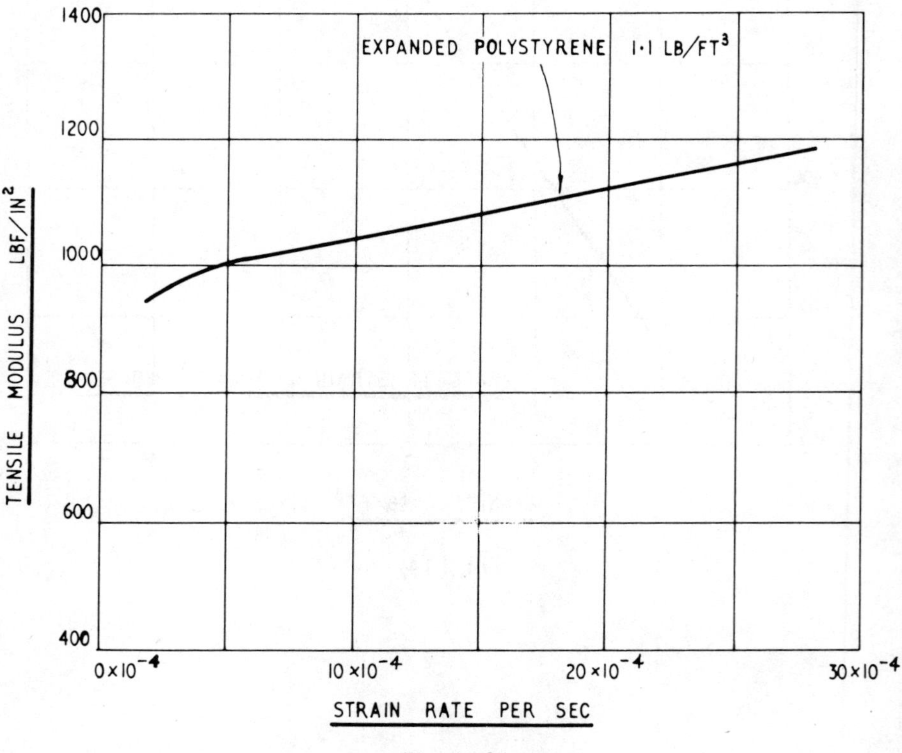

Fig. 116

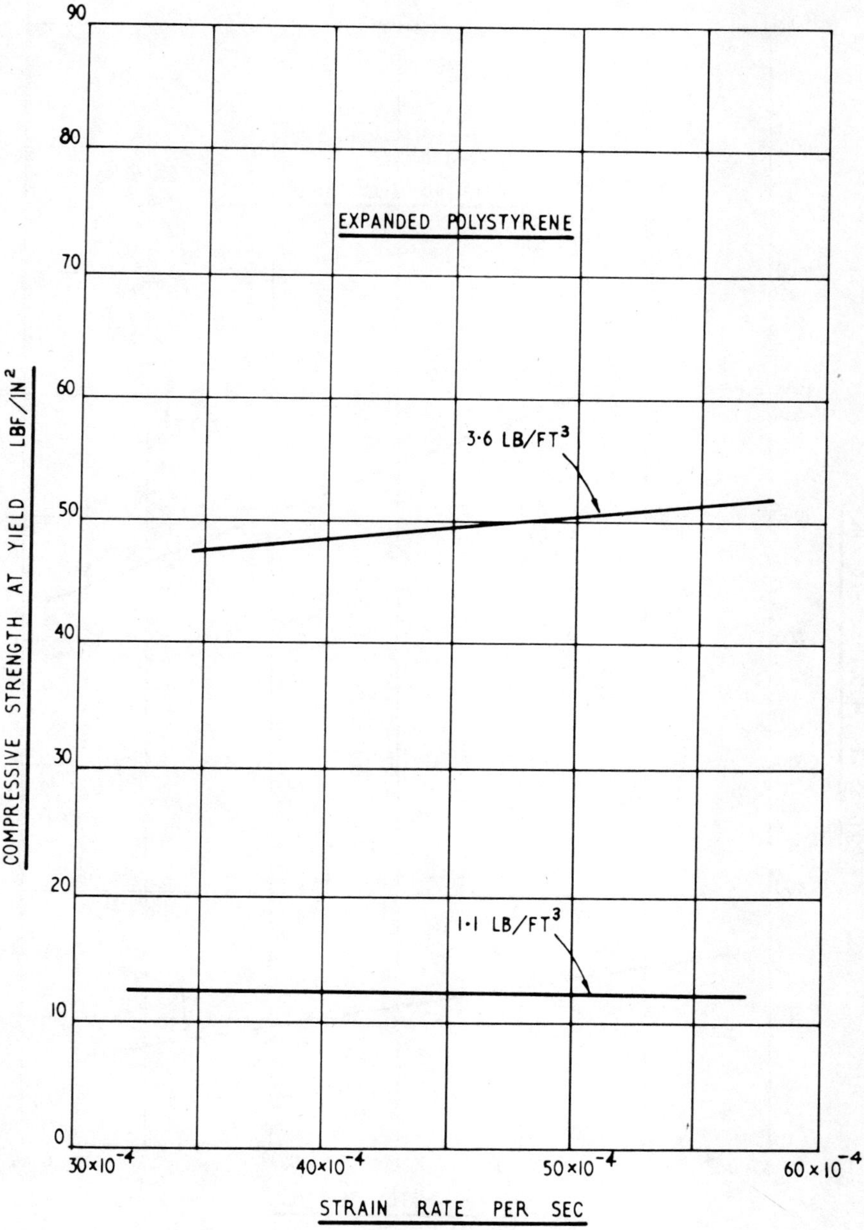

EXPANDED POLYSTYRENE
3·6 LB/FT³
1·1 LB/FT³
COMPRESSIVE STRENGTH AT YIELD LBF/IN²
0
10
20
30
40
50
60
70
80
90
30×10⁻⁴
40×10⁻⁴
50×10⁻⁴
60×10⁻⁴
STRAIN RATE PER SEC

Fig. 117

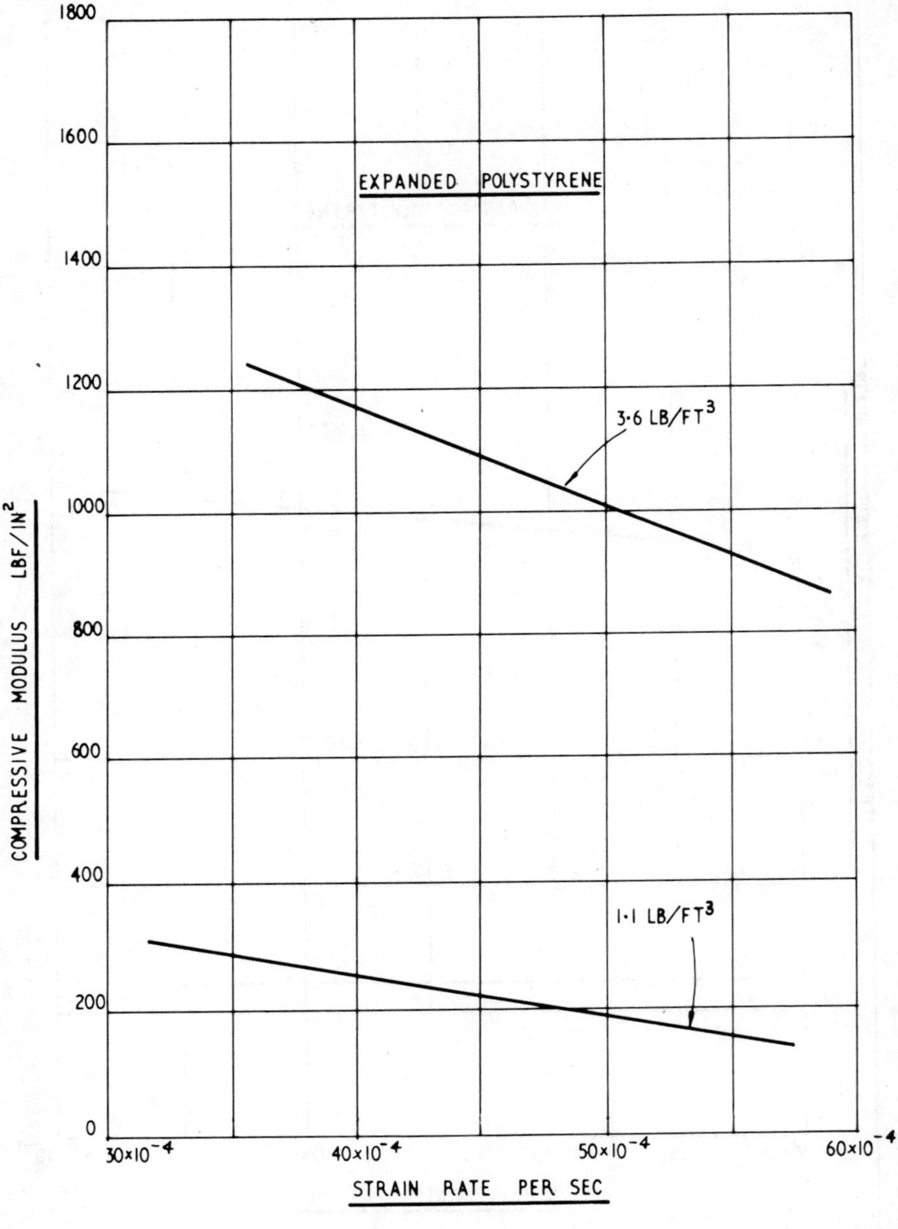
EXPANDED POLYSTYRENE
3·6 LB/FT³
1·1 LB/FT³
COMPRESSIVE MODULUS LBF/IN²
1800
1600
1400
1200
1000
800
600
400
200
0
30×10⁻⁴
40×10⁻⁴
50×10⁻⁴
60×10⁻⁴
STRAIN RATE PER SEC

Fig. 118

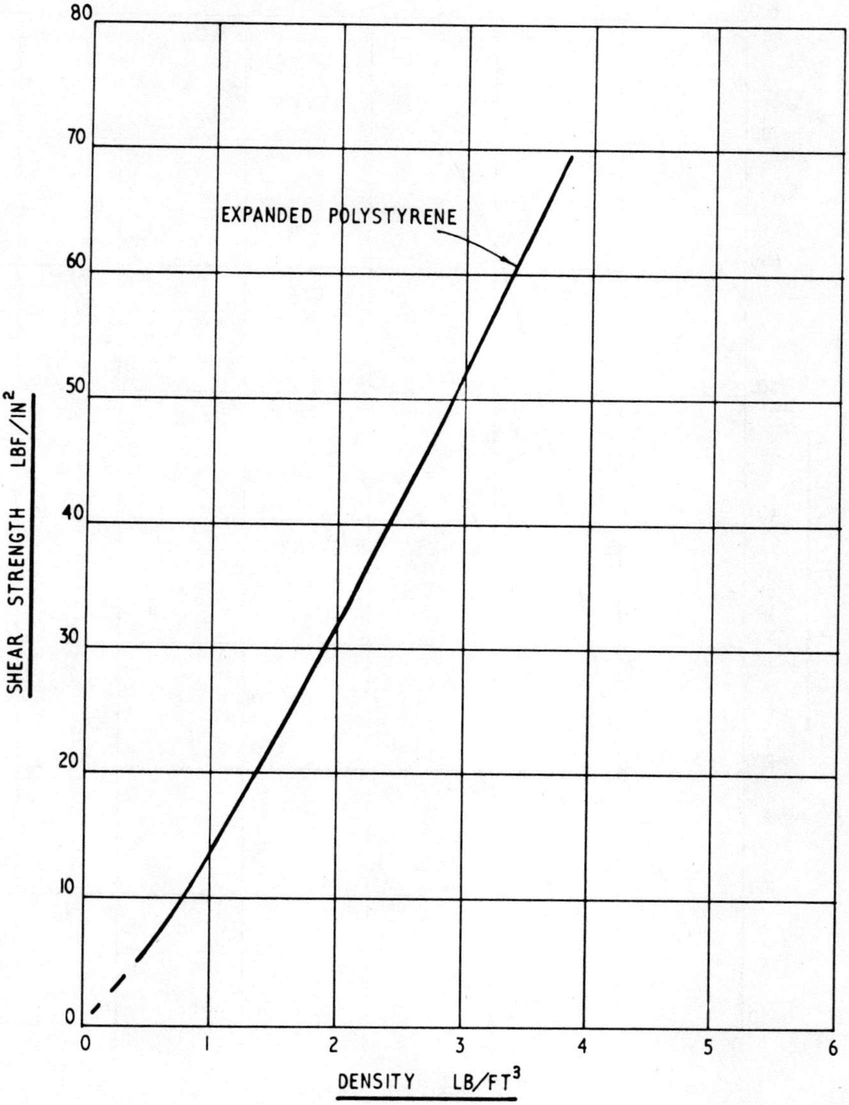
80
70
60
50
40
30
20
10
0
0
1
2
3
4
5
6
EXPANDED POLYSTYRENE
SHEAR STRENGTH LBF/IN²
DENSITY LB/FT³

Fig. 119

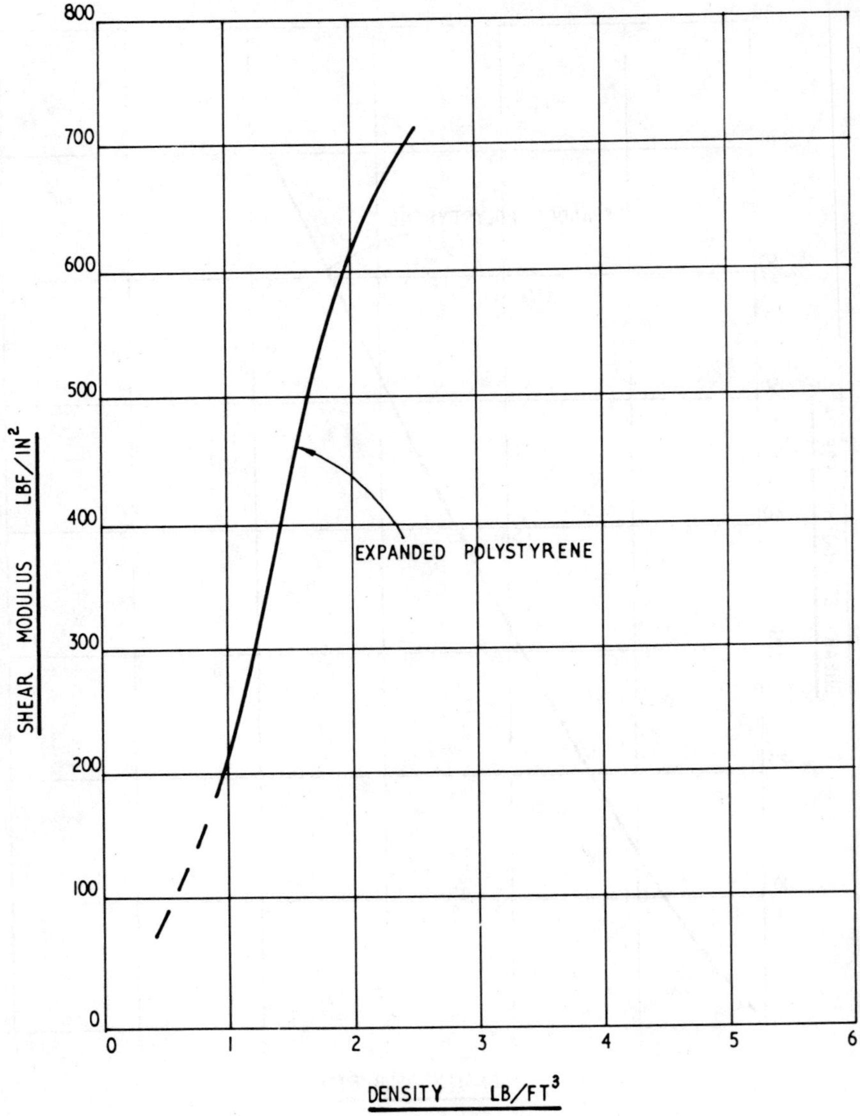
800
700
600
500
400
300
200
100
0
SHEAR MODULUS LBF/IN²
EXPANDED POLYSTYRENE
0
1
2
3
4
5
6
DENSITY LB/FT³

Fig. 120

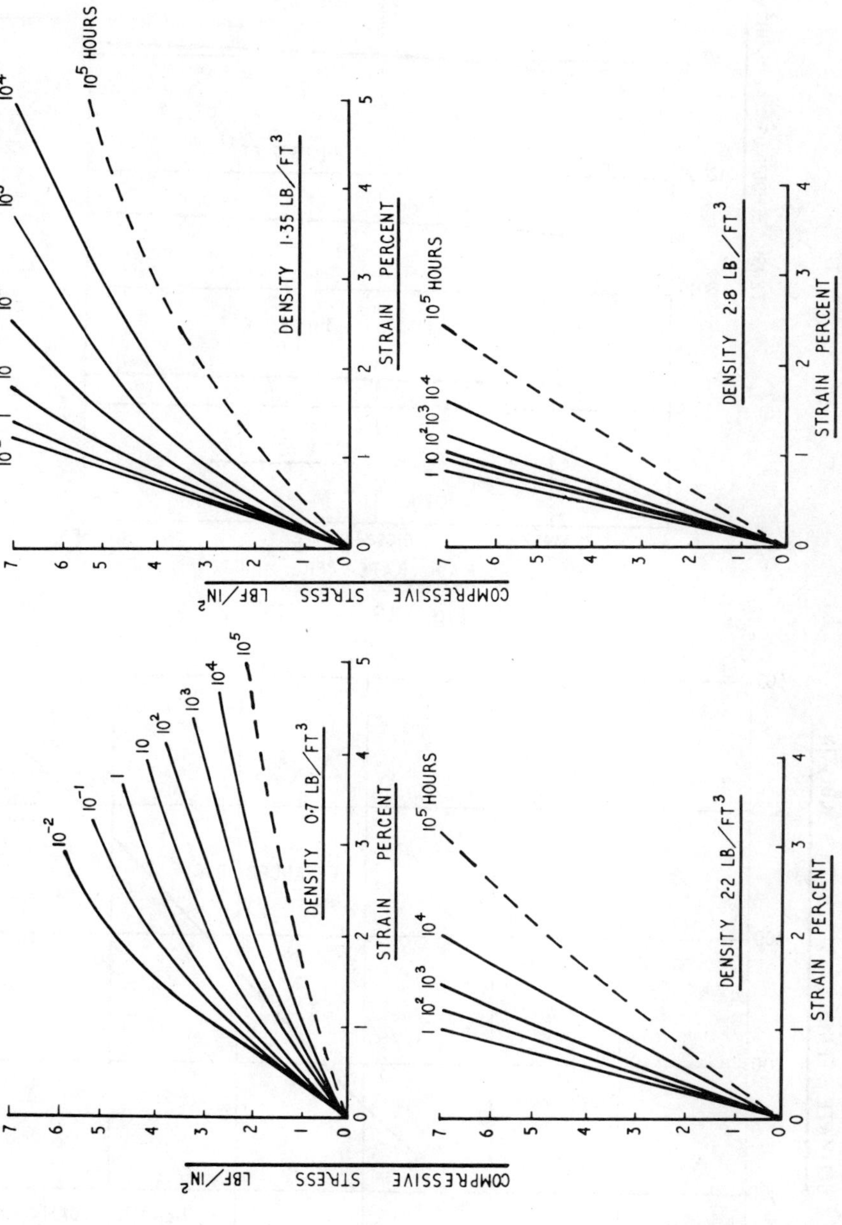
COMPRESSIVE STRESS LBF/IN²
STRAIN PERCENT
DENSITY 0·7 LB/FT³
DENSITY 1·35 LB/FT³
DENSITY 2·2 LB/FT³
DENSITY 2·8 LB/FT³
10⁵ HOURS

Fig. 121

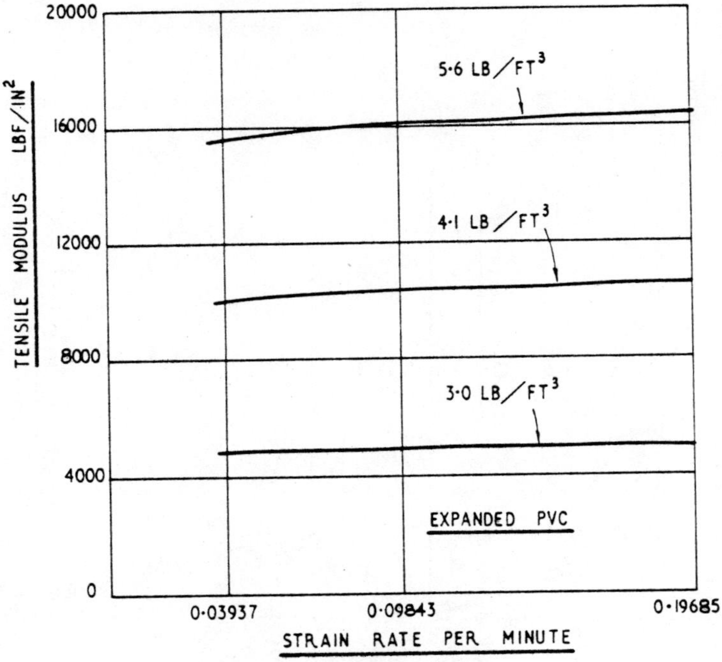

Fig. 122

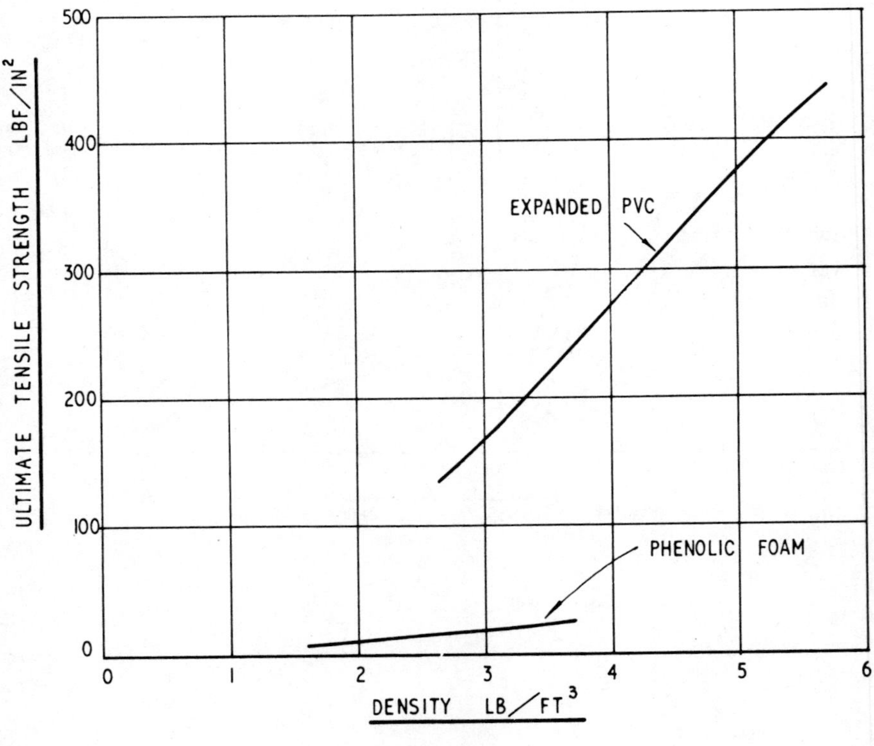

Fig. 123

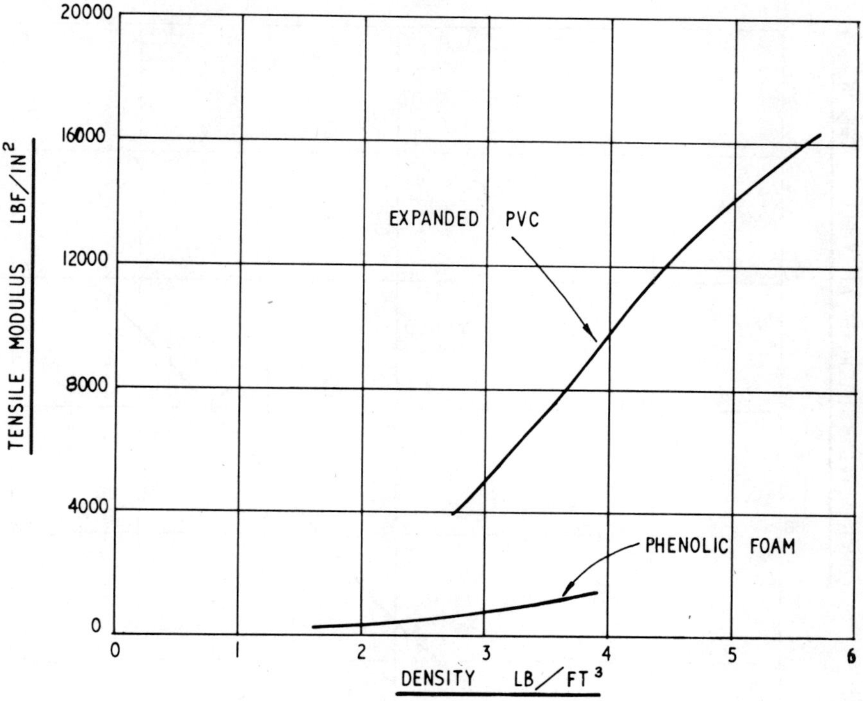

Fig. 124

shows much the same behavior as rigid polyurethane foam. Phenolic foam is very friable, with a large cell structure. It may be necessary to seal the surface by a thixotropic resin before use. The variation of the tensile, compressive, and shear strengths and moduli with density is shown in Figs. 123–128.

3.6 POLYESTER FOAM

Using foaming agents, the unsaturated polyesters,[4] and other resins, can also be foamed to densities that range widely, from 7 to 35 lb/ft³ (112–561 kg/m³) and even more. Gallagher et al.[124] have reported on the use of Lucidol's foaming agents with the unsaturated polyesters. The variation of the compressive strength and modulus with density is shown in Fig. 129. It should be noted that the foam can also be filled to improve the properties even further.

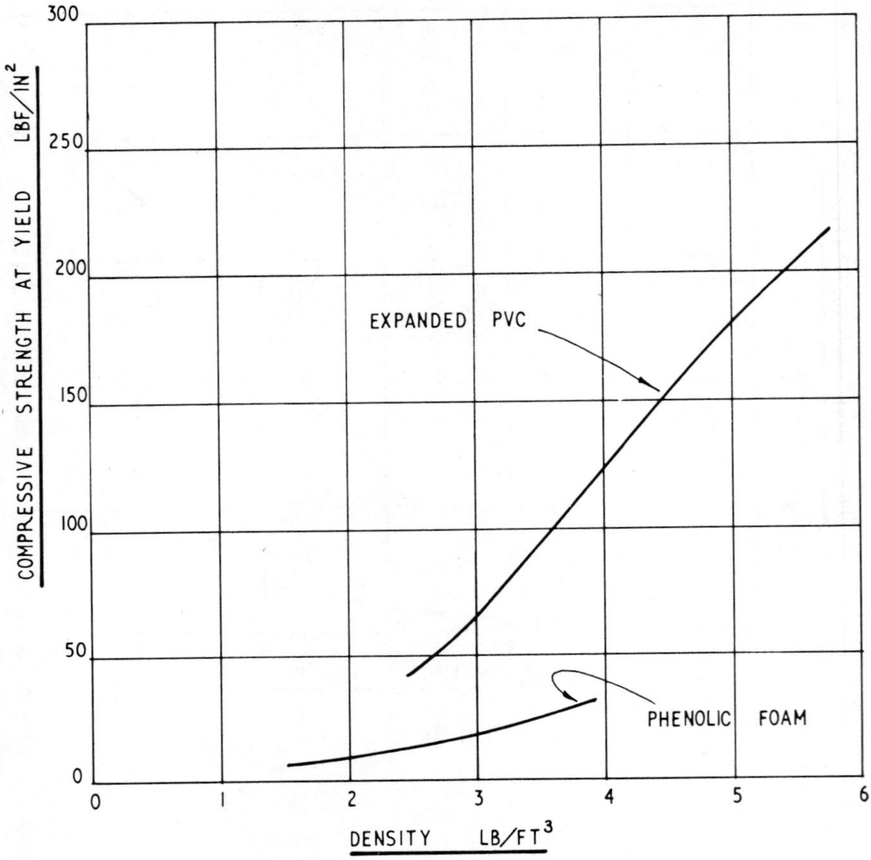

Fig. 125

3.7 POLYSULFONE FOAM

Smarook[5] has reported on a low-temperature polysulfone molding process to produce foam densities of 6–31 lb/ft³ (96–497 kg/m³). As for all the other foams, the properties are a function of the density, and the variation of some of these properties with density are shown in Figs. 130–134. The tensile, compressive, and shear strengths of the unfilled foam are shown in Fig. 130. The foam can also be filled with glass fiber, graphite fibers, and the like. Figures 131–134 show the variation of both flexural and tensile strengths and moduli with various percentages of carbon fiber by weight.

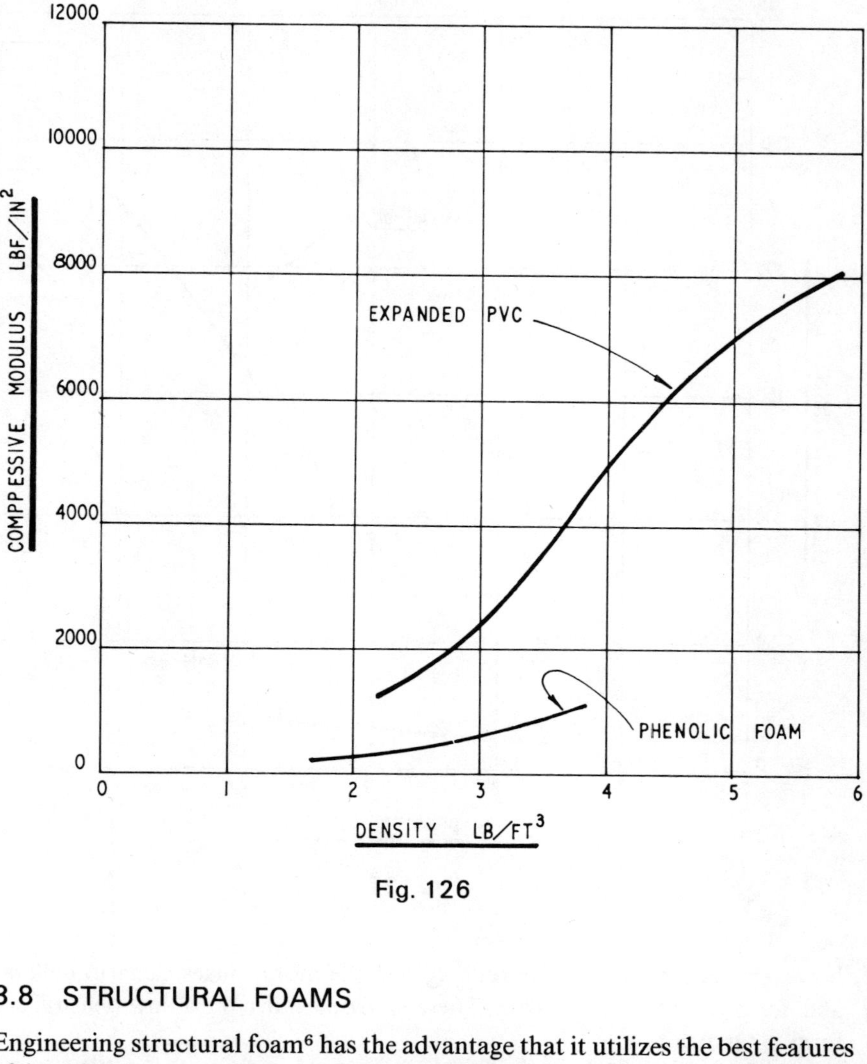

Fig. 126

3.8 STRUCTURAL FOAMS

Engineering structural foam[6] has the advantage that it utilizes the best features of low-density core materials, together with the formation of tough, solid, integral skins in a unitized sandwich construction. With the use of injection molding, there are further advantages in the manufacturing process.

To form the low-density core and the solid skin, an inert gas is dispersed in a polymer melt during processing. It would be possible to use a blowing agent that decomposes when subjected to the heat of processing. Structural foam is produced by release into a closed mold. As the gas expands, the mold is filled with foam, having a cellular structure. Against the mold surface,

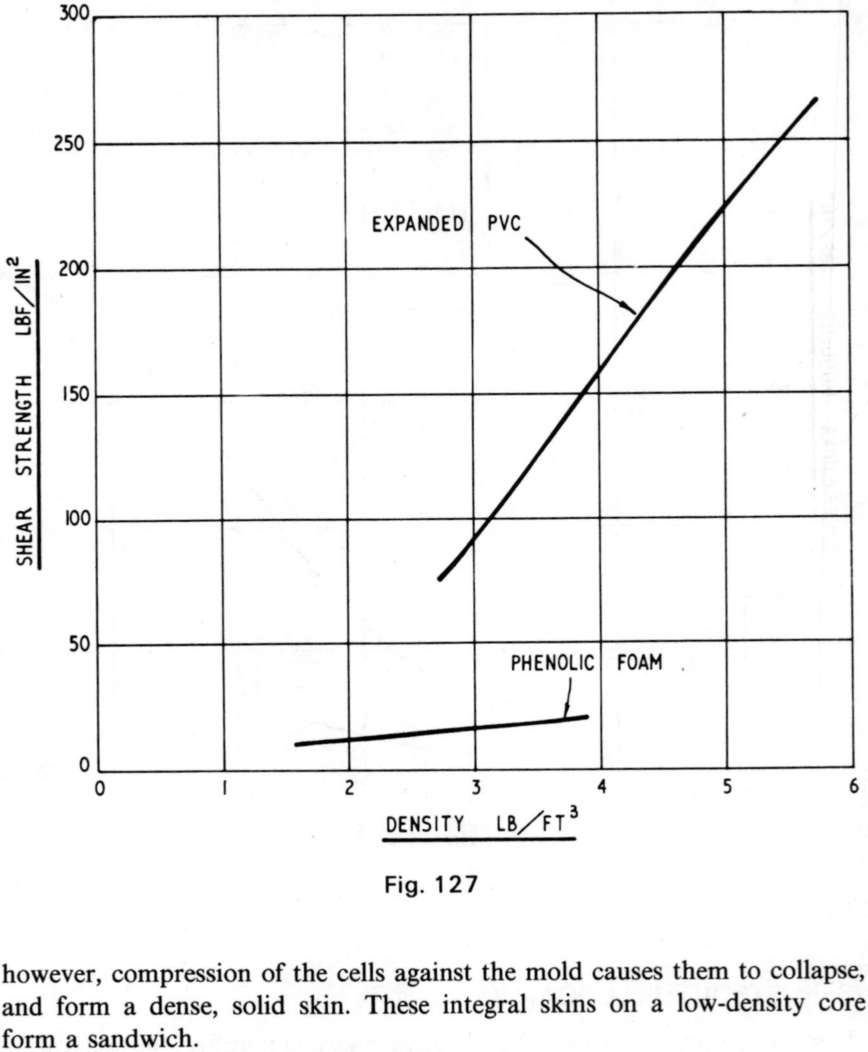

Fig. 127

however, compression of the cells against the mold causes them to collapse, and form a dense, solid skin. These integral skins on a low-density core form a sandwich.

A large variety of thermoplastic resins can be used, such as ABS, nylon, polycarbonate, phenylene-oxide based resins, and the like. Thermosets can also be used, such as polyurethane. Structural foam can also be filled for even greater strength and stiffness. In the design of parts subjected to high temperatures, the heat distortion temperatures should be kept in mind. Thermoplastic polyester, glass reinforced nylon, polycarbonate, and polysulfone all have good heat distortion temperatures.

The mechanical properties of the structural foams will depend on the resin used, and the properties of some of these materials will be considered here.

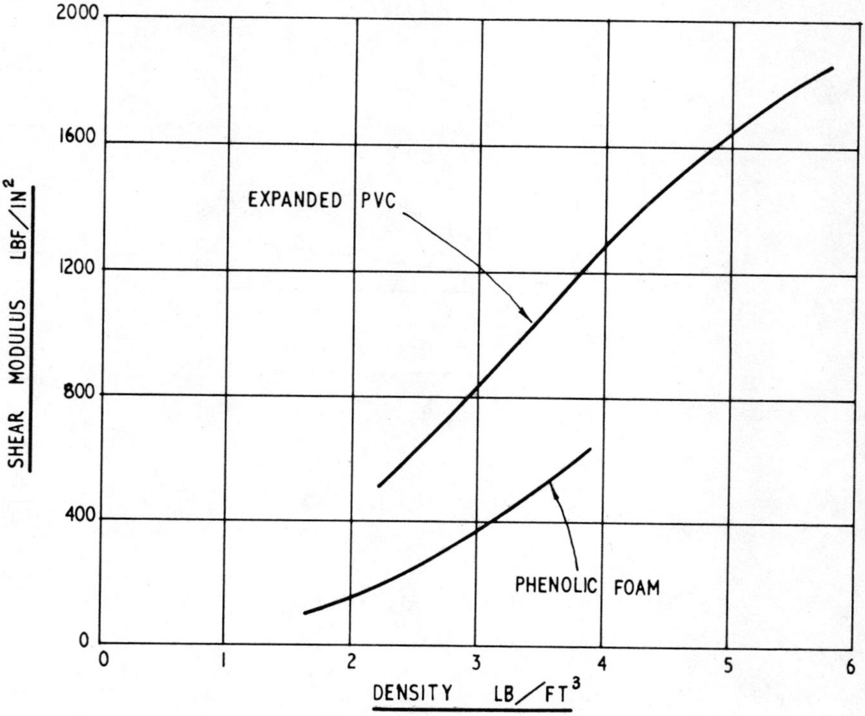

Fig. 128

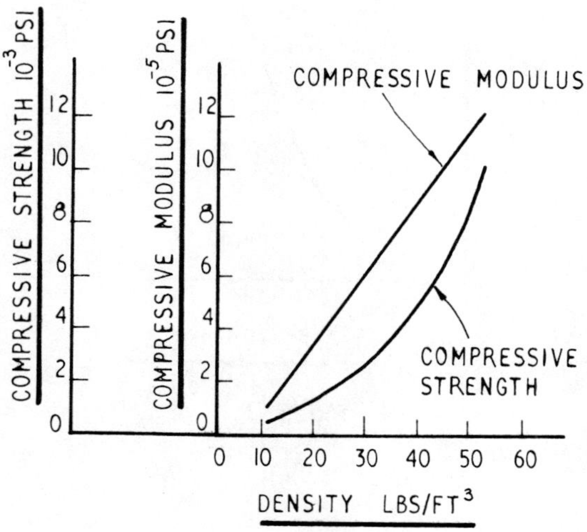

Fig. 129

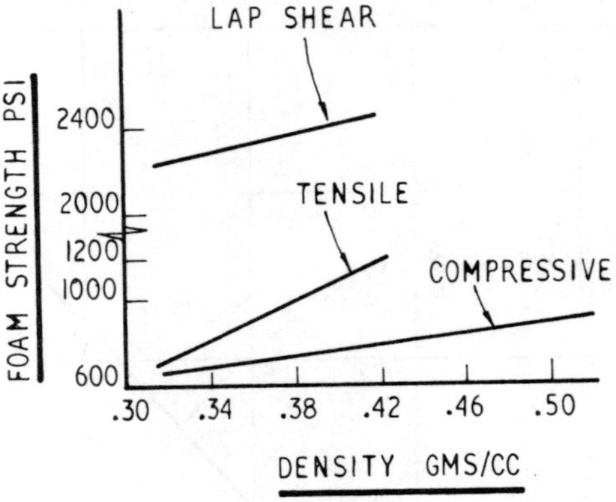

Fig. 130

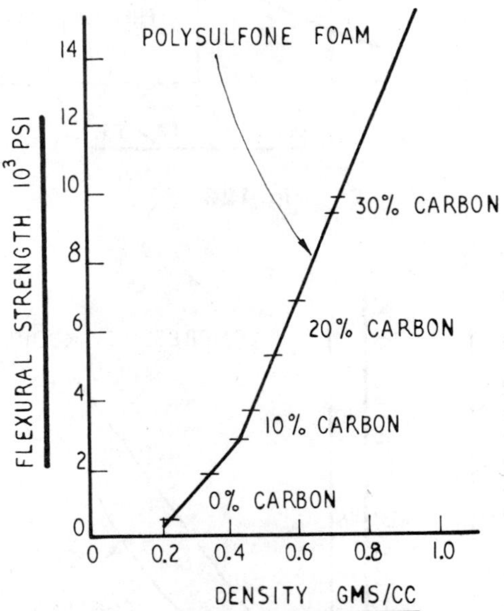

Fig. 131

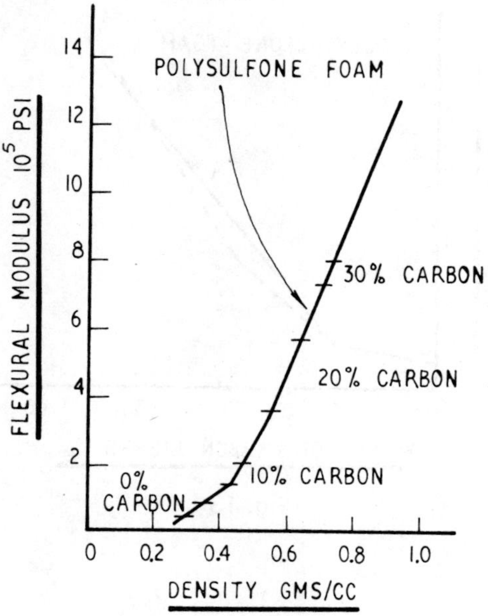

Fig. 132

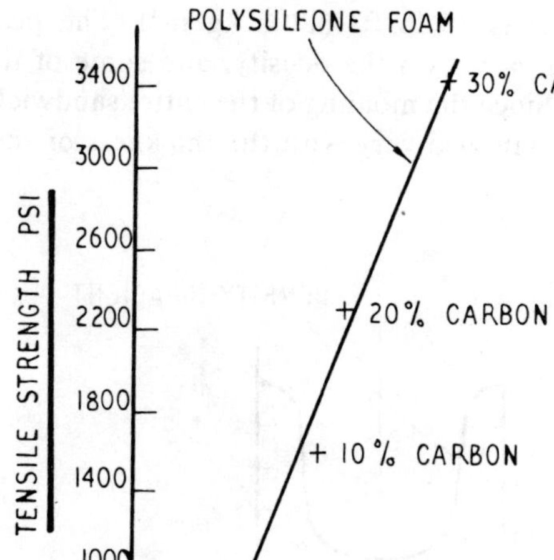

Fig. 133

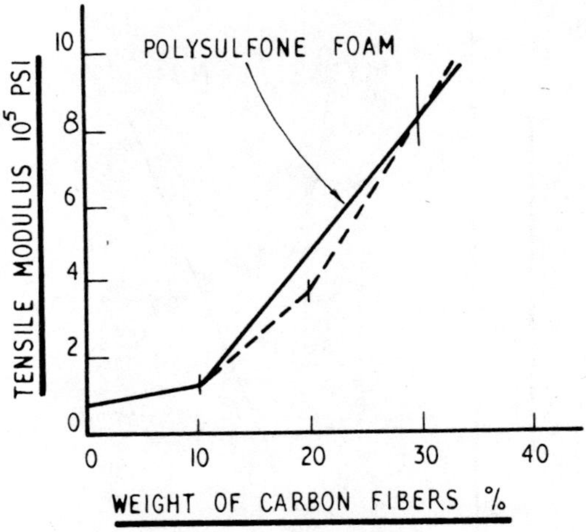

Fig. 134

Eakin[125] has studied the properties of polyurethane structural foams. The density gradient through a typical ⅜ in. (10 mm) molded part is shown in Fig. 135. The density of the core is only about 25 lb/ft³ (400 kg/m³) while the density of the skins is 75 lb/ft³ (1202 kg/m³). The properties of the polyurethane foam will vary with the density, and some of these properties are shown in Table 58. Since the molding of the entire sandwich is an integral process, the properties will also vary with the thickness of the part.

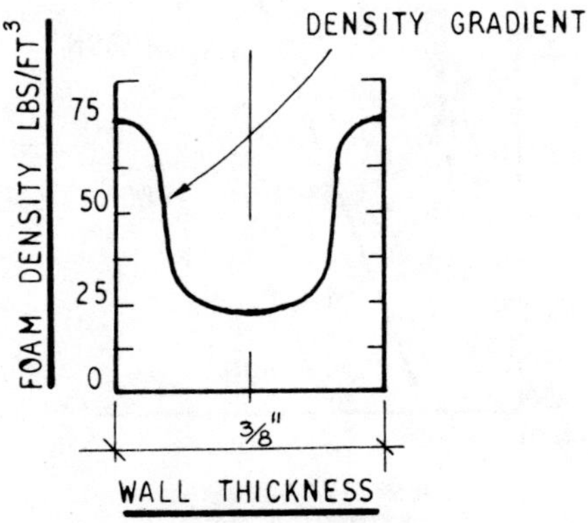

Fig. 135

Table 58

Property	Sp. gr. = 0.4 (25 lb/ft³)	Sp. gr. = 0.5 (31 lb/ft³)	Sp. gr. = 0.6 (37.5 lb/ft³)
Flexural strength, psi	3,200	4,600	6,000
Flexural modulus, psi × 10^5	1	1.3	1.6
Compressive strength, psi	800	1,800	2,800
Tensile strength, psi	1,600	2,500	3,400
Heat deflection temperature at 66 psi, °F	180	192	205

With thermoplastic materials, General Electric[6] has given a great deal of information on the structural foams of polyester, polycarbonate, and phenylene-oxide based resin. The comparative properties of each of these materials is given in Table 59. Matonic[126] has also studied the properties of polycarbonate structural foam and these are given in Table 60.

The long term creep behavior of structural foam is also of some importance, and General Electric[6] has given the reduction in the elastic modulus with time for polycarbonate structural foam, as shown in Fig. 136.

3.9 POISSON'S RATIO OF THE LOW-DENSITY CORE MATERIALS

The Poisson's ratio is very difficult to determine for all of these materials. It is of interest to examine the values of ν for these materials given in the published literature. O'Dell and Graham[127] give the value of ν for extruded polystyrene foam as less than 0.1. The Marine Design Manual[128] gives the value of ν for a "foamed plastic core" as 0.20. Panshin et al.[129] use a value of $\nu = 0.30$ when referring to cellular plastics. Oberdick[130] gives a value of $\nu = 0.64$ for flexible polyurethane foam. Hughes[123] has determined the value of the Poisson's ratio for expanded PVC as 0.35, for densities greater than 3 lb/ft³ (48 kg/m³). However, for very low densities, he maintains that the Poisson's ratio would be small, approaching zero.

3.10 REINFORCED PLASTICS HONEYCOMB CORES

This type of core material is, for specific strength, the strongest plastics core material that can be used in sandwich construction. It consists of very small reinforced plastics laminates that are built up into a honeycomb pattern, as shown in Fig. 137. The dimensions of the individual cells, as well as the depth of the core, can be varied to suit specific requirements. The cells of the honeycomb are generally hexagonal, and vary in size from about 3⁄16

Table 59

Property	Polycarbonate	Noryl	Thermoplastic Polyesters	Polysulfone	ABS	Polyethylene	Polystyrene	Polypropylene
Specific gravity	0.80	0.80	1.10	0.80	0.80	0.60	0.70	0.80
Flexural strength, psi $\times 10^3$	10	6	20	6.3	3.7	2.73	4.5	3.2
Flexural modulus, psi $\times 10^5$	3.0	2.4	10.0	2.13	1.25	1.20	2.10	1.20
Compressive strength, psi $\times 10^3$	7.5	5.5	12.5	—	1.0	1.3	3.3	1.6
Tensile strength, psi $\times 10^3$	5.5	3.3	11.8	4.55	2.7	1.3	1.3	2.0
Heat distortion temperature at 264 psi, °F	270	205	380	341	180	94	160	132

Table 60

Property	Polycarbonate Structural Foam
Part thickness, in.	0.25
Density, g/cc	0.85
Tensile strength, psi	4500–5500
Flexural strength, psi	9000–10000
Flexural modulus, psi $\times 10^5$	2.5–3.0
Heat deflection temperature °F	265–275

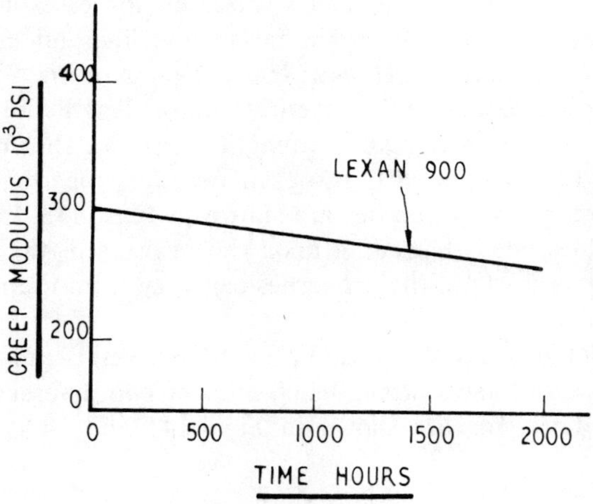

Fig. 136

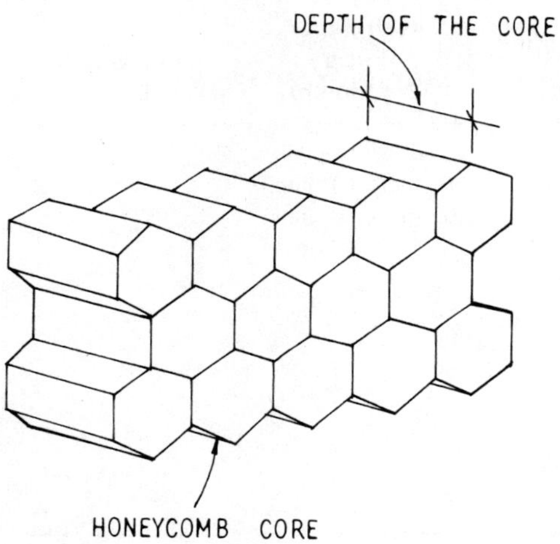

Fig. 137

in. (5 mm) to 1 in. (25 mm) across the flats. Very naturally the strength of the honeycomb will depend on the dimensions of the comb, the thickness of the walls, and the materials used. In general, however, for optimum designs, the strength of the honeycomb can be related, as in foams, to the density of the core as a whole. This quantity then forms a reasonable yardstick to which the properties of the materials can be related for effective comparison.

Honeycomb cores are available in densities ranging from 1 lb/ft^3 (16 kg/m^3) to about 9 lb/ft^3 (144 kg/m^3). The materials most commonly used are glass cloth, paper, cotton, or asbestos felt impregnated with polyester or phenolic resins. Epoxy or silicone resins may also be used. Paper honeycombs may be corrugated or flat web, or hexagonal cell with unexpanded core, where the core is in a flattened condition for ease of shipping, but this type requires an expander at the fabrication line and may be difficult to use with slow curing resins. Hauser, Hauser, and Norling[131] give the structure of various paper honeycombs, together with the lengthwise shear strength and modulus, and the compressive strength; these are shown in Tables 61 and 62. Holland and Forsberg[132] also examined the properties of glass cloth honeycomb cores; these properties are shown in Figs. 138–140. The curves show mean values, and, depending upon the resin used, the dimensions of the honeycomb, and so on, the properties can vary from the mean by about ±20%.

For elevated temperatures, phenolic or NCNS resins are more suitable. The compressive and shear strengths of cores of two different types of heat resistant phenolic resins are shown in Figs. 141 and 142. Holloway and

Table 61

Type of Honeycomb	Kraft Paper Basis Weight (lb)	Cell Size (in.)	Phenolic Resin (%)
Verticel			
A	60 (flat), 60 (flute)	¼	20
B	60 (flat), 80 (flute)	⅜	15
C	60 (flat), 60 (flute)	½	15
Union Camp			
A	80	½	18
B	99	¾	11
C	99	1	11
Hexcel			
A	80	½	18
B	99	¾	11
C	99	1	11

Table 62

Sandwich Core Material	Density (lb/ft^3)	Lengthwise Shear Strength (psi)	Lengthwise Shear Modulus (psi)	Compressive Strength (psi)
Polystyrene foam	1–2	40–80	—	20–40
Urethane foam	2	36	—	16–43
Paper honeycomb				
Verticel				
A	3.45	162	20,000	140
B	2.52	100	11,900	82
C	1.43	56	6,790	44
Union Camp				
A	2.25	40	—	170
B	1.73	24	—	91
C	1.30	16	—	53
Hexcel				
A	2.2	41	4,400	140
B	1.7	31	3,200	104
C	1.2	22	2,100	64

Note: In the case of the hexagonal cores, the shear strength and shear modulus values are perpendicular to the ribbon direction.

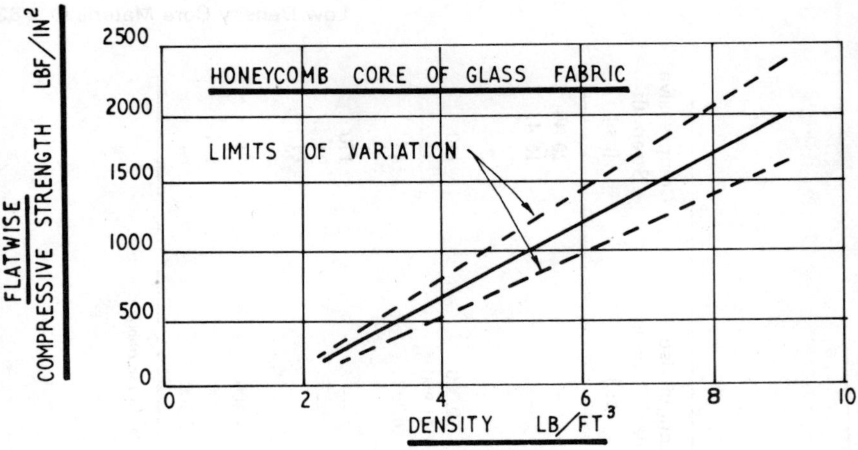

Fig. 138

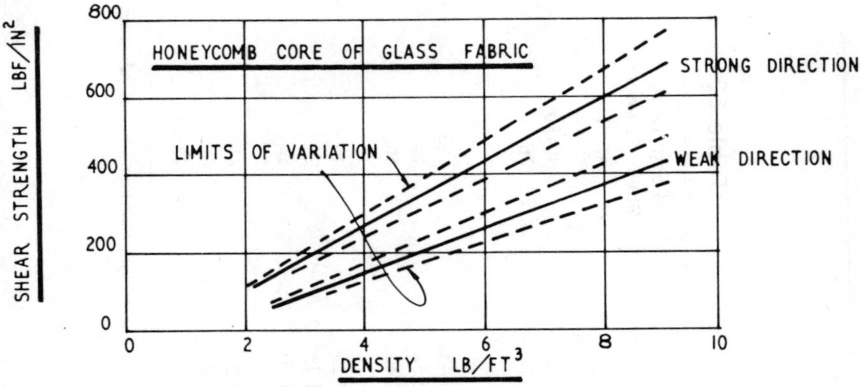

Fig. 139

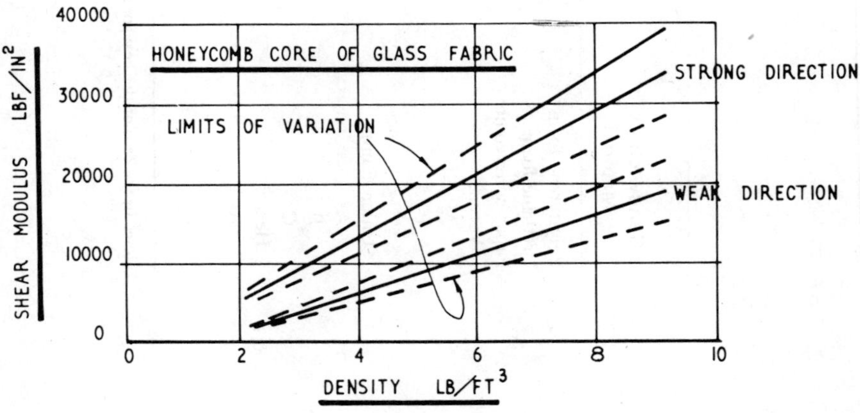

Fig. 140

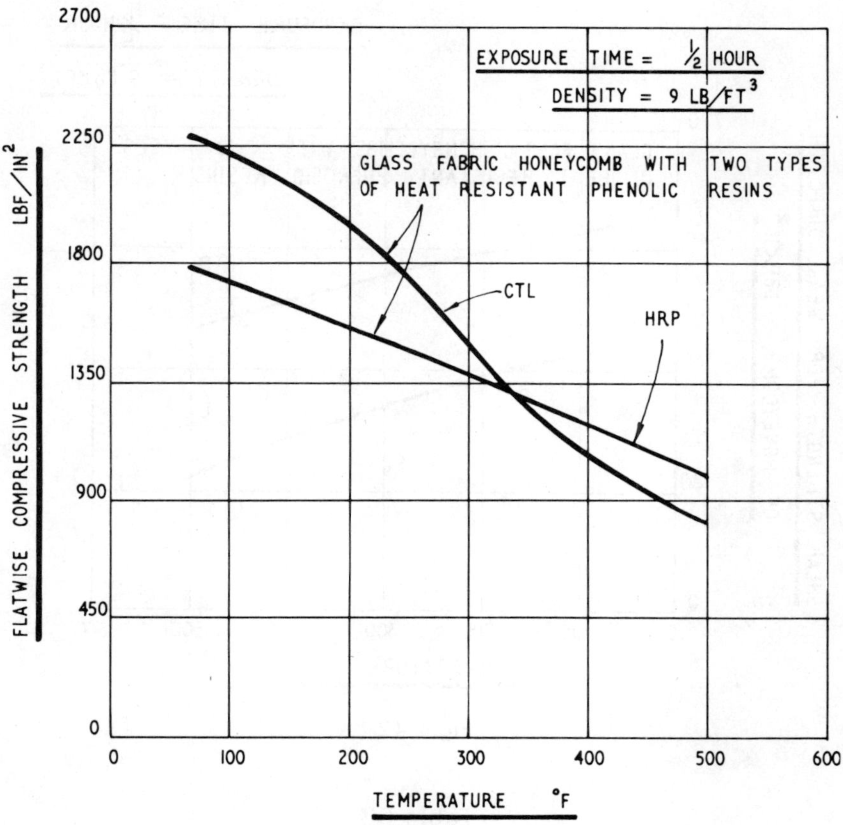

Fig. 141

MacDonald[133] have given the compressive strength of paper honeycomb impregnated with phenolic resin as 270 psi (1862 kPa). The honeycomb had ¼ in. (6 mm) cells, and the density of the honeycomb core was 3.1 lb/ft³ (50 kg/m³). Kray[134] has given the compressive properties of NCNS/Polyamide honeycomb cores; these are shown in Table 63. The core had $\frac{3}{16}$ in. (5 mm) hexagonal cells. They were then saturated with NCNS resin, a thermosetting polymer. Much of this information has been published by the Society of the Plastics Industry (SPI) and the Society of Plastics Engineers (SPE) of the United States, to whom grateful acknowledgement is made.

It may be noted that the thermal insulation of honeycomb cores could be considerably improved by foaming into the cells. This would also improve the mechanical properties, since the core would now consist of composite materials acting together.

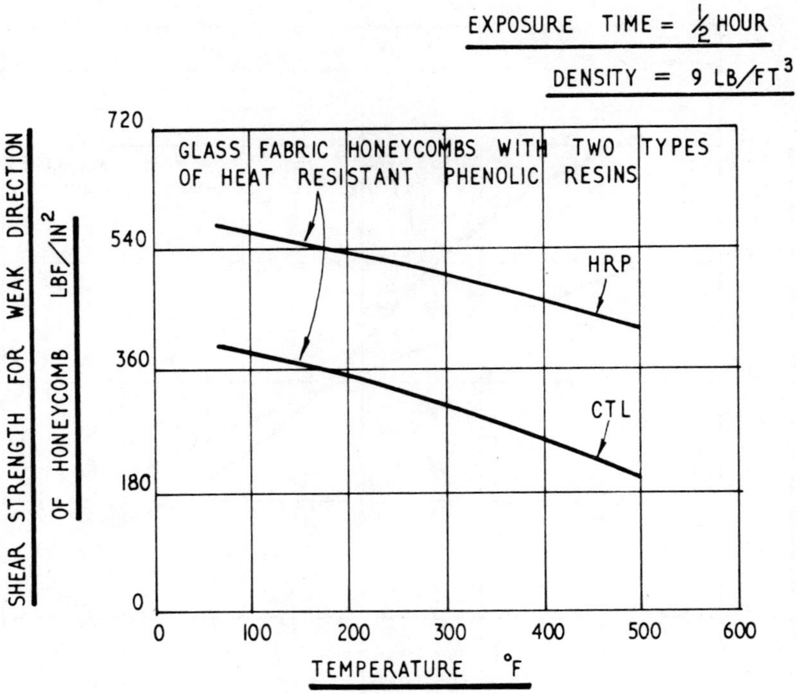

Fig. 142

Table 63

Core Density (lb/ft³)	Temperature (°F)	Compressive Strength (psi)	Compressive Modulus (psi)
4.64	77	218	1751
3.64	77	167	1496
5.35	400	194	1217
3.33	400	127	1134

Note: The properties are for NCNS/Polyamide (Nomex) Honeycomb cores.

Plastics honeycomb cores are very suitable for use in ground radomes of sandwich construction. Shearer[135] has reported on a 140 ft (42.7 m) diameter space frame radome. The structure, over the base plan area, consisted of over 1600 hexagonal panels. Each panel had a press molded perifery channel, 6 in. (152 mm) thick paper honeycomb core, and preimpregnated fabric for the skins. The entire panel was molded in one operation.

CHAPTER 4

SANDWICH CONSTRUCTION

4.1 INTRODUCTION

The analysis and design of various structures in plastics is considered in later chapters. Equally important, however, are the actual forms of construction. For small spans, single skin construction with FRP composites could be used. For larger spans, it is necessary to use sandwich construction. This features a low-density core of the material types considered in Chapter 3, and thin, strong, reinforced plastics inner and outer skins. The methods of analysis would determine the bending moments, axial forces, and shearing forces at any cross section. It is then necessary to determine the stresses in the facings and the core of the sandwich at that section. These problems, and related aspects such as the large shearing deformations of the core, are considered in this chapter. It may be noted here that there is already a great deal of literature on sandwich construction, though not necessarily in plastics materials, which could be easily applied to structural sandwiches in plastics.

4.2 SANDWICH BEHAVIOR UNDER TRANSVERSE LOADING

Consider a typical cross section of a sandwich, as shown in Fig. 143. The sandwich under transverse loading is usually designed on the assumption[122,136,137] that the bending moment at the cross section is taken by the facings whereas the shear is taken by the core. On this assumption, the flexural rigidity of the sandwich is given by

$$D_f = \frac{E_f b(h^3 - c^3)}{12(1 - \nu_f^2)} \tag{37}$$

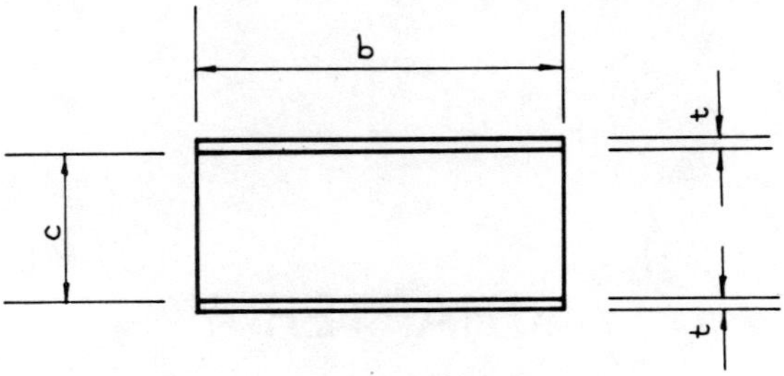

Fig. 143

where

t = thickness of the facings
b = width of the sandwich
c = core thickness
h = total thickness $= c + 2t$
E_f = elastic modulus of the facings
ν_f = Poisson's ratio of the facings.

The stresses in the facings are given by

$$\sigma_f = \frac{2M}{(h+c)bt} \tag{38}$$

where

σ_f = the stress in the facing
M = the bending moment at the cross section.

With thin facings, the shearing rigidity of the sandwich can be written as

$$N = bcG_c$$

where

G_c = modulus of rigidity of the core.

The shearing stress τ can be written as

$$\tau = \frac{2V}{b(h+c)} \tag{39}$$

where

V = shearing force at the cross section.

The deflection of the sandwich is obtained by superimposing the shearing deflections of the core on the bending deflections. In plastics sandwich construction, with relatively weak cores, the shearing deflections can be several times the bending deflections, due to the very low value of G_c for the low-density core materials.

4.2.1 Assumptions on Which the Analysis is Based

The application of plate theory to sandwich construction is based on three very important assumptions, none of which is strictly valid. These assumptions are the following[138]:

1. That the stresses in the thin facings are not affected by the compressive stresses in the core,
2. That the analysis of the panel as determined by the plate theory is unaffected by the large shearing deformations, and
3. That there is no compressive failure of the core under the applied loads.

The first of these assumptions neglects the localized bending of the thin facings about their own neutral axis. It assumes that the facings are always held at a constant distance apart and are in tension or compression. In the case of cores with very small elastic moduli, the facings are, in fact, plates on elastic foundations, and the compressive stresses in the core do affect the facings. The second assumption is even less valid than the first. The theory of plate bending is based on the assumption that points of the plate lying initially on a normal to the middle plane of the plate remain on that normal after bending. If the shearing deformations are large, this assumption is not valid. Further, large shearing deformations, in the case of sandwich panels, can affect the load distributions as given by the plate theory.

Consider the two strips shown in Fig. 144, loaded by a vertical point load W at 0. The load W_1 carried by the short span l_1, and the load W_2 carried by the long span l_2 are given by

$$W_1 = \frac{W}{1+(l_1/l_2)^3}$$
$$W_2 = \frac{W}{1+(l_2/l_1)^3} \qquad (40)$$

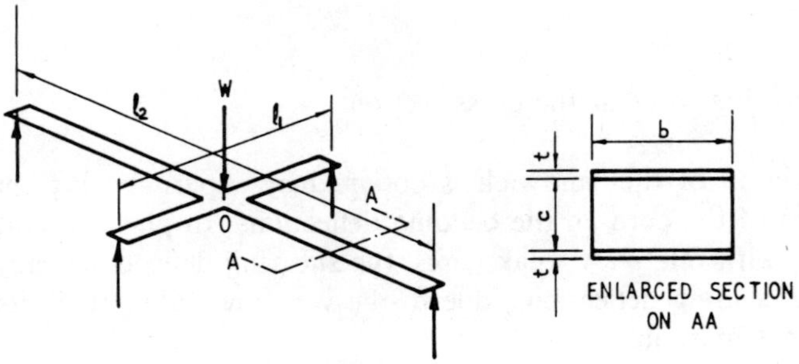

Fig. 144

If the section of both strips is a sandwich, and if the shearing deformations of the core are large, then the loads W_1 and W_2 are given by

$$W_1 = \frac{W}{1 + \left(\dfrac{Nl_1^3 + 12D_f l_1}{Nl_2^3 + 12D_f l_2}\right)}$$

$$W_2 = \frac{W}{1 + \left(\dfrac{Nl_2^3 + 12D_f l_2}{Nl_1^3 + 12D_f l_1}\right)} \qquad (41)$$

where

D_f = flexural rigidity of the sandwich
N = shearing rigidity of the sandwich.

The distribution of the loads, and hence the moments in the strips, given by Equation 40 is not necessarily the same as that given by Equation 41. The example given above, however, is rather an extreme case. In cases of absolute symmetry (such as a circular sandwich plate under symmetrical loading), the maximum transverse moments and shearing forces are unaltered by the large shearing deflections of the core, due to this effect. More exact theories of sandwich panel behavior are considered in Sections 4.4–4.6.

The author has experimentally investigated the third assumption by using expanded polystyrene and rigid polyurethane foam as cores.[112] Bending tests were run on sandwich beams under 3-point and 4-point loading. The specimens were made by bonding 1 in. (25 mm) and 1½ in. (38 mm) thick cores, 2 in. (51 mm) wide, to 1⁄32 in. (0.8 mm) and 1⁄16 in. (1.6 mm) thick FRP composites reinforced with glass cloth. In the case of the 3-point loading tests, failure was found in all cases to be a local compressive failure under the central point load. Progressive compressive facing failure then occurred. The theoretical facing stresses can be calculated on the assumption, previously stated, that all the bending moment is taken by the facings. These theoretical facing stresses at failure are plotted against the density of the core for 1 in. (25 mm) and 1½ in. (38 mm) thick cores in Figs. 145 and 146. In the case of the 4-point loading tests, failure occurred at the lower densities in shear, but, at the highest density, local compressive failure occurred. The theoretical facing stresses at failure have been plotted against the density of the core for 1 in. (25 mm) and 1½ in. (38 mm) thick cores in Figs. 147 and 148. A study of Figs. 145–148 shows that the facing stresses at failure vary almost linearly with the density for both types of loadings. The values of the facing stresses at failure are well below the ultimate strengths of the

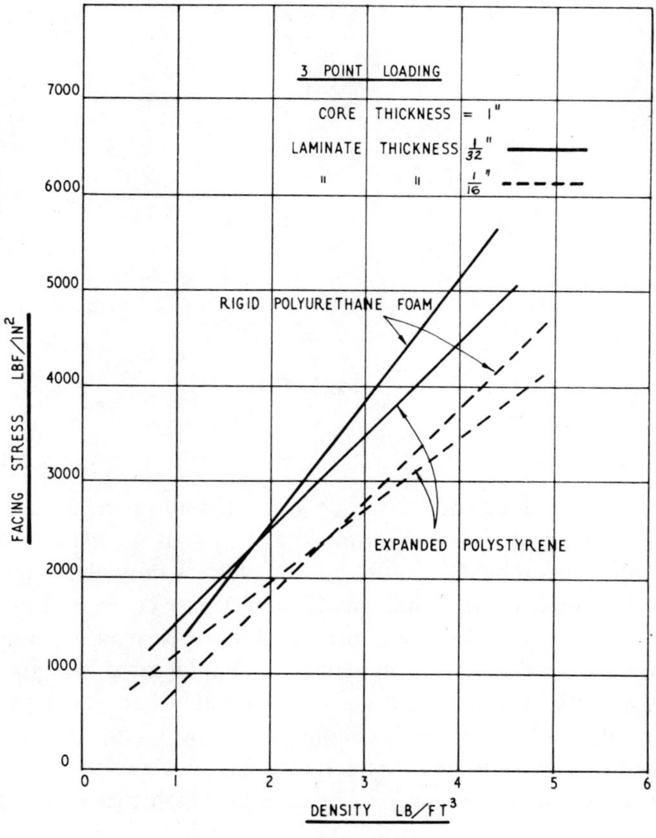

Fig. 145

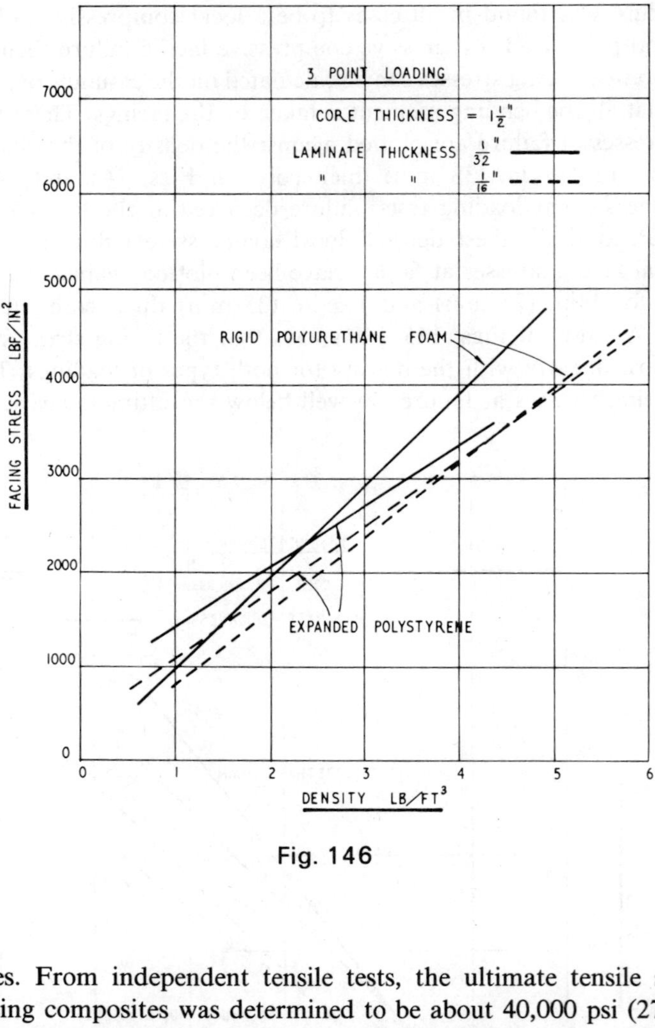

Fig. 146

composites. From independent tensile tests, the ultimate tensile strengths of the facing composites was determined to be about 40,000 psi (276 MPa). The maximum theoretical facing stress at failure is only about 6000 psi (41 MPa). With thinner facings, and smaller and denser cores, and by spreading the load evenly so that local compressive failure is avoided, higher facing stresses can be reached. In commercially manufactured composites, large variations in both thickness and glass fiber content can occur (particularly in hand lay-up) and, in general, composites below $\frac{1}{32}$ in. (0.8 mm) should not be used. From a thermal insulation point of view, the core thickness should be increased, depending on the area in which the sandwich panel is to be used.

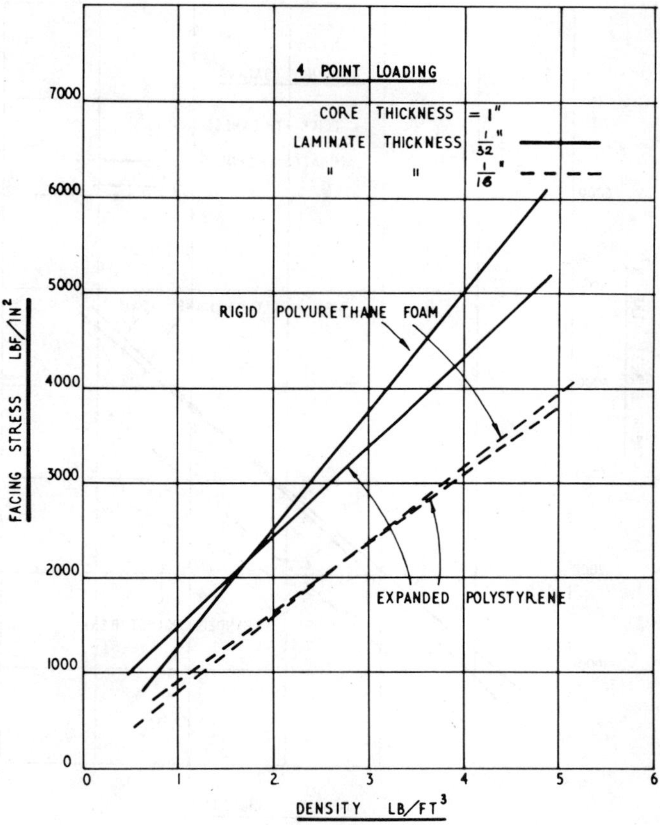

Fig. 147

4.2.2 Types of Sandwich Failures under Transverse Loads

A sandwich panel under transverse loading can fail in one of four ways:

1. Tensile or compressive failure of the facings
2. Shear failure of the core
3. Shear failure of the adhesive layer (bond failure)
4. Local compressive failure of the core

It is possible to guard against failure of the first two types by limiting the stresses given by Equations 38 and 39, respectively. Failure of type 3 can be prevented by proper fabrication techniques and good workmanship.

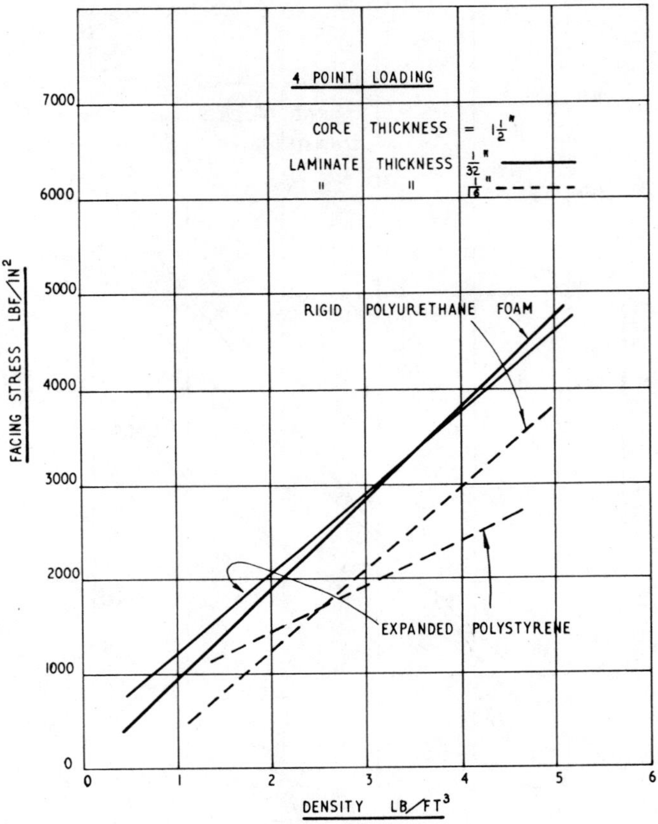

Fig. 148

Failure of type 4 is caused by the application of point loading as shown in Fig. 149. One design data sheet[139] gives the empirical formula for the compressive stress in the core under a point load P as

$$\sigma_c = 0.192P\left(\frac{D_f'}{k_c}\right)^{2/3} \tag{42}$$

where

$$D_f' = \frac{E_f t^3}{12(1-\nu_f^2)}$$

and

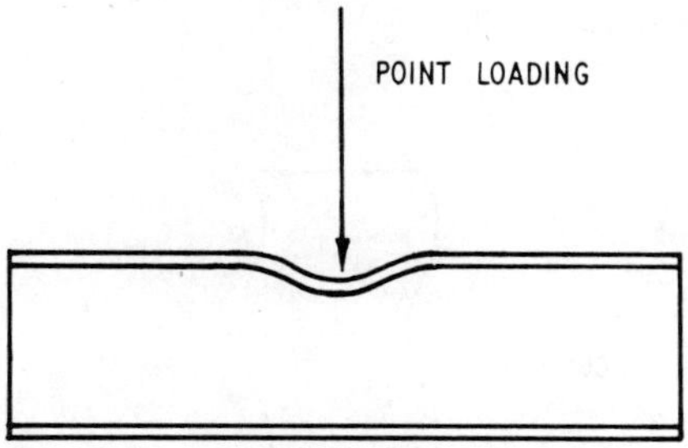

Fig. 149

$$k_c = \frac{E_c}{2(1 - \nu_c^2)}$$

Equation 42 is based on the assumption that the facing is an infinitely large plate on an elastic foundation. The stress in the facing is then given by

$$\sigma_f = 0.366(1 + \nu_f) + \frac{P}{t^2}\left(\log_{10} \frac{E_f}{0.0343 k_c} - 0.266\right) \tag{43}$$

4.3 SANDWICH FAILURE UNDER COMPRESSIVE END LOADS

When subjected to compressive end loads, a sandwich panel of length l, can fail in one of four ways:

1. Overall buckling failure
2. Wrinkling of the facings
3. Facing failure
4. Core failures

4.3.1 Overall Buckling Failure

The critical buckling load of a sandwich strip can be written as[122,123]

$$P_{cr} = \frac{P_E}{1 + \dfrac{P_E}{G_c b c}} \tag{44}$$

where

$$P_E = \frac{\pi^2 D_f}{l^2}$$

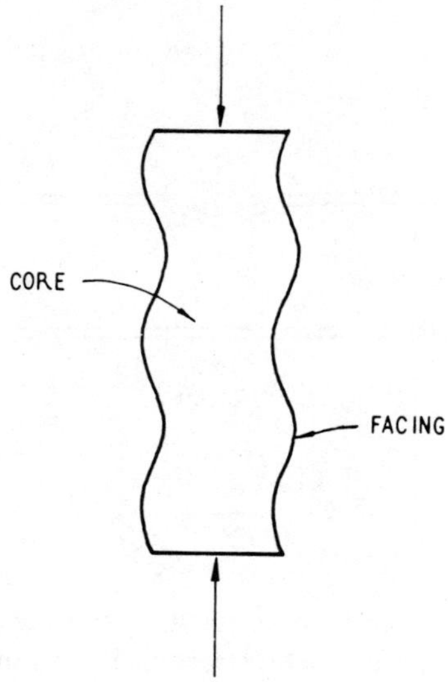

Fig. 150

4.3.2 Wrinkling of the Facings

This type of failure is shown in Fig. 150. The stress σ_f at which wrinkling occurs is given as[122,140]

$$\sigma_f = 0.50(E_c E_f G_c)^{1/3} \tag{45}$$

This wrinkling stress has also been given as[139]

$$\sigma_f = 0.63 E_f^{1/3} E_c^{2/3} \tag{46}$$

4.3.3 Facing Failure

This type of failure occurs only in very short columns with very rigid cores. The facing stress is then given by

$$\sigma_f = \frac{P}{2bt} \tag{47}$$

4.3.4 Core Failures

The types of core failures that can occur are shown in Figs. 151a–151c. Figure 151a shows shear crimping failure. Figure 151b shows core crushing

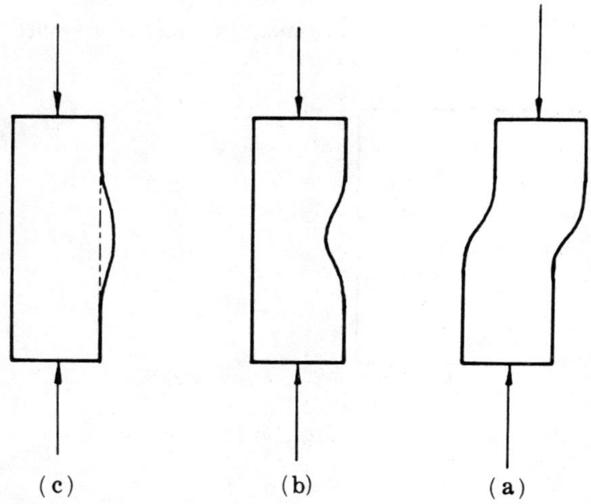

Fig. 151

failure, and Fig. 151c shows localized bond failure between the facing and the core.

4.4 RECTANGULAR SANDWICH PANEL UNDER TRANSVERSE LOADING

The behavior of flat rectangular sandwich panels with orthotropic facings and cores has been studied by March,[141] using energy methods. The analysis has been performed, assuming that any line in the core that is initially straight and normal to the middle surface of the core will remain straight after the deformation of the panel but will no longer be normal to the deformed middle surface. The method further assumes that the deflected middle surface can be expressed by the equation

$$w = f(x, y) \tag{48}$$

where

w = deflection in the Z direction.

The directions of the axes are shown in Fig. 152. The results obtained on the application of the theory to specific edge conditions are considered below. It may be noted that the formulas given are easily applicable to isotropic facings or core materials.

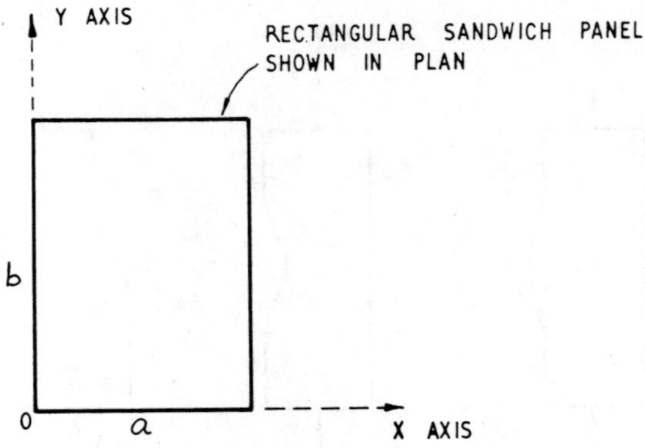

Fig. 152

4.4.1 Panel with Simply Supported Edges

The form of the middle deflected surface is assumed to be

$$w = w_0 \sin\alpha x \sin\beta y \tag{49}$$

where

w_0 = maximum deflection of the panel
$\alpha = \pi/a$
$\beta = \pi/b$
a, b = length of the sides of the panel as shown in Fig. 152.

March[141] has shown that if the shear deformations of the core are taken into account, then the maximum deflection of the panel w_0 can be written as

$$w_0 = w_0'(1+\eta) \tag{50a}$$

where

w_0' = maximum deflection of the panel neglecting shear deformations in the core
η = a correction factor.

This correction factor η is given by

$$\eta = \frac{ctR}{k'} \tag{50b}$$

where R and k' are constants that vary with each individual case considered, and where

c = core thickness
t = facing thickness.

In this particular case,

$$R = \frac{\pi^2}{2\lambda a^2}\left(E_x \frac{b^2}{a^2} + E_y \frac{a^2}{b^2} + 2A\right) \tag{51}$$

and

$$k' = G'_{yz} + G'_{zx} \frac{b^2}{a^2} \tag{52}$$

where

E_x, E_y = elastic moduli for the facings in the X and Y directions, respectively, and

$$\lambda = (1 - \mu_{xy}\mu_{yx}) \tag{53}$$

where

μ_{xy}, μ_{yx} = Poisson's ratio for the facings

and

$$A = E_x \mu_{yx} + 2\lambda G_{xy} \tag{54}$$

where

G_{xy} = shear modulus for the facings
G'_{yz}, G'_{zx} = shear modulus for the core.

In addition the maximum deflection w'_0 of the panel neglecting shear deformations in the core can be obtained as

$$w'_0 = \frac{96\text{p}}{\pi^2 \beta^2 R (h^3 - c^3)} \tag{55}$$

The actual maximum deflection of the panel can therefore be determined from Equation 50a. The bending moments, twisting moments, shears, and reactions can be approximately determined by neglecting the effects of shear deformations and using ordinary orthotropic plate theory. This has been proved valid for specific cases by Reissner,[142] Ericksen,[143] and Hughes.[123]

4.4.2 Panel with All Edges Clamped

The form of the middle deflected surface is taken as

$$w = w_0 \sin^2 \alpha x \sin^2 \beta y \tag{56}$$

With all notations as listed in Section 4.4.1, March[141] has shown once again that

$$w_0 = w_0'(1+\eta) \tag{50a}$$

where

$$w_0' = \frac{8p}{\beta^2 R(h^3 - c^3)} \tag{57}$$

and

$$\eta = \frac{ctR}{k'} \tag{50b}$$

However, the constants for this case are

$$R = \frac{2\pi^2}{3\lambda a^2}\left(3E_x\frac{b^2}{a^2} + 3E_y\frac{a^2}{b^2} + 2A\right) \tag{58}$$

and

$$k' = G'_{yz} + G'_{zx}\frac{b^2}{a^2} \tag{59}$$

The actual deflection of the panel taking shear deformations into account can, therefore, be determined.

4.5 BUCKLING OF A RECTANGULAR SANDWICH PANEL UNDER COMPRESSIVE END LOADS

The assumptions made in the applications of energy methods to sandwich panels under transverse loading, given in Section 4.4, are equally applicable to the buckling behavior of sandwich panels. Once again, the facings and the core are considered as orthotropic, and it is again assumed that the deflected middle surface can be expressed in the form

$$w = f(x,y) \tag{48}$$

In applying the formulas to isotropic facings stressed beyond the proportional limit, March[141] suggests that better correlation with experimental work is obtained by using a reduced modulus E_r given as

$$E_r = \frac{2E_f E_t}{E_f + E_t} \tag{60}$$

where

E_f = elastic modulus of the isotropic facings below the proportional limit
E_t = tangent modulus of the isotropic facings at the stress in question.

The application to specific edge conditions is considered below.

4.5.1 Panel with All Edges Simply Supported

The deflected surface is once again assumed to be

$$w = w_0 \sin\alpha x \sin\beta y \tag{49}$$

This is based on the assumption that the panel buckles into a single half wave. If there appears to be some possibility that the panel will buckle into more than one half wave, then β is replaced by $n\beta$, where n is an integer whose value makes the critical buckling load an absolute minimum. This aspect of the problem is considered in Sections 4.5.3 and 4.5.4. The critical load P_{cr} taking shear deformations into account can then be written as

$$P_{cr} = \frac{P'_{cr}}{1+\eta} \tag{61}$$

where

P'_{cr} = critical buckling load neglecting shear deformations of the core
η = a correction factor.

For buckling in the form of one half sine wave and, with the notation given in Section 4.4.1,

$$P'_{cr} = \left[\frac{3t(c+t)^2 + t^3}{3}\right] R \tag{62}$$

and

$$\eta = \frac{ctR}{k'} \tag{50b}$$

where

$$R = \frac{\pi^2}{2\lambda a^2}\left[E_x \frac{b^2}{a^2} + E_y \frac{a^2}{b^2} + 2A\right] \tag{51}$$

and

$$k' = G'_{yz} + G'_{zx}\frac{b^2}{a^2} \tag{52}$$

4.5.2 Panel with the Loaded Edges Simply Supported and the Remaining Edges Clamped

For a panel deflecting in a single half wave, the form of the deflected middle surface can be written as

$$w = w_0 \sin^2\alpha x \sin\beta y \tag{63}$$

Then

$$P_{cr} = \frac{P'_{cr}}{1+\eta} \tag{61}$$

where

$$P'_{cr} = \left[\frac{3t(c+t)^2 + t^3}{9}\right] R \tag{64}$$

and

$$\eta = \frac{ctR}{k'} \tag{50b}$$

In this case,

$$R = \frac{8\pi^2}{\lambda a^2}\left[E_x \frac{b^2}{a^2} + \frac{3}{16} E_y \frac{a^2}{b^2} + \frac{A}{2}\right] \tag{65}$$

and

$$k' = 3G'_{yz} + 4G'_{zx}\frac{b^2}{a^2} \tag{66}$$

4.5.3 Panel with the Loaded Edges Clamped and the Remaining Edges Simply Supported

The critical buckling load P_{cr} is given by

$$P_{cr} = \frac{P'_{cr}}{1+\eta} \tag{61}$$

where

P'_{cr} = critical buckling load, which depends on the number of longitudinal half waves into which the panel buckles.

With one longitudinal half wave,

$$w = w_0 \sin\alpha x \sin^2\beta y \tag{67}$$

$$k' = 4G'_{yz} + 3G'_{zx}\frac{b^2}{a^2} \tag{68}$$

$$R = \frac{8\pi^2}{\lambda a^2}\left[\frac{3}{16} E_x \frac{b^2}{a^2} + E_y \frac{a^2}{b^2} + \frac{A}{2}\right] \tag{69}$$

and

$$P'_{cr} = \frac{(h^3 - c^3)R}{24} \tag{70}$$

With two longitudinal half waves,

$$w = w_0 \sin\alpha x \sin\beta y \sin 2\beta y \tag{71}$$

$$k' = 5G'_{yz} + G'_{zx}\frac{a^2}{b^2} \tag{72}$$

$$R = \frac{\pi^2}{2\lambda a^2}\left[E_x\frac{b^2}{a^2} + 41E_y\frac{a^2}{b^2} + 10A\right] \tag{73}$$

and

$$P'_{cr} = \frac{(h^3 - c^3)R}{30} \tag{74}$$

With three longitudinal half waves,

$$w = w_0 \sin\alpha x \sin\beta y \sin 3\beta y \tag{75}$$

$$k' = 10G'_{yz} + G'_{zx}\frac{b^2}{a^2} \tag{76}$$

$$R = \frac{\pi^2}{2\lambda a^2}\left[E_x\frac{b^2}{a^2} + 136E_y\frac{a^2}{b^2} + 20A\right] \tag{77}$$

and

$$P'_{cr} = \frac{(h^3 - c^3)R}{60} \tag{78}$$

4.5.4 Panel with All Edges Clamped

$$P_{cr} = \frac{P'_{cr}}{1+\eta} \tag{61}$$

With one longitudinal half wave,

$$w = w_0 \sin^2\alpha x \sin^2\beta y \tag{56}$$

$$k' = G'_{yz} + G'_{zx}\frac{b^2}{a^2} \tag{59}$$

$$R = \frac{2\pi^2}{3\lambda a^2}\left[3E_x\frac{b^2}{a^2} + 3E_y\frac{a^2}{b^2} + 2A\right] \tag{58}$$

$$P'_{cr} = \frac{(h^3 - c^3)R}{6} \tag{79}$$

With two longitudinal half waves,

$$w = w_0 \sin^2\alpha x \sin\beta y \sin 2\beta y \tag{80}$$

$$k' = 15G'_{yz} + 4G'_{zx}\frac{b^2}{a^2} \tag{81}$$

$$R = \frac{\pi^2}{2\lambda a^2}\left[16E_x\frac{b^2}{a^2} + 123E_y\frac{a^2}{b^2} + 40A\right] \tag{82}$$

$$P'_{cr} = \frac{(h^3 - c^3)R}{90} \tag{83}$$

With three longitudinal half waves,

$$w = w_0 \sin^2\alpha x \sin\beta y \sin 3\beta y \tag{84}$$

$$k' = \frac{15}{4}G'_{yz} + \frac{1}{2}G'_{zx}\frac{b^2}{a^2} \tag{85}$$

$$R = \frac{\pi^2}{2\lambda a^2}\left[2E_x\frac{b^2}{a^2} + 51E_y\frac{a^2}{b^2} + 10A\right] \tag{86}$$

$$P'_{cr} = \frac{2(h^3 - c^3)R}{45} \tag{87}$$

4.6 ANALYSIS OF A RECTANGULAR SANDWICH PANEL IN BENDING

As an example consider a rectangular sandwich panel with simply supported edges under transverse loading, with details given below:

$b = 10$ ft (3.05 m)
$a = 6$ ft (1.83 m)
$t = 1/8$ in. (3 mm)
$c = 1.00$ in. (25 mm)
$h = 1.25$ in. (32 mm)
$E_f = 1.2 \times 10^6$ psi (8.27 GPa)

$E_c = 1000$ psi (6.89 MPa)
$G_c = 400$ psi (2.76 Mpa)
$\mu = 0.30$
$p = 30$ lb/ft^2 (146.5 kg/m^2)

$$G_f = \frac{E_f}{2(1+\mu)} = 4.6 \times 10^5 \text{ psi (3.17 GPa).}$$

Using the formulas given in Section 4.4.1, the values of the various constants can be easily obtained. It should be noted that $E_x = E_y = E_f$ and that $G'_{yz} = G'_{zx} = G_c$. We therefore have the following:

$$\lambda = (1 - 0.3 \times 0.3) = 0.91$$

$$k' = 400 + 400 \frac{120^2}{72^2} = 1512 \text{ psi (10.42 MPa)}$$

$$A = 1.2 \times 10^6 \times 0.30 + 2 \times 0.91 \times 4.6 \times 10^5 = 1.2 \times 10^6 \text{ psi (8.27 GPa)}$$

The value of R can then be obtained by the use of Equation 51 as $R = 6.45 \times 10^3$ psi/in.^{2} (68.9 kPa/mm^2) and the value of η can be obtained by the use of Equation 50b as

$$\eta = \frac{ctR}{k'} = \frac{1 \times 1/8 \times 6.45 \times 10^3}{1512} = 0.53$$

The value of w'_0 is obtained by the use of Equation 55 as

$$w'_0 = \frac{96 \times 30/144}{\pi^2 \times (\pi^2/120^2) \times 6.45 \times 10^3 (1.25^3 - 1^3)} = 0.48 \text{ in. (12 mm)}$$

Thus, finally, the actual maximum deflection of the panel taking shear deformations in the core into account can be obtained by the use of Equation 50a as

$$w_0 = w'_0 (1 + \eta) = 0.48(1 + 0.53) = 0.73 \text{ in. (18.5 mm)}$$

If the same panel has all its edges clamped, then the analysis can be carried out in a similar manner but with the use of equations given in Section 4.4.2. With the same panel data given above, the values of the various constants can be determined as

$$k = 1512 \text{ psi (10.42 MPa)}$$

and

$$R = 1.91 \times 10^4 \text{ psi/in.}^2 \text{ (204 kPa/mm}^2\text{)}$$

This gives the value of η as

$$\eta = \frac{1 \times 1/8 \times 1.91 \times 10^4}{1512} = 1.58$$

The value of w_0' can be obtained by the use of Equation 57 as

$$w_0' = \frac{8 \times 30/144}{(\pi^2/120^2) \times 1.91 \times 10^4 \, (1.25^3 - 1^3)} = 0.13 \text{ in. (3.3 mm)}$$

As can be seen, the bending deflections are much smaller with clamped edges than with simply supported edges. However, the shearing deflections are not materially affected by the clamping and so the value of η is considerably higher in the clamped case. This gives the value of w_0 the maximum total deflection taking shearing deformations into account as,

$$w_0 = 0.13\,(1 + 1.58) = 0.34 \text{ in. (8.6 mm)}$$

4.7 ANALYSIS OF A RECTANGULAR SANDWICH PANEL IN BUCKLING

The buckling of a simple rectangular sandwich panel subjected to compressive end loads can be considered by the use of the formulas given in Sections 4.5.1, for the case with all edges simply supported.

With the same data for the panel discussed in Section 4.6, the value of the critical buckling load neglecting shear deformations can be obtained by the use of Equation 62, where the value of R is obtained from Equation 51 as

$$R = \frac{\pi^2}{2 \times 0.91 \times 72^2}\left[1.2 \times 10^6 \times \frac{120^2}{72^2} + 1.2 \times 10^6 \times \frac{72^2}{120^2} + 2 \times 1.2 \times 10^6\right]$$

$$= 6.45 \times 10^3 \text{ psi/in.}^2 \text{ (68.9 kPa/mm}^2\text{)}$$

This gives the value of P_{cr}' as

$$P'_{cr} = \left[\frac{3 \times 1/8\,(1 + 1/8)^2 + (1/8)^3}{3}\right] 6.45 \times 10^3 = 1025 \text{ lb/in. } (18.3 \text{ kg/mm})$$

The value of η is obtained from Equation 50b as 0.53 and this gives the critical buckling load of the panel with shear deformations considered as

$$P_{cr} = \frac{1025}{1 + 0.53} = 670 \text{ lb/in. } (12 \text{ kg/mm}).$$

CHAPTER 5

ARCHITECTURAL FORMS FOR USE WITH PLASTICS

5.1 INTRODUCTION

It was shown in Section 1.3.3 that the very low modulus of elasticity of plastics makes them unsuitable for use in conventional structural forms, such as, for instance, beam and slab construction. It is therefore very important to utilize such structural forms as folded plates, synclastic and anticlastic shells, stressed skin systems, tension structures, and the like, which give adequate stiffness by virtue of their shape.[7-9,138] All such structural forms are not necessarily efficient. A structural form that is unsuitable in single skin construction, where plate buckling has to be prevented, may be quite suitable for the same structure in sandwich construction, where the problem of plate buckling does not arise. However, sandwich construction is more expensive than single skin construction, and it may turn out that a relatively inefficient structure in single skin construction, with a thickness larger than is absolutely necessary, is more economical than a very efficient sandwich structure. The choice would be dictated by the particular circumstances prevailing. A quick preliminary design, using several likely structural forms, can easily be made; the most economical structural form can then be selected.

5.2 FOLDED PLATE STRUCTURES

In general, folded plate structures consist of two or more plates joined together in such a manner that the external loading is resolved at the folds into components acting in the planes of the plates forming the structure. These components can then be resisted very efficiently, as they act in the direction of the greatest stiffness of the plates. Folded plates, being composed of flat elements, are very amenable to prefabrication. Further, the joints can be very simple.

Folded plate structures are basically of two types—prismatic and nonprismatic. Prismatic structures are formed from rectangular plates, and are the most common form of folded plate structures. They are assumed to be supported on rigid end diaphragms. Figures 153a and 153b are examples of prismatic roof structures. Figure 153a uses a two-plate unit, whereas Fig. 153b uses a basic five-plate unit.

Nonprismatic folded plate structures can be of three types—pyramidal, prismoidal, and composite. Pyramidal structures,[144] such as pavilion roofs, are rather special forms of folded plate structures, and are shown in Fig. 154. They can be based on any regular polygon. These may lie between the triangle at one end of the scale and the cone at the other, as shown in Fig. 154. The cone, in fact, represents a pyramid based on a polygon with an infinite number of sides.

The disadvantage of pyramidal structures is that, if used as single pyramids, the overall dimensions of the individual plates becomes excessive. At very low loads, therefore, the plates buckle, shedding load onto the stiffly jointed and hence straight corners. This buckling can be prevented in a number of ways. The plates can be ribbed, as shown in Fig. 155a. However, ribbing is expensive and the same effects can be achieved by corrugations, as shown in Fig. 155b. Ribbing or corrugations make these structures more expensive. If the large deflections and the buckled form of the plates are not objectionable from an architectural point of view, it is advisable to treat the central parts of the plates as being structurally ineffective. They act merely as covering membranes, as shown in Fig. 156, and transfer all the load to the corners. The corners have to be appropriately stiffened to carry all the load from two adjacent membranes. This can be done either by increasing the thickness of the plates at the corners or by chamfering, as shown in Fig. 157. A pyramidal folded plate conical dome has been built and tested by the author,[145] of the type shown in Fig. 268. Some of the test results are discussed in Section 10.7.

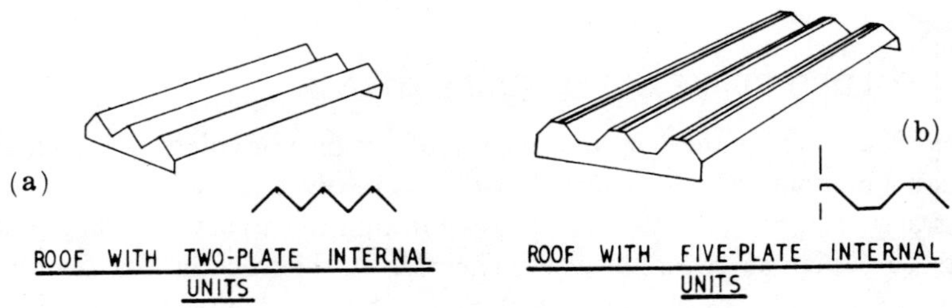

Fig. 153

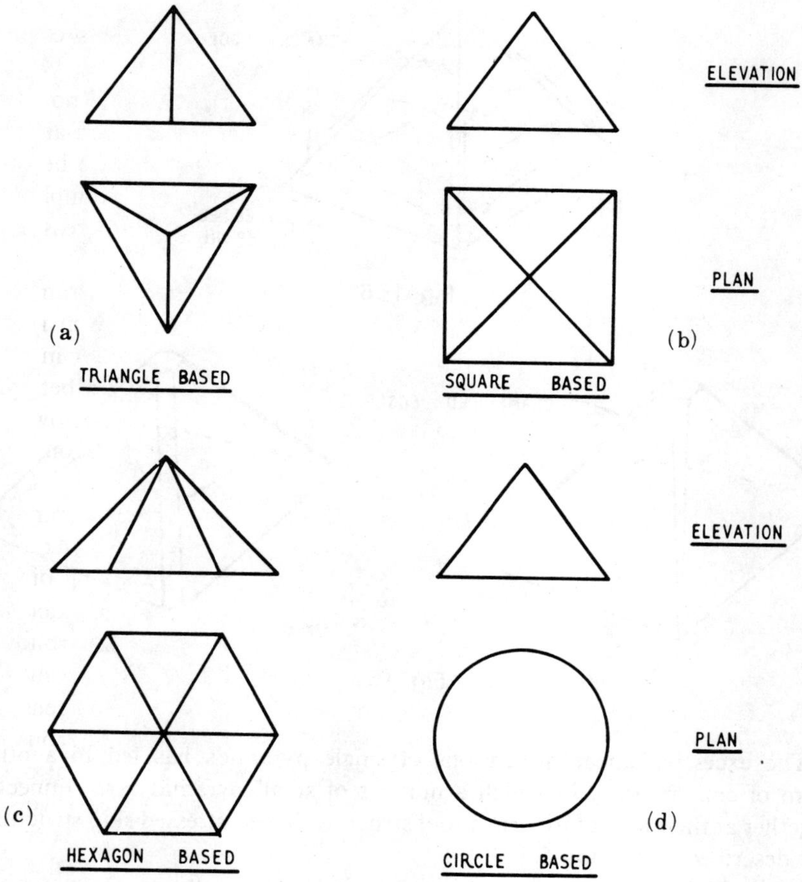

Fig. 154

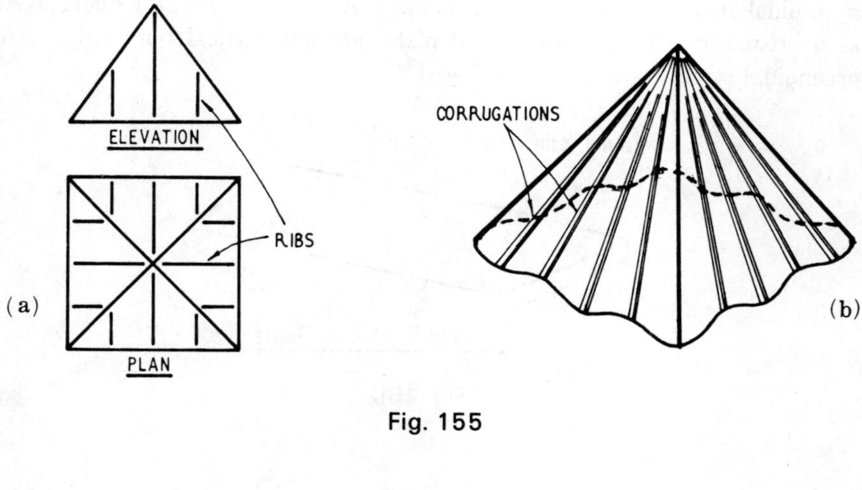

Fig. 155

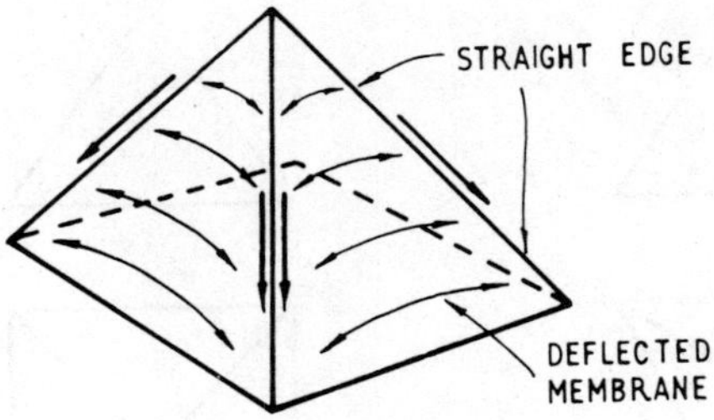

Fig. 156

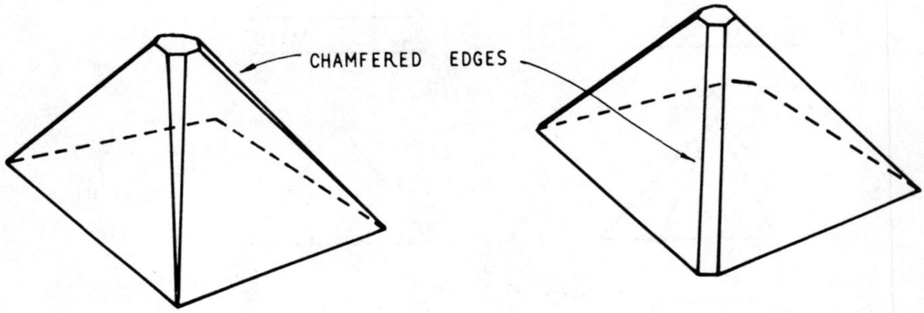

Fig. 157

The excessive linear dimensions of single pyramids has led to another form of construction, in which a number of small pyramids are connected together at their apexes to form a roof structure. These stressed skin structures are described in Section 5.4.

Prismoidal structures[144] are an intermediate form between prismatic and single pyramid structures. A prismoidal structure can be derived from a pyramidal structure by cutting off its apex. Alternately, it can be conceived as a prismatic structure whose end plates are not vertical but inclined. A prismoidal structure is shown in Fig. 158.

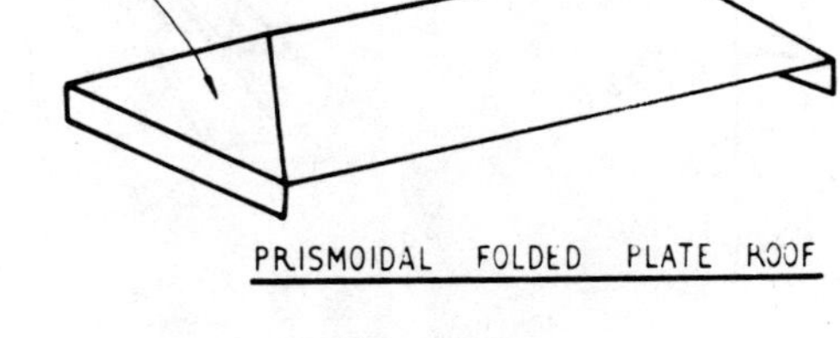

Fig. 158

Composite folded plate structures[138] are the most important type, and are quite distinct from the types discussed earlier. In composite folded plate structures, the folded plate action of the plates is tied in with the elastic behavior of the structure. The distinguishing feature of composite folded plate construction is that, taken as a whole, the structure is folded not only laterally but longitudinally as well. Examples of composite folded plate structures are shown in Fig. 159. Figure 159a shows a three-pinned composite folded plate portal frame or rigid frame, using the two-plate internal unit. The beam and column units have to be end and edge shaped in order to mate with each other. The flat triangular panels between the columns, marked A, may be omitted to provide suitable openings. Figure 159b shows the two-pinned composite folded plate portal or rigid frame. Figure 159c shows a composite folded plate barrel vault, while Fig. 159d shows a composite folded plate dome.

Another interesting composite folded plate structure is shown in Fig. 160. The entire structure can be built up from only one type of unit. This unit

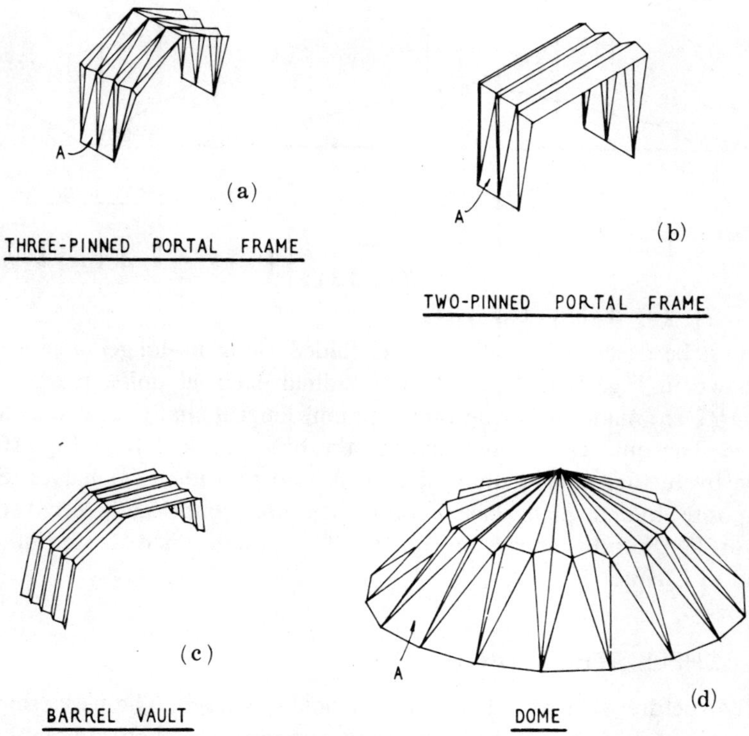

Fig. 159

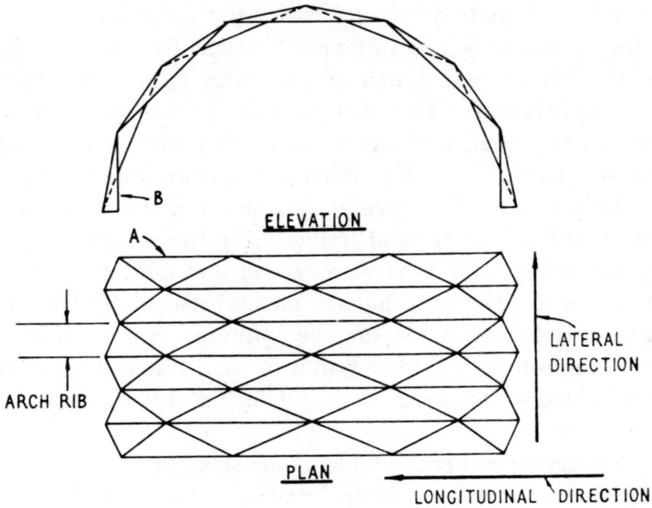

Fig. 160

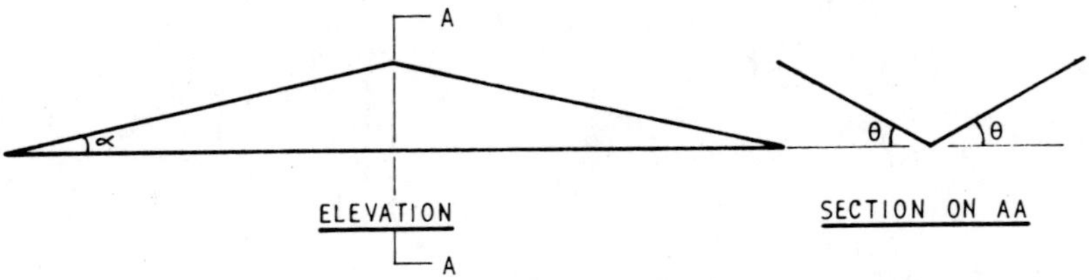

Fig. 161

can best be described as a rhomboid folded along its longer diagonal, and is shown in Fig. 161. The half longitudinal external units, marked A, in Fig. 160, are made by slicing the basic unit longitudinally into two identical halves. The quarter external units at the base, marked B in Fig. 160, are made by further slicing units of type A into two identical halves. Such a composite folded plate barrel vault has been built and tested by the author[7,138,146] and is shown in Fig. 196. The structure is discussed in greater detail in Chapter 8.

5.3 SHELL STRUCTURES

Shell structures can be either singly or doubly curved. The best example of the singly curved shell is the barrel vault, supported on rigid end diaphragms, as shown in Fig. 162. The barrel can have various types of cross sections.

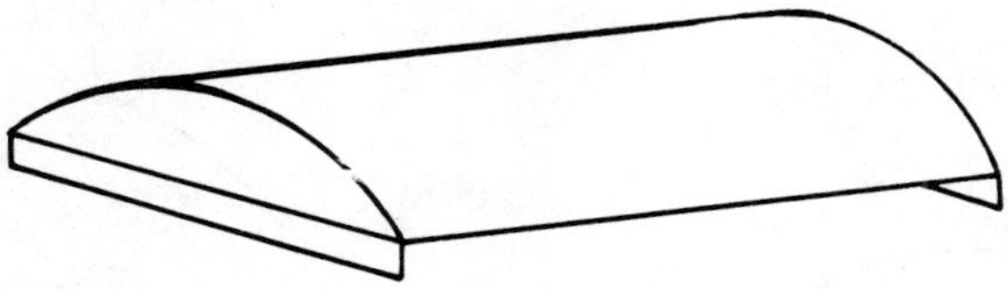

Fig. 162

Fig. 163

They can circular, parabolic, elliptical, and so on. The cross section can even be corrugated to prevent local buckling of the skin, as shown in Fig. 163.

The behavior of such a shell is very largely influenced by its span/width ratio. A long shell with a large span/width ratio tends to behave as a beam with a curved cross section. A short shell does not. Another example of the singly curved shell is the cone, which has already been mentioned as a folded plate pyramid with an infinite number of sides. The cone can of course be used by itself, as in the nose of a rocket but, for civil engineering structures, it has to be reduced to a pyramid with a finite number of sides for ease of connection to other units. The analysis and design of a barrel shell in plastics are considered in Chapter 9.

Doubly curved shells offer the most exciting possibilities. They can be of two types—synclastic and anticlastic. Synclastic surfaces have their principal curvatures in the same direction, while anticlastic surfaces have their principal curvatures in opposite directions, as shown in Fig. 164. The spherical dome is the simplest example of the doubly curved synclastic shell. The dome is principally in compression, and this may lead to overall instability. This instability can be prevented by ribbing or by use of sandwich construction. The analysis and design of a single skin FRP spherical dome are considered in detail in Chapter 9.

Any structural form, however, that puts the material wholly or principally in tension is better suited to single skin construction. The synclastic tension membrane, shown in Fig. 165, is an excellent structural form for this type

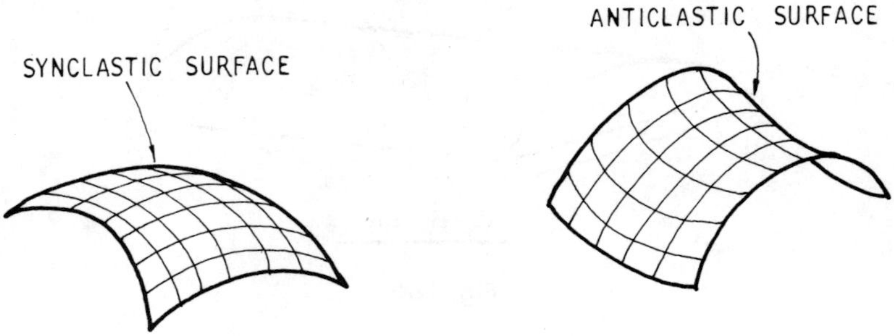

Fig. 164

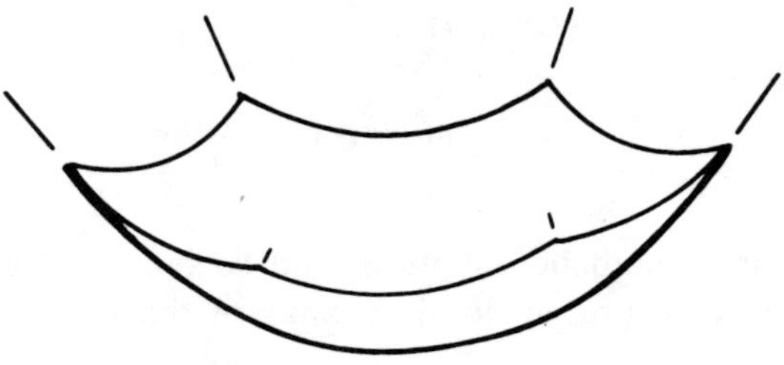

Fig. 165

of construction. Such membranes are almost entirely in tension, except at the edges where bending moments do arise. In synclastic shells, whether in tension or in compression, the principal curvatures need not necessarily be equal. A number of combinations are possible to yield a number of different shapes, not all perhaps, of optimum structural efficiency, but all nevertheless entirely functional in that they serve specific purposes. The elliptic paraboloid,[147] the ellipsoid, and the toroid are further examples of synclastic shells that could be used for plastics sandwich construction.

Anticlastic surfaces are of great importance for single skin construction. The conoid is shown in Fig. 166. Since the principal curvatures are in opposite directions, the buckling of the thin skin in one direction is effectively prevented by the tension in the other direction.

There are some limitations on the use of anticlastic surfaces dictated by the structure supporting the membrane. The structure eventually has to transfer all the load to the foundations and may become very uneconomical. Excellent examples of anticlastic surfaces are the soap film membranes shown in

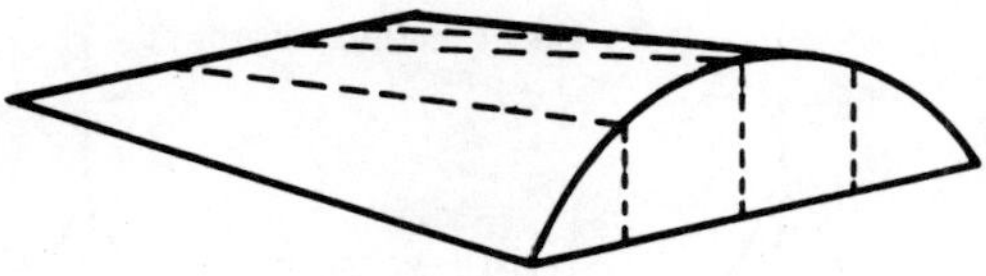

Fig. 166

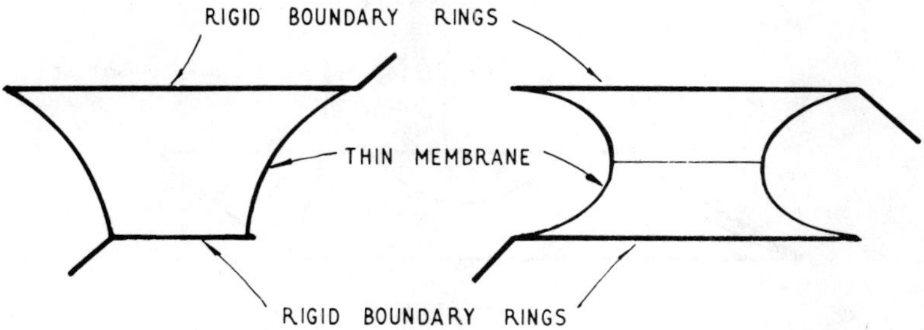

Fig. 167

Fig. 167. The membrane is under constant tension over the entire surface bounded by the circular edge members. In fact, the surface is a hyperbolic paraboloid and is obtained by moving a parabola on another parabola of opposite curvature, as shown in Fig. 164b. The surface can also be generated by the movement of a straight line on two skew lines, as shown in Fig. 168.

Under the particular loading condition of a uniform load on a horizontal projection, such as snow load, the hyperbolic paraboloid is under the action of only membrane stresses with absence of moments. However, under less favorable loading conditions, such as unsymmetrical loading or point loading,

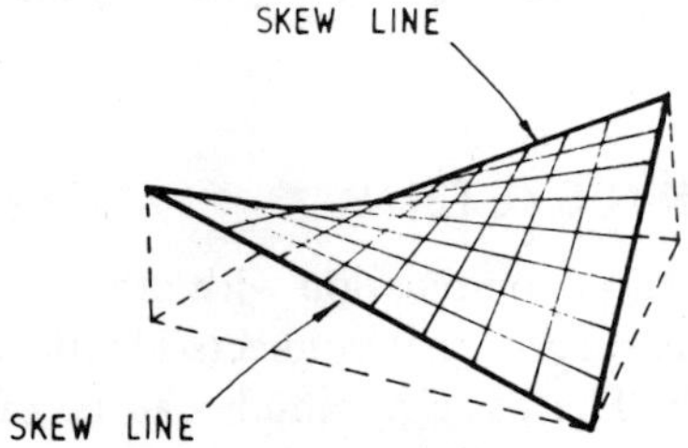

Fig. 168

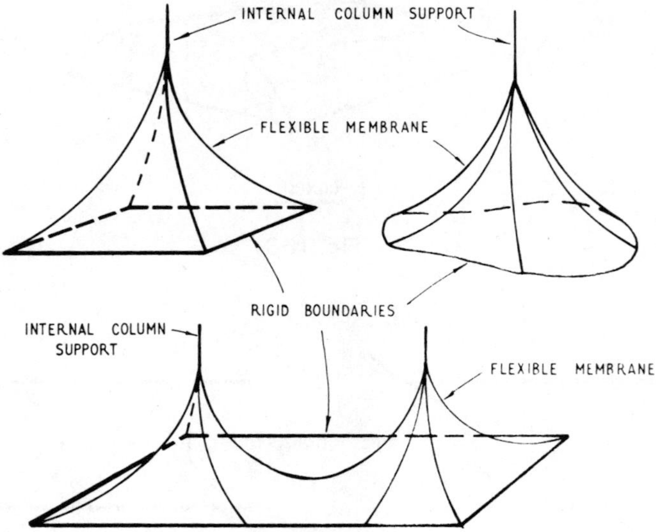

Fig. 169

bending moments do arise, particularly at the edges. The analysis and design of a hypar in single skin construction are considered in Chapter 9.

There are several other forms of anticlastic surfaces that have no strict mathematical basis but that are nevertheless efficient structural forms, particularly for plastics. These forms are dependent, not only upon the shape of the boundary, but on the loading itself. They are so flexible that they alter their geometry, under the external load, in an effort to keep the membrane under tension. This is, however, the ideal state when applied to a material that has no flexural rigidity at all. Such tension membranes with anticlastic surfaces are shown in Fig. 169. The rigid boundaries can be regular or irregular, as shown in the Fig. 169. Several such tentlike structures have been built using plastics covering membranes. The fabric, which can be fiberglass, is coated with teflon and has excellent flammability and weathering characteristics.

5.4 STRESSED SKIN SYSTEMS

Plastics can also be used in combination with other materials, such as aluminum, in a type of construction that connects identical pyramids together to form a structure.[148-150] The basic pyramids can be of the type discussed in Section 5.2 and shown in Fig. 154. As these pyramids have to be connected together, however, practical considerations narrow the choice to the square-

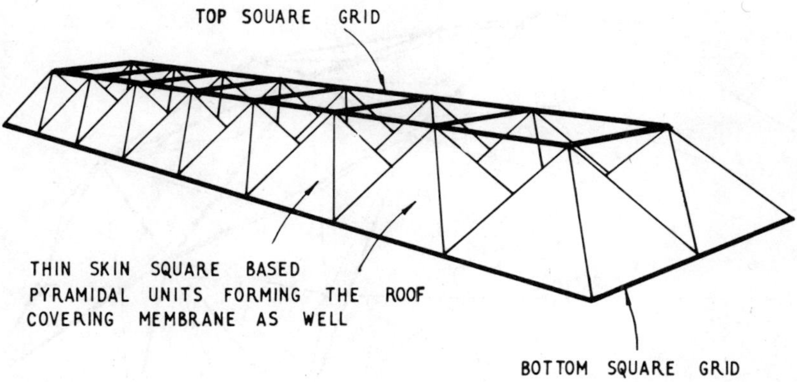

Fig. 170

based or the hexagonal-based pyramid. The square-based pyramids are connected together by square aluminum grids to yield a stable structure, as shown in Fig. 170. The aluminum grids take the tension and the compression, and the pyramids take all the shearing forces in much the same way as the core in sandwich construction. However, if the connecting flanges of the pyramids are made sufficiently stiff, they can form the bottom grid of the structure as well. The pyramids can also be used with their apexes downwards, as shown in Fig. 171. In this case, however, a continuous, nonstructural sheeting is necessary. Though the pyramid continuum takes all the shearing forces, the individual pyramids are acted upon by vertical and horizontal forces at the apex and along the boundaries. They show, therefore, much the same behavior as the larger pyramidal structures discussed in 5.2. Plate

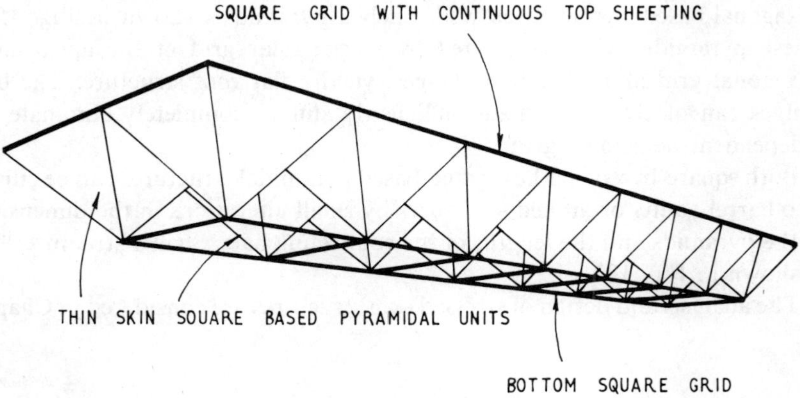

Fig. 171

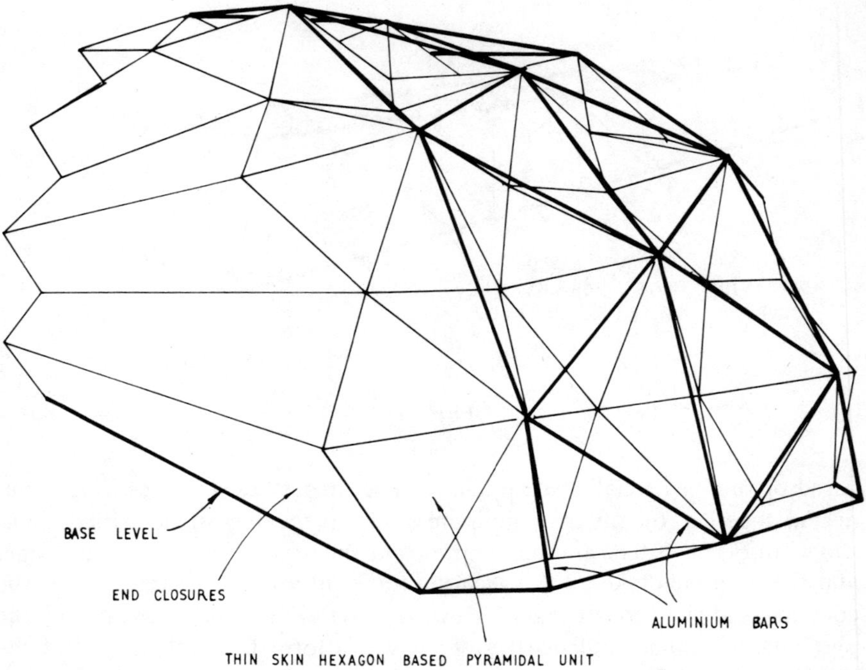

Fig. 172

buckling can occur at very low loads, and this can, once again, be prevented by ribbing or corrugations, as shown in Fig. 155. In this case, however, the buckling is never severe and may be permitted. The corners then need to be strengthened, as shown in Fig. 157. Much of the foregoing applies to hexagonal-based pyramids as well. Such a pyramid is shown in Fig. 154c. These pyramids, when connected by a triangular grid at the apex and a hexagonal grid along the base flanges, yield a flat roof structure. The base flanges can of course be made sufficiently stiff to completely eliminate the independent hexagonal grid.

Both square-based and hexagonal-based pyramidal structures can be curved into barrel vaults or arched structures by small alterations in the dimensions of the pyramids and the lengths of the connecting bars. Such a structure[149,151] is shown in Fig. 172.

The analysis and design of stressed skin structures are considered in Chapter 10.

CHAPTER 6

ANALYSIS OF FOLDED PLATE STRUCTURES

6.1 EXISTING CLASSICAL METHODS FOR THE ANALYSIS OF FOLDED PLATES

The analysis of folded plate structures has been studied by several authors, among whom Whitney et al.,[152] Simpson,[153] Parme,[154] Yitzhaki,[155] and Goldberg and Leve[156] deserve special mention. A summary of the methods of the first four authors shows that there are only small differences between them. Consider the folded plate roof, shown in Fig. 173, spanning longitudi-

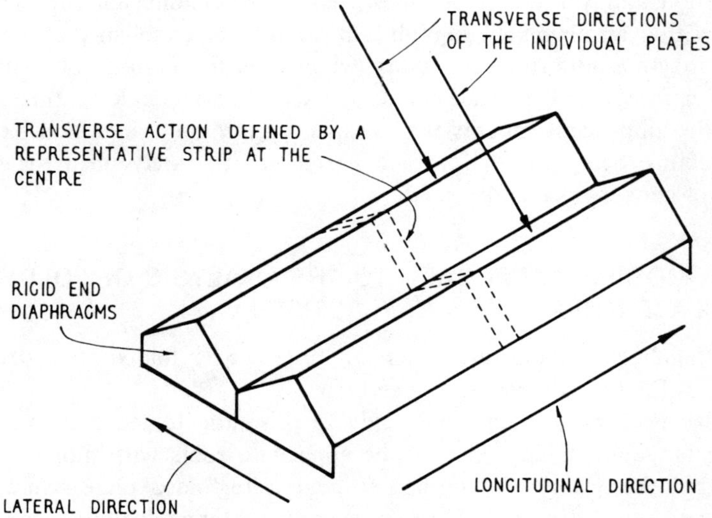

Fig. 173

nally between rigid end diaphragms. A strip at the center is considered as spanning between, or being continuous over, the ridges and the valleys. This defines the transverse strip action. The strip then transfers its reactions as loads into the planes of the plates, which span as beams in the longitudinal direction between the end diaphragms. The beams deflect under the load and thus cause a settlement of the supports of the strip. This modifies the transverse moments and therefore the reactions on the plates. It is only in the manner in which compatibility is established between the transverse strip action and the longitudinal beam action, and in the choice of unknowns, that the four methods differ.

The assumption made in all four methods is that the transverse moments can be determined by analyzing a representative strip at the center, these moments then being assumed to exist over the entire length of the folded plate. For long folded plates, where the span/plate width ratio is greater than 2, this assumption gives the correct moments over the center, where the maximum moments do occur, but gives incorrect moments over the supports, where the transverse moments should be zero. For short folded plates, where the span/plate width ratio is less than 2, it is necessary to consider a transverse plate action with the entire plate in bending. This approach also determines those smaller transverse moments in the longitudinal direction that are neglected by the simple approach.

Goldberg and Leve[156] recognized the shortcomings of the simple methods. For prismatic folded plates, they replaced the transverse strip action by a plate action, and considered the simultaneous plate bending and the membrane action of the various plates. Equilibrium can then be established at the joints, leading to $4n$ equilibrium equations, where n is the number of joints with unknown forces and displacements. The assumption made in this method is that the joint displacements can be expanded into half range Fourier series, the structure being solved for each harmonic separately until the desired degree of convergence is obtained.

6.2 A MODIFIED APPROACH TO THE ANALYSIS OF FOLDED PLATE STRUCTURES IN PLASTICS

All of the above methods are unsuitable for use with folded plate structures in plastics for the two reasons given below:

1. The methods are applicable only to prismatic folded plate structures with rectangular plates. They can be applied to roofs with more than two plates, but they cannot be applied to composite folded plate structures of the type shown in Fig. 159. They are even strictly inapplicable to the simple roof structure of the type shown in Fig. 174, where the transverse plate action requires the analysis of trapezoidal plates.

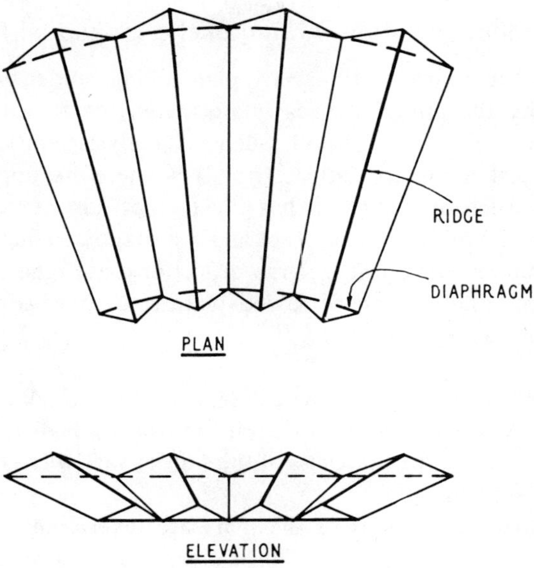

Fig. 174

2. All the existing methods make the assumptions that the transverse deflections of the plate are small in comparison to the thickness of the plate. If these deflections are large, as could very easily occur in single skin construction in plastics, the small deflection theories no longer apply. Large deflection theories that require the stretching of the middle surface to be taken into account must then be applied in all cases in which the plate is bent to a nondevelopable surface. This leads to nonlinear equations, making the problem considerably more involved. In this case, additional boundary conditions, due to movable or immovable edges, must also be taken into account. If the transverse deflections of the plate are very much larger than the thickness of the plate, the flexural rigidity of the plate can be neglected and the plate treated as a flexible membrane. However, if the deflections are very large, and comparable to the overall dimensions of the plate, the simplification in the formula for the curvature may not hold good either, making the problem even more involved.

For folded plate structures in plastics, the author[7,138] gives a very general and flexible approach that is applicable to any shape or type of folded plate structure, in single skin or sandwich construction. The structure is first divided into internal and external units.[157] The internal or external unit is then analyzed for a transverse plate action of the individual plates, followed by an analysis for the overall longitudinal action of the structure.

6.2.1 The Transverse Plate Action of the Individual Plates

The analysis of the plates for transverse plate action, under normal loading, is carried out by the use of classical plate theory, or by a matrix method such as the finite elements method. Such an analysis considers the entire plate and not just a representative strip. It is therefore applicable to any shape of plate. Such an approach has the further advantage that, if plate theory is used, a large number of exact and approximate solutions for plates of various shapes, even subject to large deflections, are readily available.

The transverse plate action of folded plate structures is determined on the following five assumptions:

1. In cases where the deflections are less than the thickness of the skin or sandwich, small deflection theories can be applied. If the transverse deflections exceed the thickness of the skin or sandwich, large deflection theories are applicable.
2. The boundary conditions of the plate are determined by the type of connection between the units. This is discussed in more detail in Section 7.1.
3. The relative displacements of the edges of the plate, due to the longitudinal deformations of the structure, can be considered as negligible.
4. Very large transverse deflections do not occur, and the simplified formula for the curvature is assumed to be valid.
5. In sandwich construction, plate theory is considered to be applicable. However, shear will be taken by the core, and the shearing deflections can be calculated separately. This assumption has been discussed in some detail in Section 4.2.1.

6.2.2 The Overall Longitudinal Action of the Structure

The overall longitudinal action of the structure can be easily determined. In structures such as rigid frames or barrel vaults, the overall longitudinal action is the frame or arch action. The structure may be analyzed on its neutral axis by any of the commonly used methods of structural engineering, with due allowance for variations in cross-sectional areas and moments of inertia over the structure. These methods are generally quite adequate but, to determine the stresses in the plates, modifications have to be made for particular cases listed below:

1. For short folded plate structures with a span/plate width ratio of less than 1.5, the straight line stress distribution is no longer valid, and deep beam theory[158] must be used.

2. In folded plate structures where the plate width/plate thickness ratio is fairly low, as might occur in sandwich construction, and where the transverse loads and deflections as well as the compressive stresses in the plane of the plate are not excessive, the entire width of the plate may be considered as effective in the analysis.

However, where the plate width/plate thickness ratio is high, as would occur in single skin construction, and the transverse loads and deflections as well as the compressive stresses in the plane of the plate are large, the buckled central portions of the plates do not contribute fully to the stiffness of the structure. The straight line stress distribution for a plate subject to bending moment, shown in Fig. 175a, then passes on the compressive side through successive stages, shown in Figs. 175b and 175c, shedding more and more load onto the stiffly jointed and hence straight ridges and valleys. In the final idealized state, shown in Fig. 175d, only a certain reduced area may be considered as effective and uniformly stressed. In effect, the two flat sheets meeting at a fold prevent each other from buckling so that the folds remain straight and unbuckled, providing a kind of space frame skeleton to support the load.[138,159,160] An attempt is made in Chapter 10 to define the area of the plate that is effective along the edge or fold of two plates. This approach, also termed the "strut analogy," transforms the structure from one formed by continuous plates to a skeletal structure which may then be analysed by matrix methods of structural analysis, such as the "stiffness" method. Using this approach, a folded plate conical dome structure was analyzed using IBM's STRUDL Coding, and the results of the analysis are also discussed in Chapter 10.

In sandwich construction, the overall section may be considered as effective, and the straight line stress distribution of Fig. 175a is quite adequate.

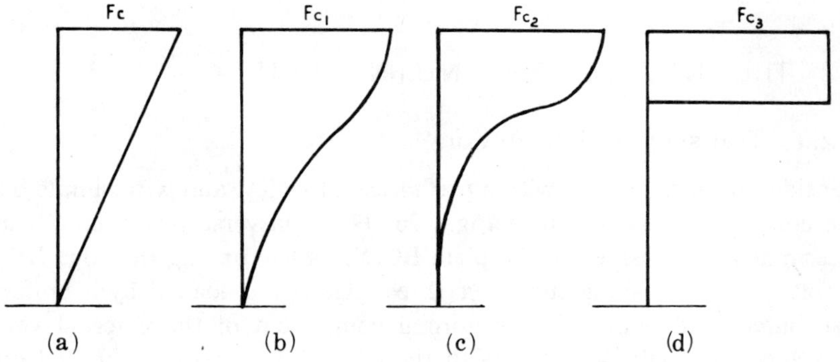

Fig. 175

CHAPTER 7

PRISMATIC FOLDED PLATE STRUCTURES

7.1 THE INFLUENCE OF CONNECTIONS

Consider the two-plate folded plate structure shown in Fig. 153a. The analysis of the structure may then be reduced to the analysis of two types of units—internal units and external units. The analysis of the units is largely dependent on the connections used. If the site connections between the units are simple bolted connections, the analysis may assume the presence of continuous hinges along the edges. The internal units may then be considered as spanning between the end diaphragms, with both longitudinal edges restrained against horizontal movement but vertical movement and angular rotation being permitted. The external units are assumed to be outer edge free. If site adhesive bonding is resorted to, the internal units may be considered as having both edges restrained against horizontal movement and angular rotation, with only vertical movement being permitted. The units can be used with a prefabricated valley or a prefabricated ridge. The former is advisable in regard to water leakage. The analysis of an internal unit is now considered in detail.

7.2 THE ANALYSIS OF AN INTERNAL UNIT

7.2.1 Transverse Plate Action

Consider an internal unit with a prefabricated valley, and with simple bolted site connections, as shown in Fig. 176. The transverse plate action can be determined by considering the plate BCDE as fixed along the edge BE and simply supported along the other three edges. It is loaded by a uniformly distributed load, which is the normal component of the external vertical load acting on half the unit. With the edge CD supported, the deflections are unlikely to be greater than half the thickness of the plate, and the small

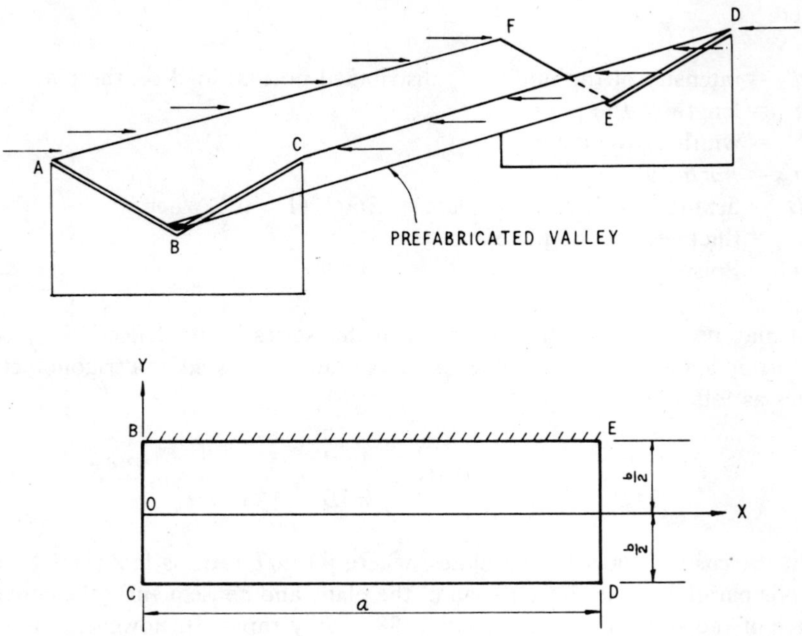

Fig. 176

deflection method of Levy, discussed in detail by Timoshenko and Woinowsky-Krieger[161] will be used. Briefly the plate is analyzed in two stages, the final results being obtained by superposition.

Stage 1: Simply supported on all four sides.
Stage 2: Subjected to moments along edge BE so that

$$\left(\frac{\partial w}{\partial y}\right)_{y=+\frac{b'}{2}}=0$$

Stage 1:
The expression for the deflection w is given by

$$w=\frac{4qa^4}{\pi^5 D}\sum_{m=1,3,5\ldots}^{\infty}\frac{1}{m^5}\left(1-\frac{\alpha_m\tanh\alpha_m+2}{2\cosh\alpha_m}\cosh\frac{2\alpha_m y}{b}\right.$$
$$\left.+\frac{\alpha_m}{2\cosh\alpha_m}\frac{2y}{b}\sinh\frac{2\alpha_m y}{b}\right)\sin\frac{m\pi x}{a}\qquad(88)$$

where

q = intensity of the uniformly distributed normal load on the plate
a = length of the plate
b = width of the plate
$\alpha_m = m\pi b/2a$
D = flexural rigidity of the plate = $Et^3/12(1-\nu^2)$, where
t = thickness of the plate and
ν = Poisson's ratio.

It may be noted that the first term in this series is the deflection w_1 of a uniformly loaded strip parallel to the X axis, and expressed as a trigonometric series as follows:

$$w_1 = \frac{q}{24D}(x^4 - 2ax^3 + a^3x) = \frac{4qa^4}{\pi^5 D} \sum_{m=1,3,5\ldots}^{\infty} \frac{1}{m^5} \sin \frac{m\pi x}{a} \tag{89}$$

In the case of short folded plates, where the a/b ratio is less than 1, this strip is parallel to the shorter span of the plate, and consequently the convergence of the series given in Equation 88 is very rapid. If, however, the a/b is greater than 1, as in Fig. 176, the strip is parallel to the longer span, and the convergence of the series given in Equation 88 is very slow. In the latter case it may be better to turn the axis around to the directions shown in Fig. 177. The expression for the deflection w is then given by

$$w = \frac{4qb^4}{\pi^5 D} \sum_{m=1,3,5\ldots}^{\infty} \frac{1}{m^5} \left(1 - \frac{\alpha_m' \tanh \alpha_m' + 2}{2 \cosh \alpha_m'} \cosh \frac{2\alpha_m' y}{a} + \frac{\alpha_m'}{2 \cosh \alpha_m'} \frac{2y}{a} \sinh \frac{2\alpha_m' y}{a}\right) \sin \frac{m\pi x}{b} \tag{90}$$

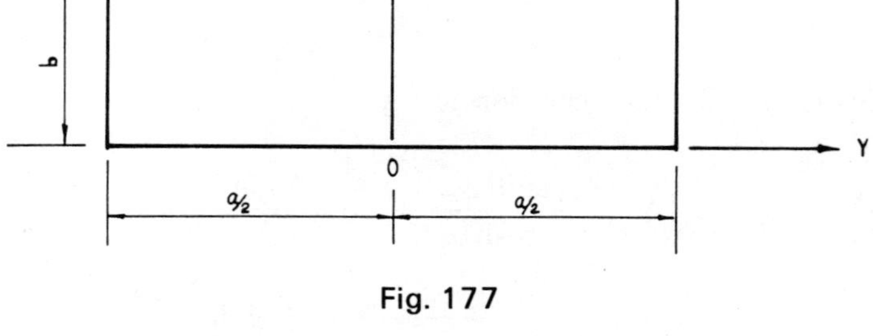

Fig. 177

where

$$\alpha_m' = \frac{m\pi a}{2b}$$

The transverse bending moments M_x and M_y can be determined from the following expressions:

$$M_x = -D\left(\frac{\partial^2 w}{\partial x^2} + \nu \frac{\partial^2 w}{\partial y^2}\right)$$
$$M_y = -D\left(\frac{\partial^2 w}{\partial y^2} + \nu \frac{\partial^2 w}{\partial x^2}\right) \tag{91}$$

Referring henceforth to the directions of the axes as in Fig. 176, the reactions on the supports V_y are determined by

$$(V_y)_{y=\pm\frac{b}{2}} = \left(Q_y + \frac{\partial M_{yx}}{\partial x}\right)_{y=\pm\frac{b}{2}} \tag{92}$$

where the shear force Q_y is given by

$$Q_y = -D\frac{\partial}{\partial y}\left(\frac{\partial^2 w}{\partial x^2} + \frac{\partial^2 w}{\partial y^2}\right) \tag{93}$$

and the twisting moment M_{yx} is given by

$$M_{yx} = -M_{xy} = -D(1-\nu)\frac{\partial^2 w}{\partial x \partial y} \tag{94}$$

Stage 2:

Moments M_y are applied along the edge BE. These moments can be expressed in the form

$$(M_y)_{y=+\frac{b}{2}} = \sum_{m=1,3,5\ldots}^{\infty} E_m \sin\frac{m\pi x}{a} \tag{95}$$

The deflection w' is then given by,

$$w' = \frac{a^2}{4\pi^2 D} \sum_{m=1,3,5\ldots}^{\infty} \frac{E_m \sin \dfrac{m\pi x}{a}}{m^2}$$

$$\times \left[\frac{1}{\cosh \alpha_m}\left(\alpha_m \tanh \alpha_m \cosh \frac{m\pi y}{a} - \frac{m\pi y}{a} \sinh \frac{m\pi y}{a}\right) + \frac{1}{\sinh \alpha_m}\left(\alpha_m \coth \alpha_m \sinh \frac{m\pi y}{a} - \frac{m\pi y}{a} \cosh \frac{m\pi y}{a}\right)\right] \quad (96)$$

The slope at the edge $y = b/2$, due to the external loading in Stage 1, is equal and opposite to the slope at the same edge due to M_y. Therefore

$$\left(\frac{\partial w}{\partial y}\right)_{y=+\frac{b}{2}} = -\left(\frac{\partial w'}{\partial y}\right)_{y=+\frac{b}{2}} \quad (97)$$

This gives the value of E_m as

$$E_m = -\frac{8q\,a^2}{\pi^3 m^3} \frac{\alpha_m - \tanh \alpha_m (1 + \alpha_m \tanh \alpha_m)}{(\alpha_m \tanh^2 \alpha_m - \tanh \alpha_m + \alpha_m \coth^2 \alpha_m - \coth \alpha_m - 2\alpha_m)} \quad (98)$$

This value of E_m when substituted in Equation 95 gives the value of the fixed end moment. The quantities M_x, M_y, $(V_y)_{y=+b/2}$ and $(V_y)_{y=-b/2}$ can then be determined as in Stage 1. The complete transverse plate action is then determined by the correct superposition of Stages 1 and 2.

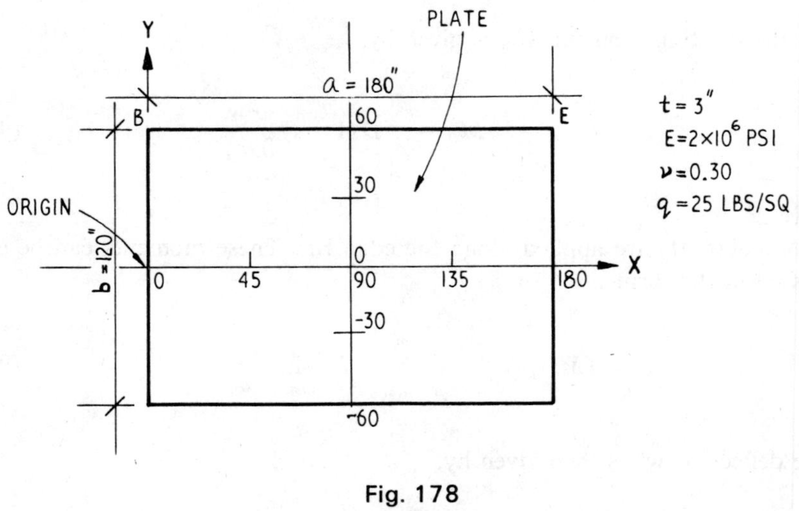

Fig. 178

In order to illustrate the use of the method, a solved example is taken below. The numerical data for the problem is shown in Fig. 178, and can be listed as follows:

$a = 15$ ft (4.57 m)
$b = 10$ ft (3.05 m)
$t = 3$ in. (76 mm)
$E = 2 \times 10^6$ psi (13.79 GPa)
$\nu = 0.30$
$q = 25$ lb/ft² (122 kg/m²).

The plate is first assumed as simply supported and with the deflected shape of Equation 88, the moments M_x and M_y can be determined by the equations given below:

$$M_x = \frac{4qa^2}{\pi^3} \sum_{m=1,3,5}^{\infty} \frac{1}{m^3} \left[1 - \frac{(1-\nu)\alpha_m \tanh\alpha_m + 2}{2\cosh\alpha_m} \cosh\frac{2\alpha_m y}{b} + \frac{(1-\nu)\alpha_m}{2\cosh\alpha_m} \frac{2y}{b} \sinh\frac{2\alpha_m y}{b} \right] \sin\frac{m\pi x}{a}$$

$$M_y = \frac{4qa^2}{\pi^3} \sum_{m=1,3,5}^{\infty} \frac{1}{m^3} \left[\nu - \frac{(\nu-1)\alpha_m \tanh\alpha_m + 2\nu}{2\cosh\alpha_m} \cosh\frac{2\alpha_m y}{b} + (\nu-1)\frac{\alpha_m}{2\cosh\alpha_m} \frac{2y}{b} \sinh\frac{2\alpha_m y}{b} \right] \sin\frac{m\pi x}{a}$$

In this case it is sufficient to take only the first two terms of the series, i.e., $m = 1, 3$, and the deflection w, and the M_x and M_y moments at quarter spans and at the center are shown plotted in Figs. 179 and 180.

The plate is now fixed along the edge BE by the application of moments M_y along the edge as given by the Equation 95, such that the slope of the plate for this second case is exactly equal and opposite to the slope of the plate in the first case, thereby leading to a fixed edge. The value of E_m has to be calculated, and for $m = 1, 3$, this is obtained as

$m = 1, \quad E_m = 299.64$
$m = 3, \quad E_m = 26.12$

The deflected shape w' for this second stage can now be obtained and is plotted in Fig. 181. The superposition of this deflected shape w' on w of the first case leads to the final deflected shape of the plate shown in Fig.

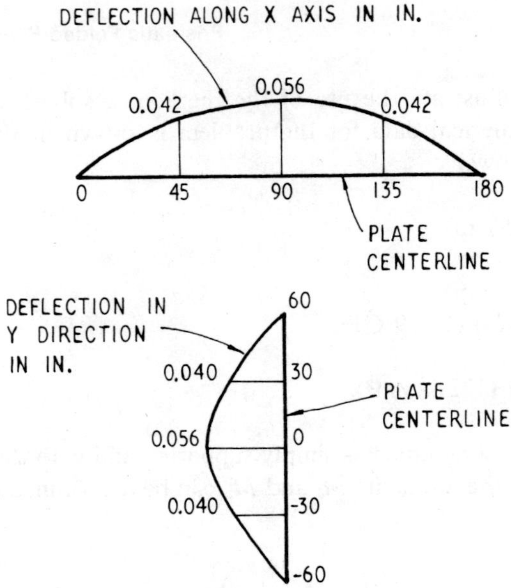

Fig. 179

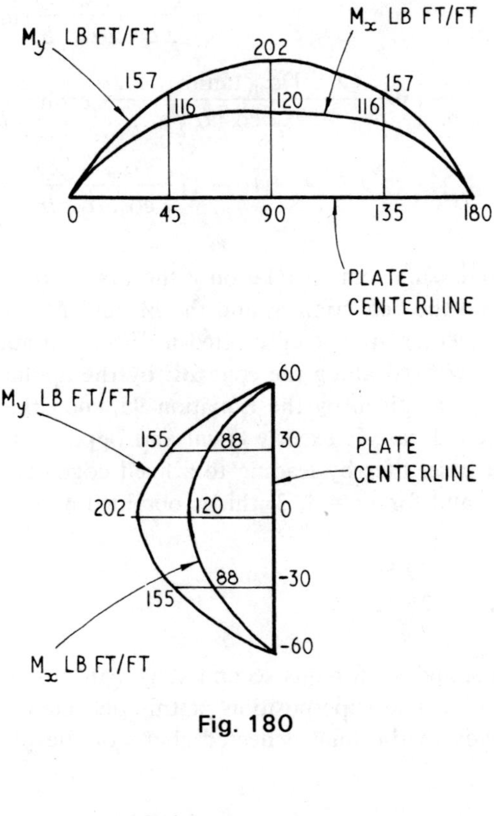

Fig. 180

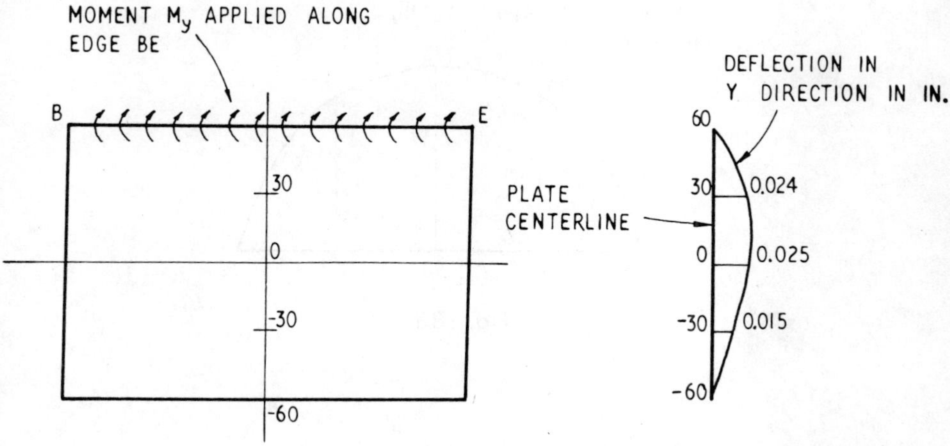

Fig. 181

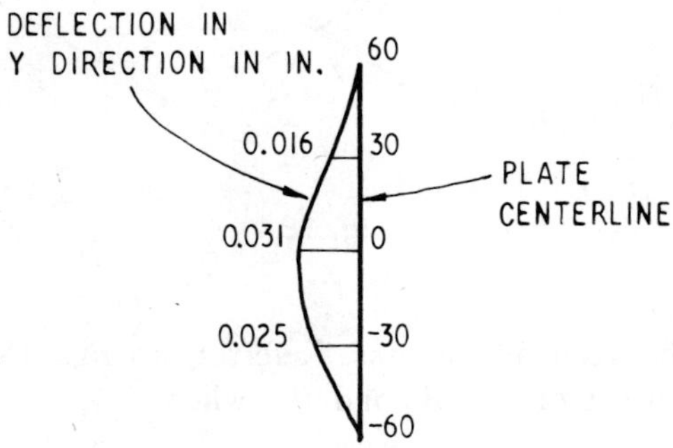

Fig. 182

182. The deflected shape has been plotted in Fig. 182 along the line $x =$ 7.5 ft (2.29 m).

The value of E_m also leads to the determination of the M_y moments applied along the edge to maintain fixity, and the distribution of these M_y moments is shown plotted in Fig. 183.

Finally the determination of the M_x and M_y moments for this second stage can be carried out in a like manner by the use of Equations 91.

7.2.2 Longitudinal Beam Action

In order to determine the longitudinal beam action of the plate, it is necessary to determine the loads transferred into the planes of the plates, which then

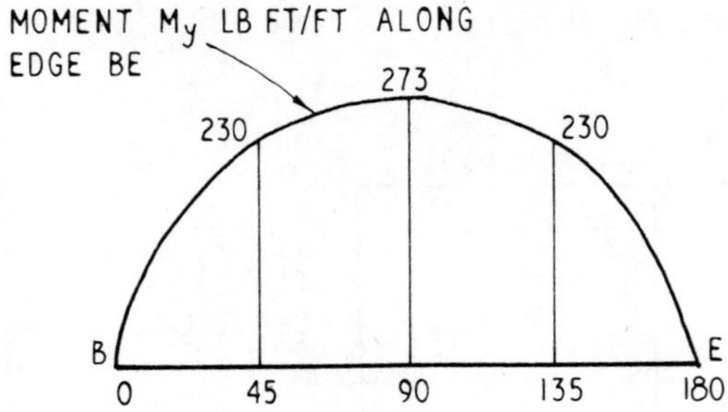

Fig. 183

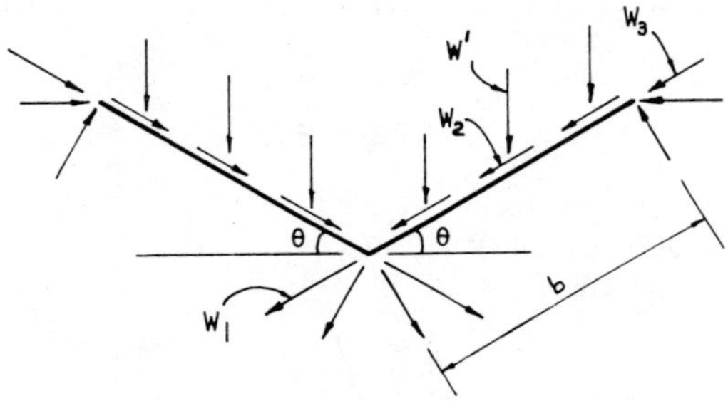

Fig. 184

span as beams between diaphragms. Referring to Fig. 184, it can be seen that this load consists of W_1, W_2 and W_3, where

W_1 = the component of the plate reaction at the valley per unit length
W_2 = the axial component of the external load per unit length
W_3 = the component of the horizontal reaction due to the supporting edge of the adjacent plate per unit length.

Let W' be the total external vertical load on the entire unit. Then, for transverse plate analysis,

$$q = \frac{W' \cos \theta}{2ab} \tag{99}$$

and, for longitudinal beam analysis,

$$W_1 = (V_y)_{y=+\frac{b}{2}} \cot \theta \tag{100}$$

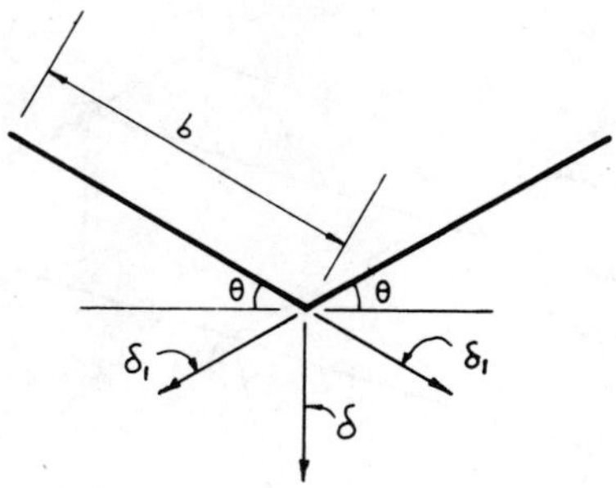

Fig. 185

$$W_2 = \frac{W' \sin\theta}{2a} \tag{101}$$

$$W_3 = (V_y)_{y=-\frac{b}{2}} \cot\theta \tag{102}$$

The vertical deflection of the folded plate unit is obtained by the resolution of the deflection of the individual beams as shown in Fig. 185. Therefore

$$\delta = \frac{\delta_1}{\sin\theta} \tag{103}$$

It may be noted that this is a more accurate approach and is easily applicable to internal units with rectangular plates. For long folded plates, the reactions $(V_y)_{y=\pm b/2}$ are very nearly constant along the edges BE and CD, except in the immediate vicinity of the diaphragms. The assumption, therefore, that the entire unit is a beam of "Vee" cross section leads to longitudinal stresses and deflections almost identical to those from the more accurate approach. In short folded plates however, the distribution of $(V_y)_{y=\pm b/2}$ along the edges BE and CD shows more variation and consequently the difference in results would be more pronounced.

The above analysis is based on the assumption that no plate buckling occurs. It cannot be applied, therefore, directly to very thin skins in single skin construction. This problem is considered in greater detail in Chapter 10.

7.3 THE ANALYSIS OF THE EXTERNAL UNIT

If the internal units have prefabricated valleys, the external unit reduces to a single plate, as shown in Fig. 186. In concrete folded plate construction, it is customary to provide an edge beam spanning between the diaphragms.

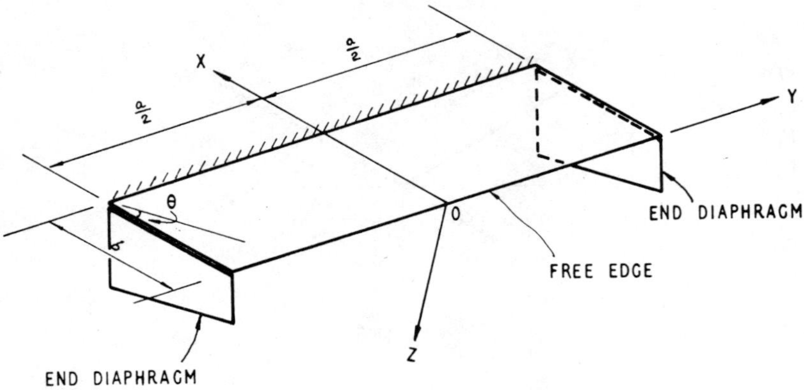

Fig. 186

This reduces the deflections and stresses in the external unit. In plastics folded plate construction, it is advisable from a constructional point of view to avoid the use of edge beams. The internal units, complete with flanges, are made in an inverted "Vee" shaped mold, and the mold and tooling costs can be a very sizable proportion of the total costs. If there are no edge beams, two external units can be made by slicing an internal unit down the valley line. The width of the external unit is less than its theoretical width by half the thickness of the cutting margin. One mold satisfies all units, which makes this method economical. As the stresses and deflections with a free edge, however, are larger than with an elastically supported edge, it may be necessary to increase the thickness of the unit. This can be done very easily on the same mold, even in sandwich construction. The thickness of the laminate can be increased, as can the thickness of the core. The core density can also be increased, if required. Were it necessary to mass produce large numbers of these units, it would be more economical to have two molds. In this chapter, the more difficult case of the external unit with a free edge is considered in detail.

The external unit has to be analyzed for a transverse plate action and a longitudinal beam action, as given in Section 6.2. The longitudinal beam action is the same as for the internal units, because the free edge of the external unit is in tension. The transverse plate action, however, is much more difficult to determine because the plate is subject to large deflections.

7.3.1 The Existing Classical Method of Analysis

The analysis of such a plate can be carried out by the energy method.[161] Let u, v, and w be the plate displacements in the X, Y, and Z directions, respectively. The total strain energy V of the system is given by

$$V = V_1 + V_2$$

where

V_1 = the strain energy due to stretching as a membrane
V_2 = the strain energy due to bending.

Then

$$\begin{aligned}
V_1 = \frac{Et}{2(1-\nu^2)} \iint \Bigg\{ & \left(\frac{\partial u}{\partial x}\right)^2 + \left(\frac{\partial u}{\partial x}\right)\left(\frac{\partial w}{\partial x}\right)^2 + \left(\frac{\partial v}{\partial y}\right)^2 + \left(\frac{\partial v}{\partial y}\right)\left(\frac{\partial w}{\partial y}\right)^2 \\
& + \frac{1}{4}\left[\left(\frac{\partial w}{\partial x}\right)^2 + \left(\frac{\partial w}{\partial y}\right)^2\right]^2 + 2\nu\left[\left(\frac{\partial u}{\partial x}\right)\left(\frac{\partial v}{\partial y}\right) + \frac{1}{2}\left(\frac{\partial v}{\partial y}\right)\left(\frac{\partial w}{\partial x}\right)^2 \right. \\
& \left. + \frac{1}{2}\left(\frac{\partial u}{\partial x}\right)\left(\frac{\partial w}{\partial y}\right)^2\right] + \frac{1-\nu}{2}\left[\left(\frac{\partial u}{\partial y}\right)^2 + 2\left(\frac{\partial u}{\partial y}\right)\left(\frac{\partial v}{\partial x}\right)\right. \\
& \left. + \left(\frac{\partial v}{\partial x}\right)^2 + 2\left(\frac{\partial u}{\partial y}\right)\left(\frac{\partial w}{\partial x}\right)\left(\frac{\partial w}{\partial y}\right) + 2\left(\frac{\partial v}{\partial x}\right)\left(\frac{\partial w}{\partial x}\right)\left(\frac{\partial w}{\partial y}\right)\right]\Bigg\} dxdy
\end{aligned} \tag{104}$$

$$\begin{aligned}
V_2 = \frac{Et^3}{24(1-\nu^2)} \iint \Bigg\{ & \left(\frac{\partial^2 w}{\partial x^2} + \frac{\partial^2 w}{\partial y^2}\right)^2 - 2(1-\nu)\left[\left(\frac{\partial^2 w}{\partial x^2}\right)\left(\frac{\partial^2 w}{\partial y^2}\right)\right. \\
& \left. - \left(\frac{\partial^2 w}{\partial x \partial y}\right)^2\right]\Bigg\} dxdy
\end{aligned} \tag{105}$$

The integrations are extended over the entire surface of the plate. In applying the energy method, it is necessary to assume suitable expressions for u, v, and w. These expressions must satisfy all the boundary conditions and will contain several parameters, the magnitudes of which are determined by the principle of virtual displacements. The boundary conditions that have to be satisfied are as follows:

Fixed edge:

$$(w)_{x=b} = 0, \quad \left(\frac{\partial w}{\partial x}\right)_{x=b} = 0$$

Simply supported edge:

$$(w)_{y=\pm\frac{a}{2}} = 0, \quad \left(\frac{\partial^2 w}{\partial y^2}\right)_{y=\pm\frac{a}{2}} = 0$$

Free edge:

$$\left[\frac{\partial^3 w}{\partial x^3}+(2-\nu)\frac{\partial^3 w}{\partial x \partial y^2}\right]_{x=0}=0, \left[\frac{\partial^2 w}{\partial x^2}+\nu\frac{\partial^2 w}{\partial y^2}\right]_{x=0}=0$$

7.3.3 The Author's Semi-Empirical Approach to the Problem

A serious disadvantage of the above method is that the accuracy of the results depends very much upon the assumed shapes for u, v, and w. A very close assumption would lead to good accuracy, but such a deflected shape might be very complicated.

The author[162] has a semi-empirical approach to the problem. The substitution of various deflection shapes into Equations 104 and 105 for V_1 and V_2, which go to make up V, shows that the final equation for the maximum deflection w_0 (at the center of the free edge) can be written in the form,

$$\left(\frac{w_0}{t}\right)+A\left(\frac{w_0}{t}\right)^3=\frac{Bqb^4}{Et^4} \tag{106}$$

where

t = thickness of the plate
q = intensity of the uniformly distributed normal loading on the plate
b = width of the plate
A, B = constants that are functions of the a/b ratio of the plate and of the Poisson's ratio ν.

The constants A and B have been determined experimentally by the author[162] for various a/b ratios, and for a Poisson's ratio of 0.35, and these are shown plotted against the a/b ratio in Fig. 187. The values of A and B for an infinitely long plate ($a/b = \infty$) are given as dotted lines. These are the limiting values for A and B. The value of A at $a/b = 1$ is very sensitive to small variations in the loading intensity q and has not been shown. Therefore the curves in Fig. 187 should not be used for $a/b < 1.5$.

7.3.3 Determination of the Transverse Moments

To determine the transverse moments in the plate, it is now necessary to assume a deflection surface in the Z direction. This deflection surface should satisfy all the boundary conditions and be expressed in the following form:

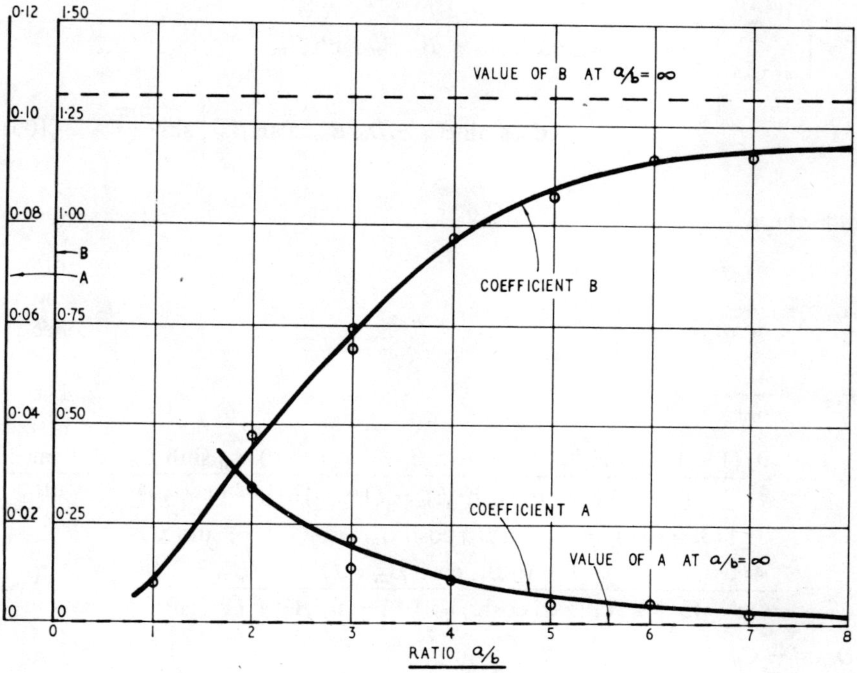

Fig. 187

$$w = w_0 f(x,b) f(y,a) \tag{107}$$

The deflection surface that rigorously satisfies all the boundary conditions is the surface determined by the small deflection theory. When expressed in the form given by Equation 107, this can be written as

$$w = \frac{w_0}{K}\left[\sum_{m=1,3,5\ldots}^{\infty}\left\{-A_m + A_m \cosh\frac{m\pi(b-x)}{a} + B_m \frac{m\pi(b-x)}{a}\sinh\frac{m\pi(b-x)}{a} + C_m \sinh\frac{m\pi(b-x)}{a} + D_m \frac{m\pi(b-x)}{a}\cosh\frac{m\pi(b-x)}{a}\right\}\sin\frac{m\pi\left(\dfrac{a}{2}+y\right)}{a}\right] \tag{108}$$

where

$$K = \left[\sum_{m=1,3,5\ldots}^{\infty} - A_m + A_m \cosh \beta_m + B_m \beta_m \sinh \beta_m + C_m \sinh \beta_m + D_m \beta_m \cosh \beta_m \right] \sin \frac{m\pi}{2} \tag{109}$$

and where

$$A_m = -\frac{4}{\pi^5 m^5}$$

$$B_m = \frac{4}{\pi^5 m^5} \frac{(3+\nu)(1-\nu)\cosh^2 \beta_m + 2\nu \cosh \beta_m - \nu(1-\nu)\beta_m \sinh \beta_m - (1-\nu^2)}{(3+\nu)(1-\nu)\cosh^2 \beta_m + (1-\nu)^2 \beta_m^2 + (1+\nu)^2}$$

$$C_m = \frac{4}{\pi^5 m^5} \frac{(3+\nu)(1-\nu)\sinh \beta_m \cosh \beta_m + \nu(1+\nu)\sinh \beta_m - \nu(1-\nu)\beta_m \cosh \beta_m - (1-\nu)^2 \beta_m}{(3+\nu)(1-\nu)\cosh^2 \beta_m + (1-\nu)^2 \beta_m^2 + (1+\nu)^2}$$

$$D_m = -C_m$$

$$\beta_m = \frac{m\pi b}{a}$$

The surface given by Equation 108 is a very accurate one. In addition to this surface, two other rather simpler surfaces[162] can also be used. These two surfaces are given by the following expressions.

1. $$w = w_0 \left(1 - \sin \frac{\pi x}{2b}\right) \cos \frac{\pi y}{a} \tag{110}$$

2. $$w = w_0 \left(1 - \frac{4x}{3b} + \frac{x^4}{3b^4}\right) \left[\frac{\left(\frac{a}{2} - y\right)^4 - 4\left(\frac{a}{2}\right)\left(\frac{a}{2} - y\right)^3 + 8\left(\frac{a}{2}\right)^3\left(\frac{a}{2} - y\right)}{5\left(\frac{a}{2}\right)^4} \right] \tag{111}$$

Both satisfy the boundary conditions at all edges except at the free edge. As the maximum deflection w_0 is now known, it is possible to calculate the deflection curves along the X axis and along the free edge as given by Equations 108, 110, and 111. The theoretical transverse moments M_x and M_y can be determined by the use of Equation 91. From the deflection surface given by Equation 108, $\partial^2 w/\partial x^2$ and $\partial^2 w/\partial y^2$ can be obtained as

$$\frac{\partial^2 w}{\partial x^2}=\frac{w_0}{K}\sum_{m=1,3,5\ldots}^{\infty}\left[\left(-\frac{m\pi}{a}\right)^2 A_m \cosh\frac{m\pi(b-x)}{a}\right.$$

$$+B_m\left(-\frac{m\pi}{a}\right)^2\frac{m\pi(b-x)}{a}\sinh\frac{m\pi(b-x)}{a}$$

$$+2B_m\left(-\frac{m\pi}{a}\right)^2\cosh\frac{m\pi(b-x)}{a}+C_m\left(-\frac{m\pi}{a}\right)^2\sinh\frac{m\pi(b-x)}{a}$$

$$+D_m\left(-\frac{m\pi}{a}\right)^2\frac{m\pi(b-x)}{a}\cosh\frac{m\pi(b-x)}{a} \tag{112}$$

$$\left.+2D_m\left(-\frac{m\pi}{a}\right)^2\sinh\frac{m\pi(b-x)}{a}\right]\sin\frac{m\pi\left(\frac{a}{2}+y\right)}{a}$$

$$\frac{\partial^2 w}{\partial y^2}=\frac{w_0}{K}\sum_{m=1,3,5\ldots}^{\infty}-\left[-A_m+A_m\cosh\frac{m\pi(b-x)}{a}\right.$$

$$+B_m\frac{m\pi(b-x)}{a}\sinh\frac{m\pi(b-x)}{a}+C_m\sinh\frac{m\pi(b-x)}{a} \tag{113}$$

$$\left.+D_m\frac{m\pi(b-x)}{a}\cosh\frac{m\pi(b-x)}{a}\right]\left(\frac{m\pi}{a}\right)^2\sin\frac{m\pi\left(\frac{a}{2}+y\right)}{a}$$

From the simple deflection surface given by Equation 111, $\partial^2 w/\partial x^2$ and $\partial^2 w/\partial y^2$ can be obtained as

$$\frac{\partial^2 w}{\partial x^2}=w_0\left[\frac{12x^2}{3b^4}\right]\left[\frac{\left(\frac{a}{2}-y\right)^4-4\left(\frac{a}{2}\right)\left(\frac{a}{2}-y\right)^3+8\left(\frac{a}{2}\right)^3\left(\frac{a}{2}-y\right)}{5\left(\frac{a}{2}\right)^4}\right] \tag{114}$$

$$\frac{\partial^2 w}{\partial y^2}=w_0\left[1-\frac{4x}{3b}+\frac{x^4}{3b^4}\right]\left[\frac{12\left(\frac{a}{2}-y\right)^2-24\left(\frac{a}{2}\right)\left(\frac{a}{2}-y\right)}{5\left(\frac{a}{2}\right)^4}\right] \tag{115}$$

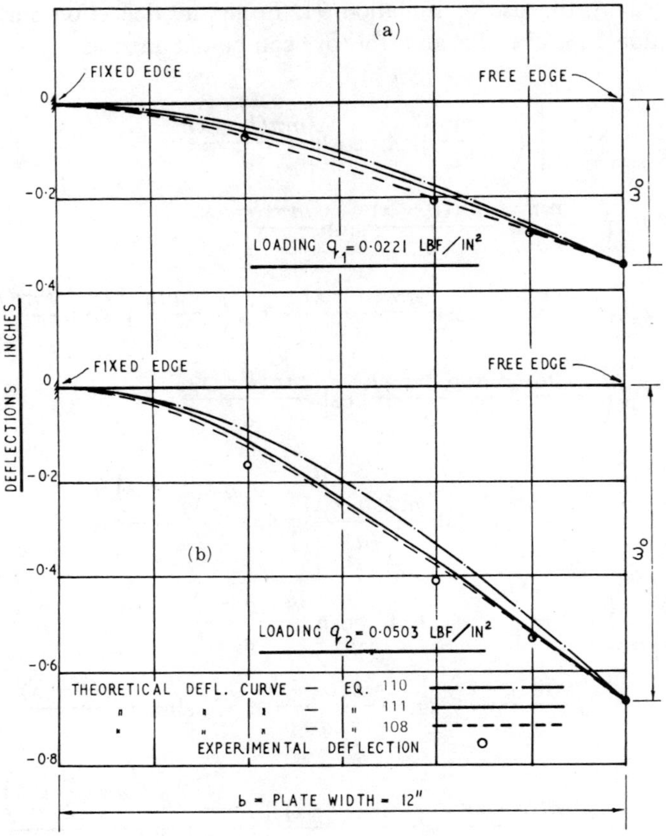

Fig. 188

In order to check the accuracy of these equations, experimental work was performed by the author on a plate with $b = 12$ in. (305 mm) and $a/b = 3$, and at load intensities $q_1 = 0.0221$ psi (0.152 kPa) and $q_2 = 0.0503$ psi (0.347 kPa). The deflection curves along the X axis as given by Equations 108, 110, and 111, together with the experimental deflection, are shown in Figs. 188a and 188b. As can be seen, Equations 108 and 111 yield curves closest to the experimental. For the same plate and load intensities, the deflection curves along the Y axis (the free edge) as given by Equations 108 and 111, together with the experimental deflection, are shown in Figs. 189a and 189b. The theoretical and experimental M_x and M_y moments are shown in Figs. 190 and 191. These moments are along a line parallel to the X axis, and 1 in. (25 mm) off the center ($y = 1$ in.). It can be seen that the moments

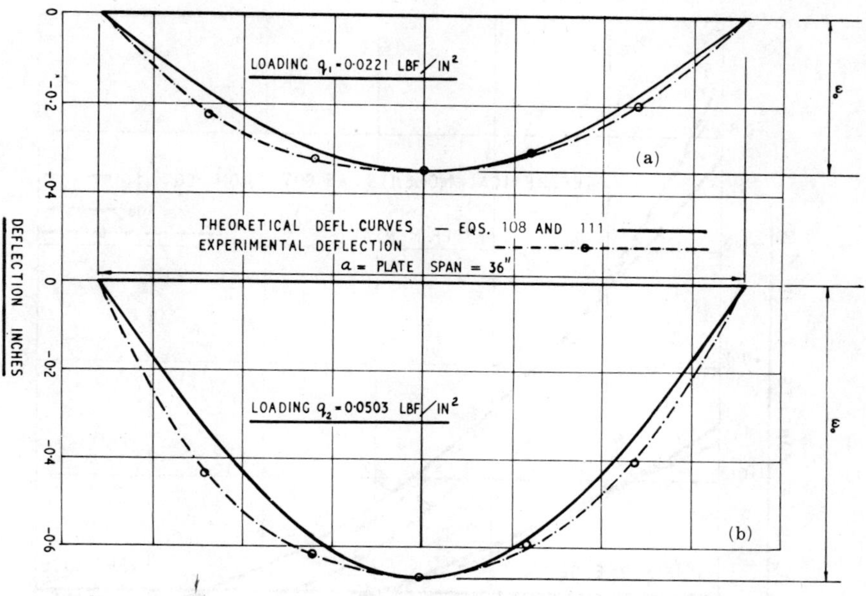

Fig. 189

predicted by the use of Equation 108 are in good agreement with the experimental. Though this deflection surface is complicated, it can be programmed very easily for a computer. All the series converge very rapidly, and it is sufficient to take the first five terms in the series, i.e., until $m = 9$. The moments predicted by the use of Equation 111 are inaccurate at the free edge. This is because the boundary conditions at the free edge have not been met. The maximum moments and the general behavior of the plate, however, can be predicted with reasonable accuracy even with the use of this simple expression. It is to be recommended for a preliminary analysis of the plate.

7.3.4 Determination of the Membrane Stresses

The theoretical membrane stresses may be determined in much the same way as the transverse moments. It is necessary to assume deflection surfaces in the X and Y directions. The strains in the X and Y directions, and the shearing strains can be written as

$$\epsilon_x = \frac{\partial u}{\partial x} + \frac{1}{2}\left(\frac{\partial w}{\partial x}\right)^2$$

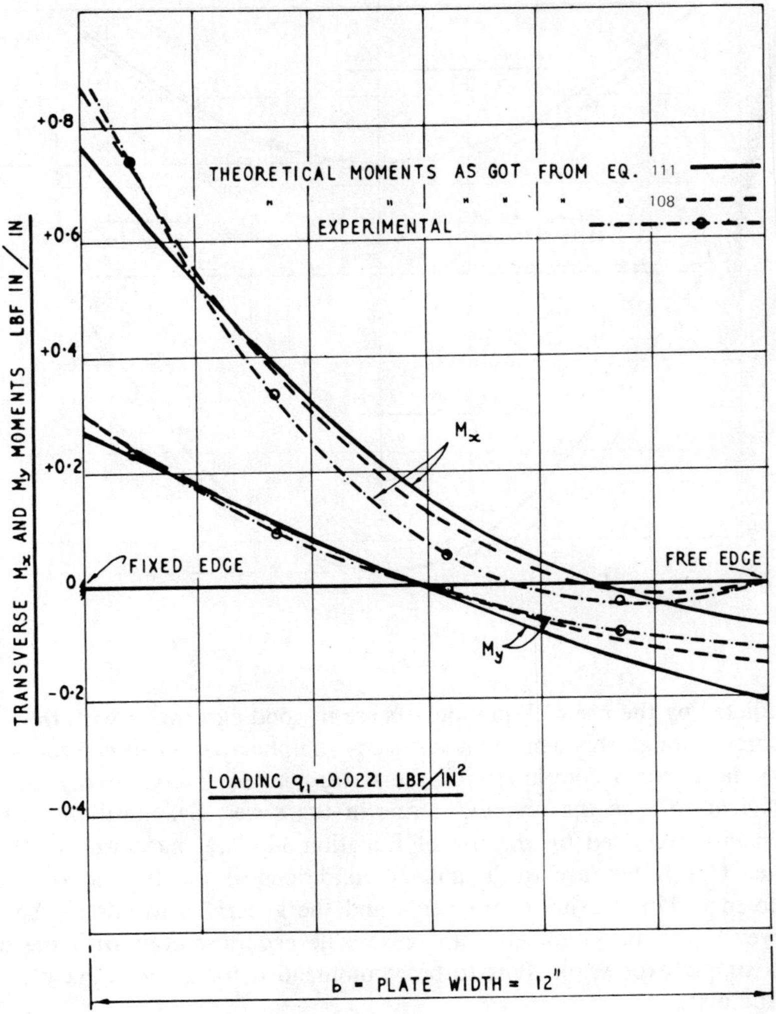

Fig. 190

$$\epsilon_y = \frac{\partial v}{\partial y} + \frac{1}{2}\left(\frac{\partial w}{\partial y}\right)^2 \tag{116}$$

$$\gamma_{xy} = \frac{\partial u}{\partial y} + \frac{\partial v}{\partial x} + \left(\frac{\partial w}{\partial x}\right)\left(\frac{\partial w}{\partial y}\right)$$

The membrane stresses are

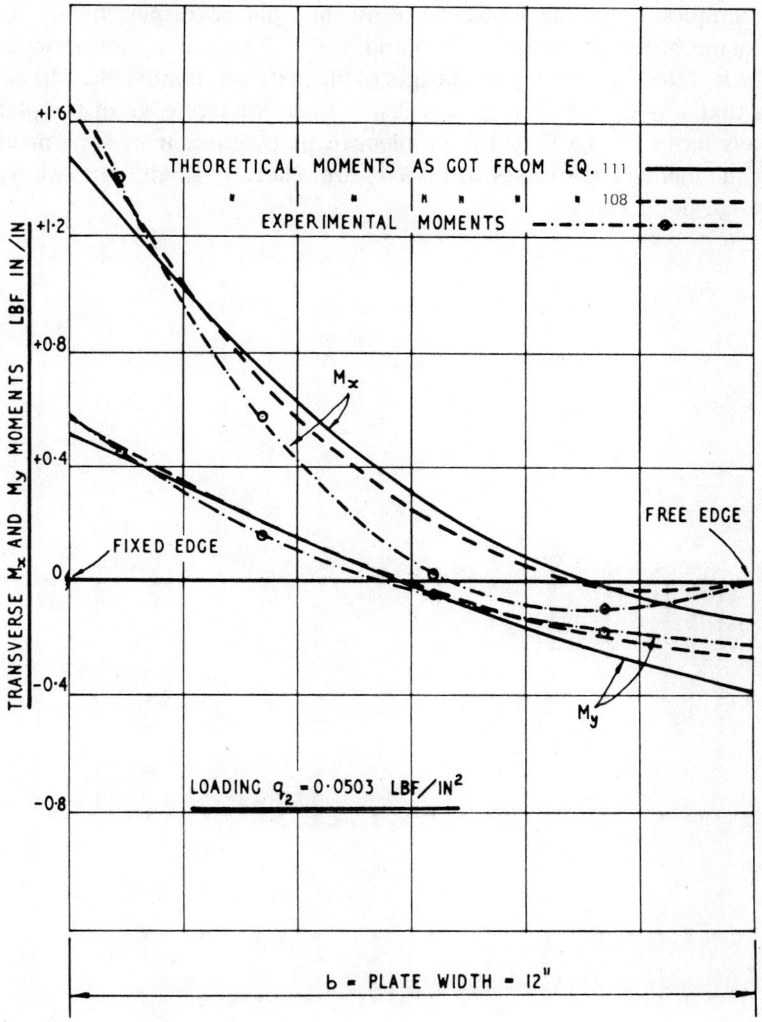

Fig. 191

$$\sigma_x = \frac{(\epsilon_x + \nu\,\epsilon_y)\,E}{(1 - \nu^2)}$$

$$\sigma_y = \frac{(\epsilon_y + \nu\,\epsilon_x)\,E}{(1 - \nu^2)} \qquad (117)$$

$$\sigma_{xy} = \gamma_{xy}\,G$$

The simplest approach would be to assume that no displacements occur in the plane of the plate, i.e., $u = 0$ and $v = 0$. The assumption that $v = 0$ is realistic if the simply supported edges of the plate are immovable. The assumption that $u = 0$, however, is unrealistic, since the free edge of the plate will always move in. To keep the problem in perpsective, it may be mentioned that the membrane stresses themselves are relatively small, particularly from a design point of view.

CHAPTER 8

COMPOSITE FOLDED PLATE STRUCTURES

8.1 THE COMPOSITE FOLDED PLATE RIGID FRAME

Composite folded plate rigid frames are shown in Fig. 159. The analysis of such structures does not present any difficulty and is only briefly considered. The approach to the analysis is the same as for the prismatic roof units, considered in Chapter 7, except that the longitudinal beam action is substituted by an overall rigid frame action.[138] The transverse plate action can be determined as before, but in this case the plate has shaped edges or skew supports. The plate is fixed on three sides and simply supported or fixed or free on the fourth side, depending on whether the unit is an internal unit or external unit, and on the connections. For long folded plates, the analysis may be simplified by neglecting the effects of edge shaping. The ridge and valley loads, as well as the tangential components of the external loads, are then transmitted to the supports by rigid frame action. The degree of rigid frame or portal action achieved, however, is largely dependent upon the column and knee stiffeners. With adequate stiffening, the rigid frame can be analyzed on its neutral axis by any of the common methods of structural engineering. Variations in cross-sectional areas and moments of inertia over the rigid frame must be taken into account.

8.2 THE COMPOSITE FOLDED PLATE BARREL VAULT

The composite folded plate barrel vault, shown in Fig. 160, is one of the most interesting shapes available for the roofing of large areas. It has the advantage that it can be built up from only one type of unit, which makes it ideal for prefabrication and hence for a material like plastics. Structures of this type, in plastics sandwich construction, could be built to spans of

200 ft (61 m). Consider the rhomboidal shaped unit, folded along its longer diagonal as shown in Fig. 161 and let the angle of the unit in elevation be α. For the units to mate, it is necessary that the angle between the valley lines of successive units in adjacent rows be 2α, and the angle between valley lines of successive units in the same row be 4α. The fold angle of the units θ may vary between practical limits of 15° and 60°. A change in θ alters the stiffness and the total length of the barrel. A shallow angle such as 15° would lead to a large area being covered by a rather flexible barrel with shallow folds. A larger angle would cause a smaller area to be covered by a stiffer barrel with deeper folds.

8.3 THE ANALYSIS OF THE BARREL VAULT

The behavior of the structure can, once again, be divided into a transverse plate action and a longitudinal arch action. Each of these will now be considered in greater detail.

8.3.1 Transverse Plate Action of the Barrel Vault

The transverse plate action is determined by the analysis of a triangular plate under normal loading. The triangular plate may be considered as fixed along all three edges if it is an internal plate and if site adhesive bonding is resorted to along the ridge lines. It may be considered as fixed along the valley line, and simply supported along the ridges, if simple bolting is used. If the valley lines too are bolted, it is necessary to insert longitudinal ties along the length of the barrel vault, to prevent an "accordian"-like effect. The external plate must be considered as free along the valley line unless an edge beam is provided, in which case it is elastically supported. The triangular plate is loaded by the component of the external load normal to the surface. For small barrel vaults, the most critical loading for the transverse plate action can be incidental live load. The transverse plate action under this loading can be determined by a rather simple approximation. Consider the triangular plate shown in Fig. 192, and subjected to a point load at the center. If the plate is long and narrow, and if the point load is applied at the center, the influence of the load does not extend into the corners but is limited to a circle of influence, as in Fig. 192. The analysis of the triangular plate may then be substituted, without much loss in accuracy, by the analysis of a circular plate loaded at the center. The analysis itself depends on whether the structure is in single skin or in sandwich construction. In single skin construction, the deflections will always be large in comparison to the thickness of the skin. In sandwich construction, the deflections will be small in comparison to the thickness of the sandwich, but the large shearing deforma-

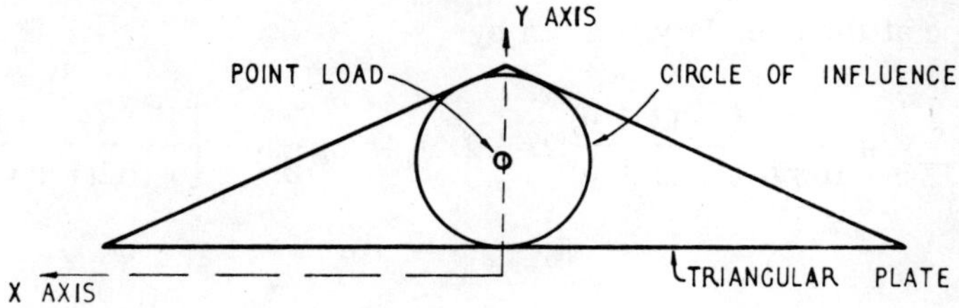

Fig. 192

tions of the core have to be taken into account. The analysis of an internal plate with fixed edges, under a point load at the center, is now considered in detail.

8.3.2 Analysis of the Plate When the Transverse Deflections Are Small

Consider a circular plate with clamped edges,[161] and let

a = radius of the plate
r = radius at any point
c = radius of the circular pad applying the load
P = the external load.

With the edges simply supported and when $r \neq 0$, the deflection w' is given by

$$(w')_{r \neq 0} = \frac{P}{16\pi D}\left\{\frac{(3+\nu)}{(1+\nu)}(a^2 - r^2) + 2r^2 \log\frac{r}{a} + c^2 \left[\log\frac{r}{a} - \frac{(1-\nu)}{2(1+\nu)}\frac{(a^2 - r^2)}{a^2}\right]\right\} \tag{118}$$

When $r = 0$,

$$(w')_{r=0} = \frac{P}{16\pi D}\left[\frac{(3+\nu)}{(1+\nu)}a^2 + c^2 \log\frac{c}{a} - \frac{(7+3\nu)}{4(1+\nu)}c^2\right] \tag{119}$$

where

ν = Poisson's ratio
D = flexural rigidity of the plate.

The slope at the boundary is given by

$$\left(\frac{\partial w'}{\partial r}\right)_{r=a} = \frac{P}{16\pi D}\left\{-\frac{(3+\nu)}{(1+\nu)}2r + 2r + 4r\log\frac{r}{a} + c^2\left[\frac{1}{r} + \frac{(1-\nu)}{2(1+\nu)}\frac{2r}{a^2}\right]\right\} \tag{120}$$

The external load is removed, and moments M are now applied at the boundary. The slope at the boundary is then given by

$$\left(\frac{\partial w''}{\partial r}\right)_{r=a} = \frac{Ma}{D(1+\nu)} \tag{121}$$

For the edges to be fully clamped, the slope as given by Equations 120 and 121 must be equal and opposite. This determines the value of M as

$$M = +\frac{P(1+\nu)}{16\pi a}\left\{\frac{(3+\nu)}{(1+\nu)}2a - 2a - c^2\left[\frac{1}{a} + \frac{(1-\nu)}{(1+\nu)}\frac{1}{a}\right]\right\} \tag{122}$$

The deflections due to M are given by

$$w'' = \frac{M(r^2 - a^2)}{2D(1+\nu)} \tag{123}$$

The total deflection, when $r \neq 0$, is hence obtained as

$$\begin{aligned}(w)_{r\neq 0} = \frac{P}{16\pi D}&\left\{\frac{(3+\nu)}{(1+\nu)}(a^2 - r^2) + 2r^2\log\frac{r}{a}\right.\\ &+ c^2\left[\log\frac{r}{a} - \frac{(1-\nu)}{2(1+\nu)}\frac{(a^2-r^2)}{a^2}\right]\\ &\left.+\left[\frac{2(3+\nu)}{(1+\nu)} - 2 - \frac{c^2}{a^2}\left(1+\frac{1-\nu}{1+\nu}\right)\right]\frac{(r^2-a^2)}{2}\right\}\end{aligned} \tag{124}$$

When $r = 0$, the total deflection can be obtained as

$$\begin{aligned}(w)_{r=0} = \frac{P}{16\pi D}&\left\{\frac{(3+\nu)}{(1+\nu)}a^2 + c^2\log\frac{c}{a} - \frac{(7+3\nu)}{4(1+\nu)}c^2\right.\\ &\left.-\left[\frac{2(3+\nu)}{(1+\nu)} - 2 - \frac{c^2}{a^2}\left(1+\frac{1-\nu}{1+\nu}\right)\right]\frac{a^2}{2}\right\}\end{aligned} \tag{125}$$

The bending moments in the plate can then be determined from the following equations.

$$
\begin{aligned}
M_r &= -D\left[\frac{d^2 w}{dr^2}+\frac{\nu}{r}\frac{dw}{dr}\right] \\
M_t &= -D\left[\frac{1}{r}\frac{dw}{dr}+\nu\frac{d^2 w}{dr^2}\right]
\end{aligned}
\tag{126}
$$

At $r \neq 0$,

$$
\begin{aligned}
(M_r)_{r\neq 0} = {} & \frac{P(1+\nu)}{4\pi}\log\frac{a}{r}+\frac{(1-\nu)Pc^2}{16\pi}\left(\frac{1}{r^2}-\frac{1}{a^2}\right) \\
& -\frac{P(1+\nu)}{16\pi}\left\{\frac{2(3+\nu)}{(1+\nu)}-2-\frac{c^2}{a^2}\left[1+\frac{1-\nu}{1+\nu}\right]\right\}
\end{aligned}
\tag{127}
$$

$$
\begin{aligned}
(M_t)_{r\neq 0} = {} & \frac{P}{4\pi}\left[(1+\nu)\log\frac{a}{r}+(1-\nu)\right]-\frac{(1-\nu)Pc^2}{16\pi}\left(\frac{1}{r^2}+\frac{1}{a^2}\right) \\
& -\frac{P(1+\nu)}{16\pi}\left\{\frac{2(3+\nu)}{(1+\nu)}-2-\frac{c^2}{a^2}\left[1+\frac{1-\nu}{1+\nu}\right]\right\}
\end{aligned}
\tag{128}
$$

At $r = 0$ it can be shown that

$$
\begin{aligned}
(M_r)_{r=0} = (M_t)_{r=0} = {} & \frac{P}{4\pi}\left[(1+\nu)\log\frac{a}{c}+1-\frac{(1-\nu)c^2}{4a^2}\right] \\
& -\frac{P(1+\nu)}{16\pi}\left\{\frac{2(3+\nu)}{(1+\nu)}-2-\frac{c^2}{a^2}\left[1+\frac{1-\nu}{1+\nu}\right]\right\}
\end{aligned}
\tag{129}
$$

Since the deflections w are small in comparison to the thickness of the plate, the membrane stresses may be considered as negligible.

In the above analysis the load has been considered as a point load at the center of the plate. This is because this analysis will be necessary for the analysis of the full scale model considered in Section 8.4. In practical construction, however, the snow load on the plate will be a uniformly distributed load over the plate, and for this loading q, the deflection can be obtained as

$$
w=\frac{q}{64D}(a^2-r^2)^2 \tag{124a}
$$

The M_r and M_t moments can also be obtained as

$$M_r = \frac{q}{16}\left[a^2(1+\nu) - r^2(3+\nu)\right] \tag{127a}$$

$$M_t = \frac{q}{16}\left[a^2(1+\nu) - r^2(1+3\nu)\right] \tag{128a}$$

As an example consider a circular plate with the following data:

radius $a = 10.27$ in. (261 mm)
uniformly distributed load $q = 27.2$ lb/ft² (133 kg/m²)
$E = 0.972 \times 10^6$ psi (6.70 GPa)
$t = 5/8$ in. (16 mm)
$\nu = 0.35$

The deflection of the plate, w, can be obtained by the use of Equation 124a. M_r and M_t can be determined by the use of Equations 127a and 128a. All are plotted in Figs. 193 and 194.

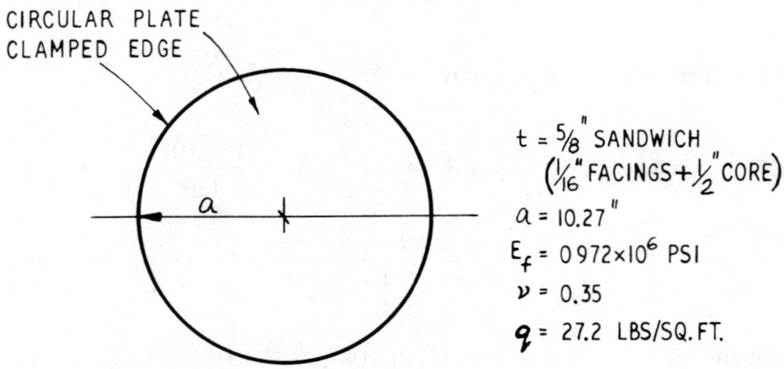

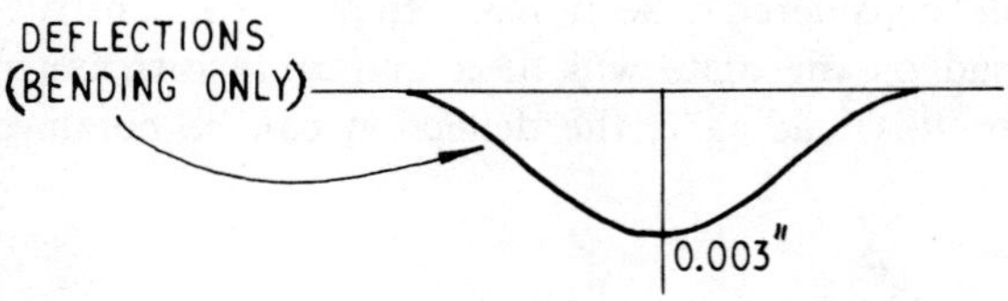

Fig. 193

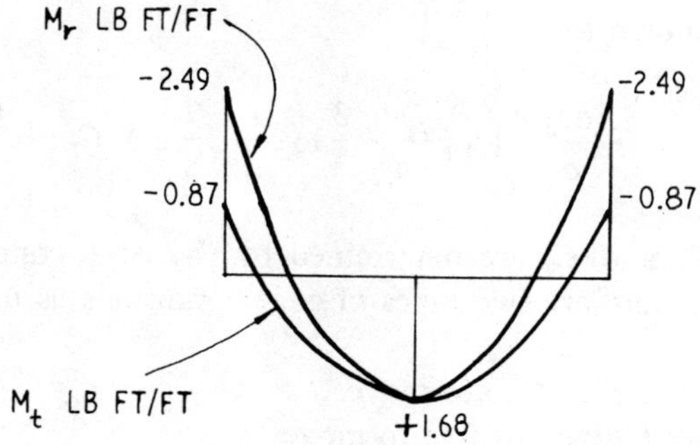

Fig. 194

8.3.3 Analysis of the Plate When the Transverse Deflections Are Large

When the transverse deflections are large, it is first assumed that the deflected surface of the plate can be expressed in the form[161]

$$w = w_0\left(1 - \frac{r^2}{a^2} + \frac{2r^2}{a^2}\log\frac{r}{a}\right) \tag{130}$$

where

w_0 = maximum deflection at the center.

It has been shown by Volmir,[163] and discussed in detail by Timoshenko and Woinowsky-Krieger,[161] that the maximum deflection w_0 can be determined by the nonlinear equation

$$\left(\frac{w_0}{t}\right) + A\left(\frac{w_0}{t}\right)^3 = \frac{BPa^2}{Et^4} \tag{131}$$

where A and B are constants that depend on the edge conditions and the Poisson's ratio of the plate. Let the membrane forces N_r and N_t be defined by a stress function f such that

$$N_r = \frac{1}{r}\frac{df}{dr}$$
$$N_t = \frac{d^2f}{dr^2} \tag{132}$$

Then it can be shown that

$$\frac{df}{dr} = -\frac{Etw_0^2 r^3}{a^4}\left(\log^2\frac{r}{a} - \frac{3}{2}\log\frac{r}{a} + \frac{7}{8}\right) + C_1 r + \frac{C_2}{r} \tag{133}$$

The constants C_1 and C_2 are determined by the edge conditions. For the plate considered there are two types of edge conditions as under:

Type 1: Clamped edge free to move.
Type 2: Clamped edge not free to move.

Type 1:
If the edge is free to move, then

$$(N_r)_{r=a} = 0 \tag{134}$$

Using Equation 132, this can be rewritten as

$$\left(\frac{1}{r}\frac{df}{dr}\right)_{r=a} = 0 \tag{135}$$

Further, as N_r must be finite in value at $r = 0$, it follows that

$$\left(\frac{df}{dr}\right)_{r=0} = 0 \tag{136}$$

The constants C_1 and C_2 can then be determined, using Equations 135 and 136, as

$$C_1 = \frac{7Etw_0^2}{8a^2} \qquad C_2 = 0 \tag{137}$$

Type 2:
Since the edge is not free to move, it follows that the radial displacement at the boundary must be zero, i.e.,

$$(u)_{r=a} = 0 \tag{138}$$

Therefore the tangential strain is

$$(\epsilon_t)_{r=a} = \frac{u}{r} = 0 \tag{139}$$

and as

$$\epsilon_t = \frac{1}{Et}(N_t - \nu N_r) \tag{140}$$

it follows that

$$(N_t - \nu N_r)_{r=a} = 0 \tag{141}$$

As the membrane forces N_r at $r = 0$ must still have a finite value, it follows that Equation 136 is still valid. The constants C_1 and C_2 may be determined by the use of Equations 136 and 141 as

$$C_1 = \frac{(9 - 7\nu)}{8(1 - \nu)} \frac{Etw_0^2}{a^2} \tag{142}$$

$$C_2 = 0$$

The stress function f for both types of edge conditions is completely determined, and the use of the stress function leads to the determination of the constants A and B of Equation 131. Volmir[163] has determined the constants A and B for $\nu = 0.30$ as

Edge free to move: $A = 0.200$, $B = 0.217$.
Edge not free to move: $A = 0.443$, $B = 0.217$.

The author has determined the same constants A and B for $\nu = 0.35$ as

Edge free to move: $A = 0.194$, $B = 0.209$.
Edge not free to move: $A = 0.453$, $B = 0.209$.

The use of Equation 130 in Equation 126 then determines the radial and tangential moments in the plate. These can be written as

$$M_r = \frac{4Dw_0}{a^2}\left[(1 + \nu)\log\frac{a}{r} - 1\right]$$
$$M_t = \frac{4Dw_0}{a^2}\left[(1 + \nu)\log\frac{a}{r} - \nu\right] \tag{143}$$

It can be seen that the use of Equation 143 leads to infinite moments at $r = 0$. If it is assumed that the load P is uniformly distributed over a small radius $r = c$, it can then be shown that

$$(M_r)_{r=0} = (M_t)_{r=0} = \frac{4Dw_0}{a^2}\left[(1+\nu)\log\frac{a}{c} - 1\right] + \frac{P}{4\pi} \tag{144}$$

The membrane forces can then be determined from the stress function f by means of Equation 132.

8.3.4 Longitudinal Arch Action of the Barrel Vault

The units of the barrel vault form a fixed arch, as may be seen in the elevation of Fig. 195. The cross-sectional area of the arch is constant, but the moment of inertia is not. Consider a small section of the arch as shown in Fig. 195. The maximum depth d of the arch rib occurs at

$$\Phi_1 = n\alpha$$

where

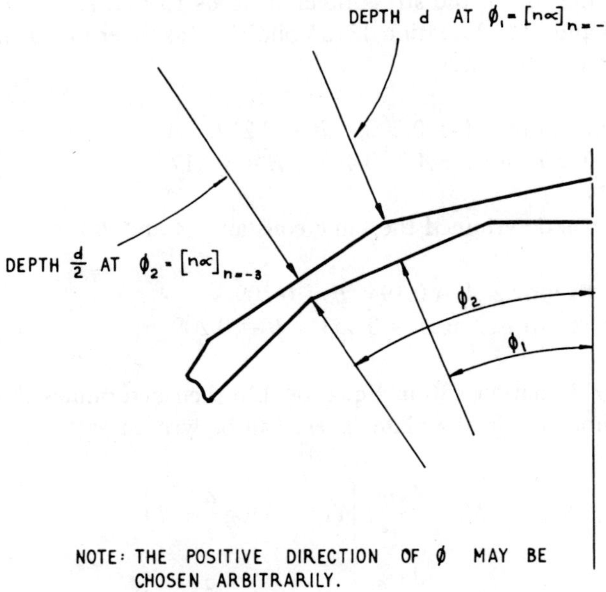

Fig. 195

Φ = the central angle
$n = \pm[0,2,4, \ldots m]$.

The minimum depth of the arch rib, $d/2$ occurs at

$$\Phi_2 = n\alpha$$

where

$n = \pm[1,3,5, \ldots, (m-1)]$

For the semicircular barrel vault shown in Fig. 160, $m = 8$ and $\alpha = 11° 15'$.

At Φ_1, the arch rib has only one plate of vertical depth d. At Φ_2, the arch rib has two plates of vertical depth $d/2$. The maximum and minimum moments of inertia can therefore be written as

$$I_{max} = kd^3$$
$$I_{min} = \frac{kd^3}{4} \qquad (145)$$

where

$$k = \frac{t}{12 \sin \theta}$$

The maximum and minimum moments of inertia given by Equation 145 can be expressed in the relationship

$$(I)_{n=\pm[0,1,2,\ldots m]} = \frac{kd^3}{8}[1-(-1)^{\frac{n+1}{2}} \sin mn\alpha][1+(-1)^{n/2} \cos mn\alpha] \qquad (146)$$

where I is a maximum when n is even and a minimum when n is odd. The analysis of the arch,[164] however, can be performed, assuming that the average moment of inertia is constant. The value of $1/I_{av}$ can be obtained by applying Simpson's rule to the rapidly fluctuating values of $1/I_{max}$ and $1/I_{min}$ over the arch. This gives

$$\frac{1}{I_{av}} = \frac{1}{6m}\left[\frac{2}{I_{max}} + \frac{4m}{I_{min}} + \frac{2(m-1)}{I_{max}}\right] \qquad (147)$$

or

$$\frac{1}{I_{av}} = \frac{1}{3}\left[\frac{1}{I_{max}} + \frac{2}{I_{min}}\right] \tag{148}$$

This value I_{av} gives the average moment of inertia of the overall section. It assumes that the entire plate width is effective and that, as in sandwich construction, no compressive buckling occurs in the central part of the plates. In single skin construction, compressive buckling causes only a reduced section towards the ridges and the valleys to be effective, causing the actual moment of inertia of the arch section to be lower than the value given by Equation 148. Zhidkov[160] hence suggests that the longitudinal arch action of the barrel vault could be determined by analyzing the structure completely as a skeletal structure. The structure then consists of three sets of arches running along the ridges and the valleys and interconnected at the nodes. This assumption is particularly realistic if large flanges are used to connect the units. The structure can then be analyzed by such matrix methods as the stiffness or equilibrium method. The analysis of a folded plate conical dome roof structure, by the use of such an approach for single skin construction, is considered in Chapter 10. The difficulty with such an approach lies in determining the exact areas that have to be assumed as concentrated along the ridge and valley lines, which then must be fed, as the member properties, into a computer program such as STRUDL. The author[7,138] has also studied the problem experimentally, and some of these findings, as well as other theoretical work on the subject, are considered in Chapter 10.

8.4 DESIGN AND TESTING OF A PROTOTYPE COMPOSITE FOLDED PLATE BARREL VAULT OF SANDWICH CONSTRUCTION

8.4.1 Description of the Structure

A full scale barrel vault in sandwich construction, of the type shown in Fig. 160, has been designed and tested by the author.[146] The general arrangement of the barrel vault with center line dimensions is shown in Fig. 196. The basic internal unit from which the structure can be built is shown in Fig. 197. The quarter external base units were made by slicing the basic unit both longitudinally and transversely into four parts. The base flanges for these units were then laid up onto the cut sections. Design specifications called for a sandwich consisting of $\frac{1}{16}$ in. (1.6 mm) thick glass fiber reinforced polyester (FRP) top and bottom facings bonded to a $\frac{1}{2}$ in. (13 mm) thick

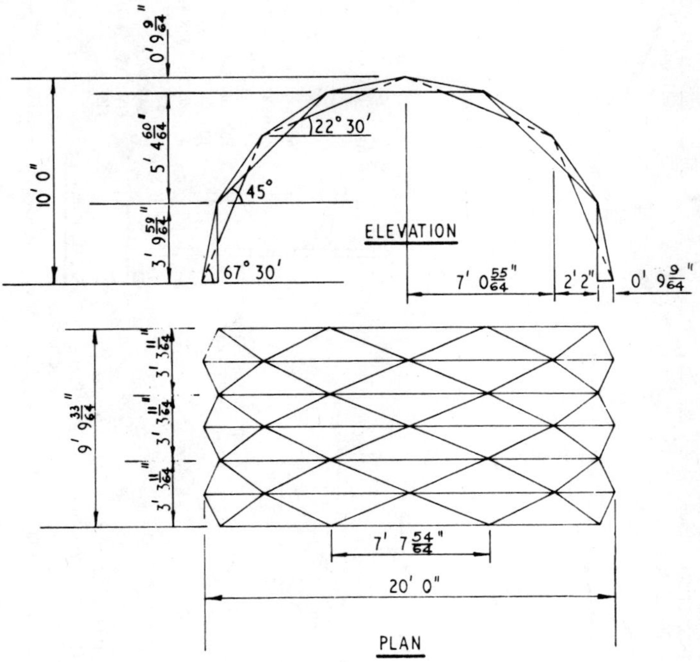

Fig. 196

rigid polyurethane foam core of a density of 2 lb/ft³ (32 kg/m³). The design was carried out according to the principles detailed in Chapter 9, by analyzing the structure according to the theory set forth in Sections 8.3.2 and 8.3.4, and it was found that, using FRP reinforced with chopped strand mat and made by a hand lay-up process, a facing composite thickness of $\frac{1}{32}$ in. (0.8 mm) was quite sufficient. However, for purposes of efficient weathering and to take environmental conditions into account (since this was a test structure and there was no guarantee as to where the structure would end up), it was decided to adopt a thickness of $\frac{1}{16}$ in. (1.6 mm). Connections were effected by $\frac{3}{8}$ in. diameter (10 mm) mild steel bolts spaced 12 in. (305 mm) on centers connecting FRP flanges 3 in. (76 mm) deep. In order to have fixity with only one line of bolts, a recess $\frac{1}{16}$ in. (1.6 mm) deep was run in each flange, as shown in Fig. 198. When the bolts were screwed tight, pressure was exerted at the top and bottom edges of the flanges, thus creating a fixed connection. Holes of $\frac{7}{16}$ in. (11 mm) diameter were provided in the units for the loading cables. These holes were suitably stiffened by FRP inserts. A single unit is shown in Fig. 197. At the nodes of the barrel a special two-piece aluminum connector was used. This is shown in Figs. 199

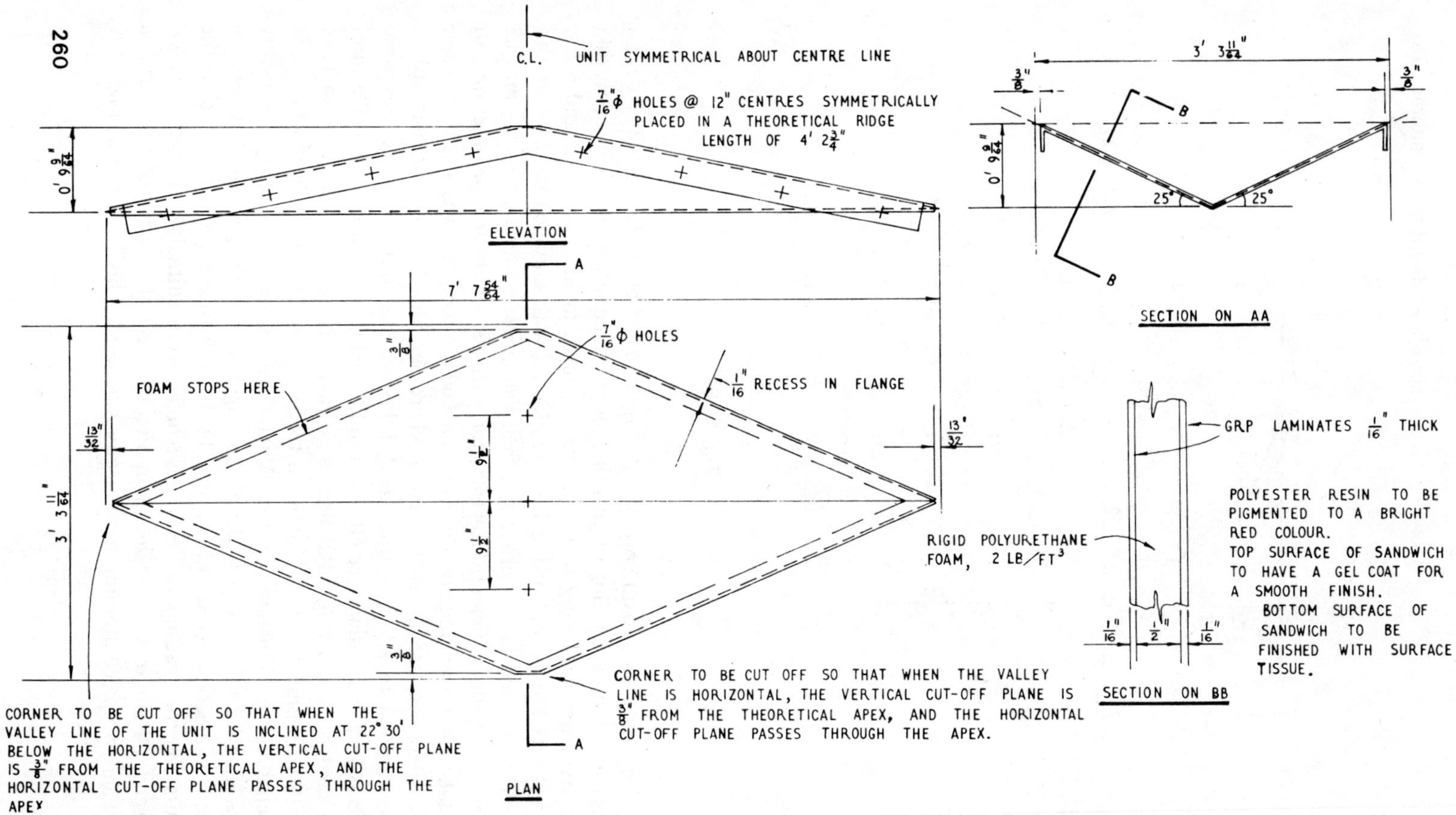
C.L.
UNIT SYMMETRICAL ABOUT CENTRE LINE
7/16" ϕ HOLES @ 12" CENTRES SYMMETRICALLY PLACED IN A THEORETICAL RIDGE LENGTH OF 4' 2 3/4"
0' 9 9/64"
ELEVATION
7' 7 54/64"
3/8"
7/16" ϕ HOLES
1/16" RECESS IN FLANGE
FOAM STOPS HERE
13/32"
13/32"
9 1/2"
9 1/2"
3' 3 11/64"
CORNER TO BE CUT OFF SO THAT WHEN THE VALLEY LINE OF THE UNIT IS INCLINED AT 22° 30' BELOW THE HORIZONTAL, THE VERTICAL CUT-OFF PLANE IS 3/8" FROM THE THEORETICAL APEX, AND THE HORIZONTAL CUT-OFF PLANE PASSES THROUGH THE APEX
CORNER TO BE CUT OFF SO THAT WHEN THE VALLEY LINE IS HORIZONTAL, THE VERTICAL CUT-OFF PLANE IS 3/8" FROM THE THEORETICAL APEX, AND THE HORIZONTAL CUT-OFF PLANE PASSES THROUGH THE APEX.
A
A
PLAN
3' 3 11/64"
3/8"
3/8"
B
B
0' 9 9/64"
25°
25°
SECTION ON AA
GRP LAMINATES 1/16" THICK
POLYESTER RESIN TO BE PIGMENTED TO A BRIGHT RED COLOUR.
TOP SURFACE OF SANDWICH TO HAVE A GEL COAT FOR A SMOOTH FINISH.
BOTTOM SURFACE OF SANDWICH TO BE FINISHED WITH SURFACE TISSUE.
RIGID POLYURETHANE FOAM, 2 LB/FT³
1/16"
1/2"
1/16"
SECTION ON BB

Fig. 197

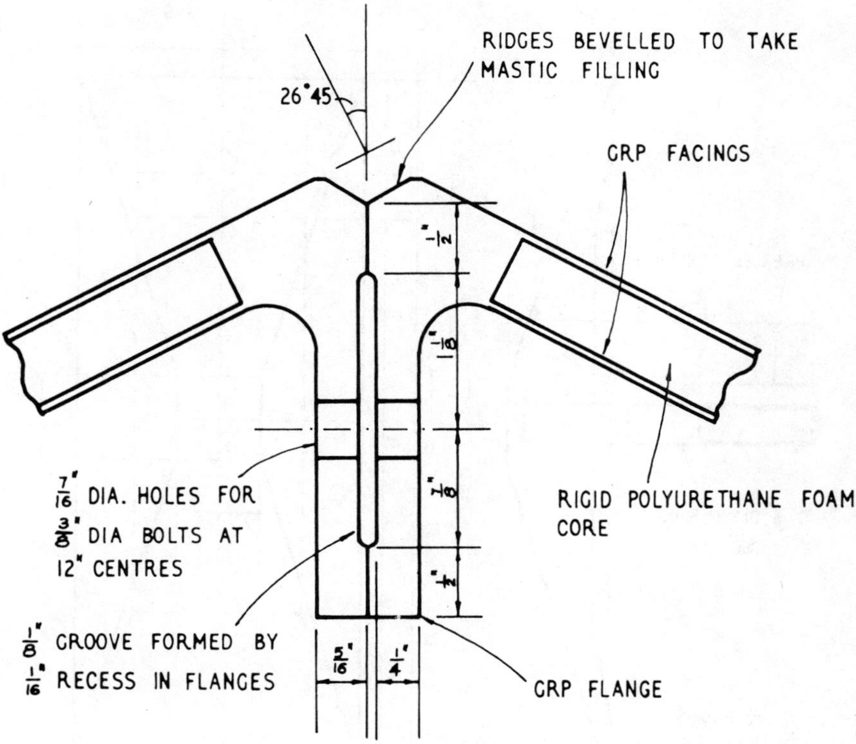

Fig. 198

and 200. The connector had the dual purpose of forming a rigid connection at the joint and preventing water leakage through it. To prevent leakage of water through the structure from the ridge lines, it was necessary to bevel the ridges, as shown in Fig. 198. A mastic filling could then be run into the ridge lines to ensure watertightness. The author and two others erected the structure in 3 hours. Both sides were worked on simultaneously. Due to the very low self weight of the units [30–35 lb (13.6–16kg)], no propping was necessary for the free cantilevers. Due to the large deflections, however, it was necessary to gently force the cantilevers apart for the central units to be fitted in.

8.4.2 Theoretical Analysis of the Structure for the Test Loadings

The analysis of the structure for the loadings to be applied during testing was performed on the basis of the theory set forth in Sections 8.3.2 and 8.3.4. The details of the structure for purposes of the analysis are as follows:

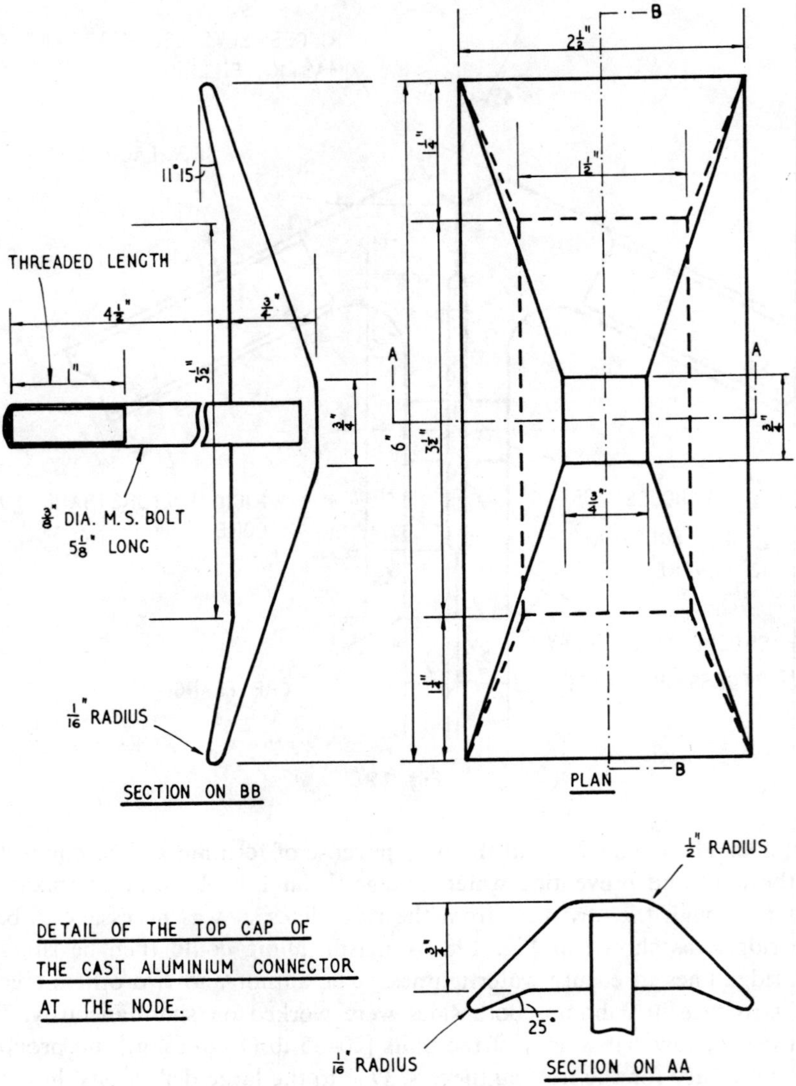

Fig. 199

Neutral axis span of the barrel = 230.86 in. (5864 mm)
Fold angle $\theta = 25°$
Angle of the unit $\alpha = 11° \ 15'$
$m = 8$.

For the transverse plate action of 8.3.2,

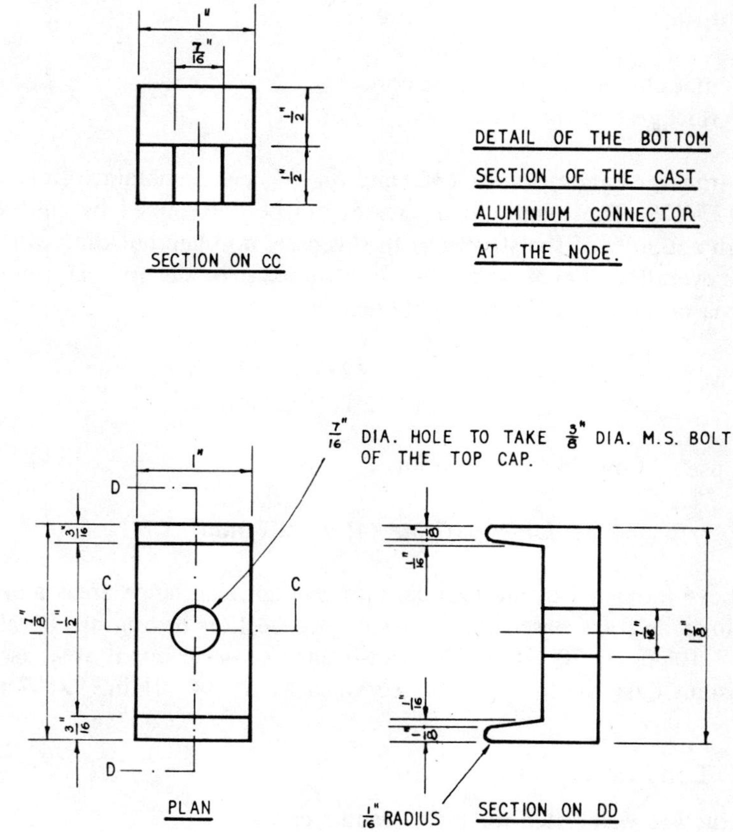

Fig. 200

radius of the plate $a = 10.266$ in. (260.8 mm)
radius of the loading pad $c = 2.00$ in. (51 mm).

The loading blocks were 4 in. (102 mm) square, but it was assumed that the load was applied on a circular area of radius 2 in. (51 mm). The transverse bending deflections can be determined by means of Equations 124 and 125. To this must be added the shearing deflections of the core, which can be approximately determined by considering a wedge shaped element of the core and are given by

$$(w_s)_{r \nless c} = \frac{P}{2\pi G_c k} \log \frac{a}{r} \tag{149}$$

where

G_c = modulus of rigidity of the core
k = thickness of the core.

The transverse moments M_r (M_y) and M_t (M_x) can be obtained from Equations 127–129. The longitudinal stresses can be determined by considering the arch action[164] of the structure. In this case, no plate buckling can occur and the overall section may therefore be assumed to be effective. The moments of inertia of an arch rib can be obtained as

$$I_{max} = 29.65 \text{ in.}^4 \ (123.4 \times 10^5 \text{ mm}^4)$$
$$I_{min} = 7.57 \text{ in.}^4 \ (31.5 \times 10^5 \text{ mm}^4)$$

The use of Equation 148 then gives

$$I_{av} = 10.07 \text{ in.}^4 \ (41.9 \times 10^5 \text{ mm}^4)$$

The above moments of inertia take into account the flange areas along the ridge lines and are based on a transformed section having an E value of 0.972×10^6 psi (6.70 GPa). The transformed cross-sectional area, assumed as constant over the arch, can be obtained as $A = 4.181$ in.^{2} (2697 mm^2).

8.4.3 Loading on the Structure

The structure was tested for three loading cases:

Case 1. An equivalent snow load of 30 lb/ft^2 (146.5 kg/m^2).
Case 2. A central line load of 326 lb/arch rib (148 kg/arch rib).
Case 3. A line load of 208 lb/arch rib (94 kg/arch rib) at $\Phi = \pm \pi/4$.

To determine the complete behavior of the barrel vault in Case 3, the load had to be applied successively at $\Phi = +\pi/4$ and $\Phi = -\pi/4$.

8.4.4 Analysis of the Test Results

The theoretical and experimental results for the loading cases in Section 8.4.3 are now discussed in detail. The longitudinal deflections of the internal arch rib under the applied loading in Cases 1–3 are shown in Figs. 201–203. It can be seen that the assumption that the entire section is effective, and that no plate buckling occurs, leads to good correlation between the theoretical and experimental deflections. The transverse deflections of the

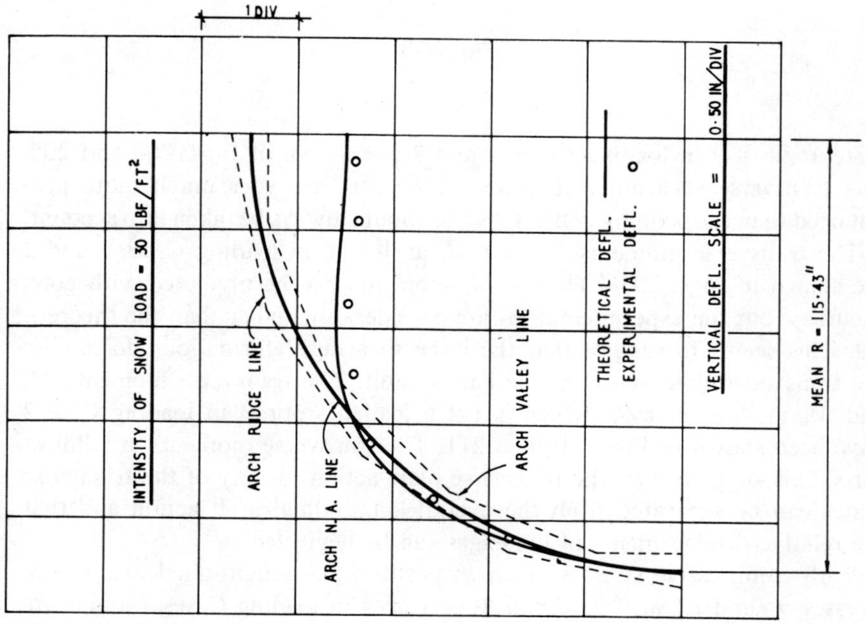

Fig. 201

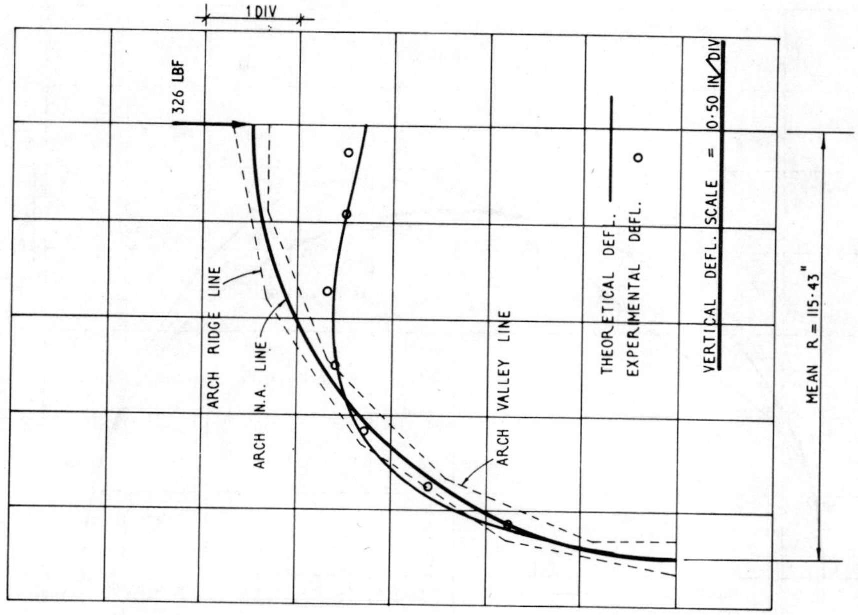

Fig. 202

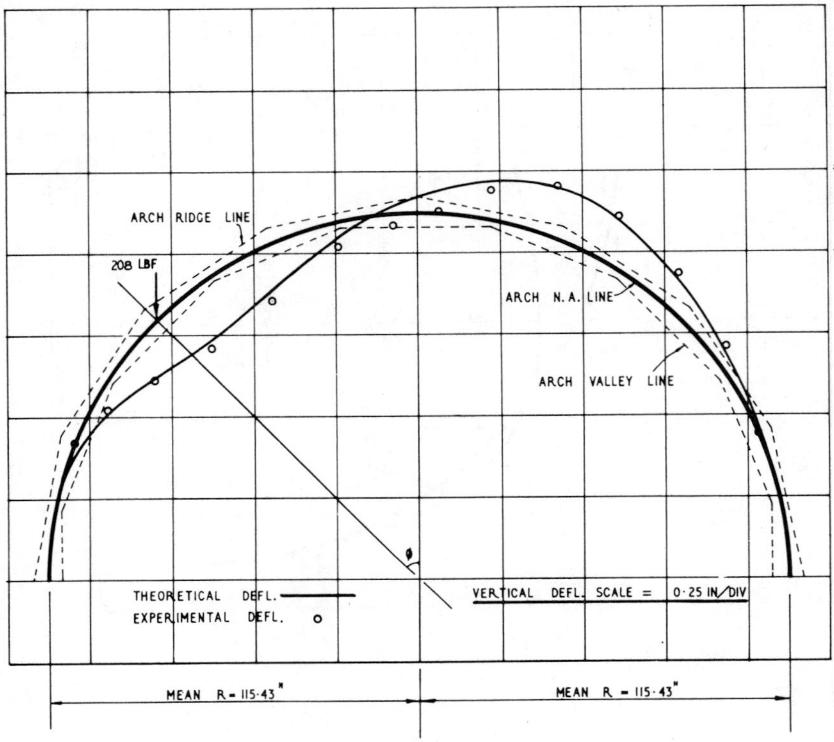

Fig. 203

plate at $\Phi = 0$ in loading Cases 1 and 2 are shown in Figs. 204 and 205. The transverse shearing deflections of the core are very much more pronounced than the bending deflections and should always be taken into account.

The transverse moments M_x and M_y at $\Phi = 0$ in loading Cases 1 and 2 are shown in Figs. 206–209. The M_y moments can be predicted with good accuracy, but the experimental M_x are considerably higher than the theoretical. This seems to suggest that the large shearing deformations do modify the transverse plate action of the barrel vault. The transverse moments M_x and M_y at $\Phi = +\pi/4$ (which is not a loaded section) in loading Case 2 have been shown in Figs. 210 and 211. The transverse moments are almost zero. This suggests that the transverse plate action for any of the triangular plates can be separated from the overall longitudinal arch action and that the relative displacements of the edges can be neglected.

With compressive stresses taken as positive, the longitudinal and lateral stresses σ_x and σ_y at $\Phi = 0$ and $\Phi = +\pi/4$ in loading Cases 1 and 2 are

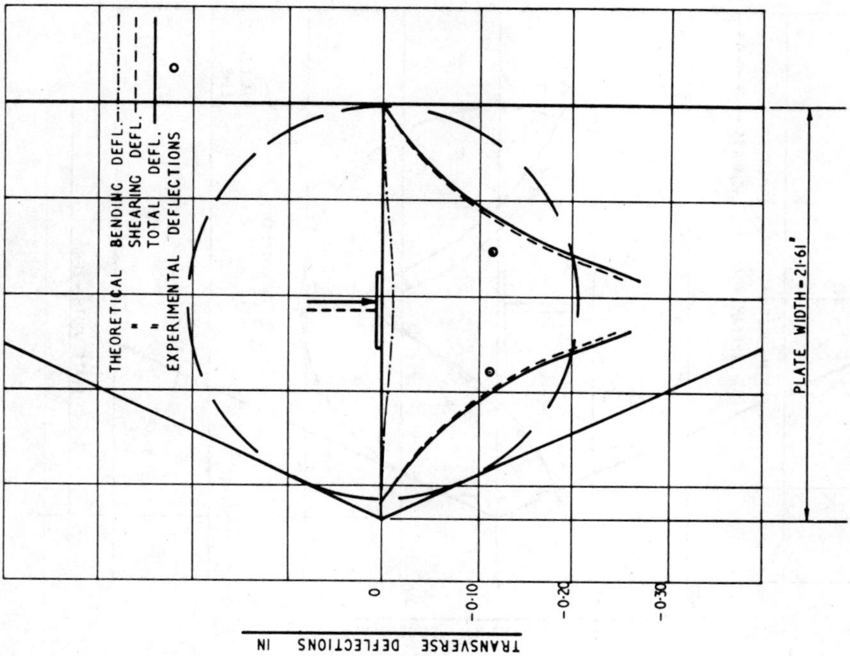

Fig. 205

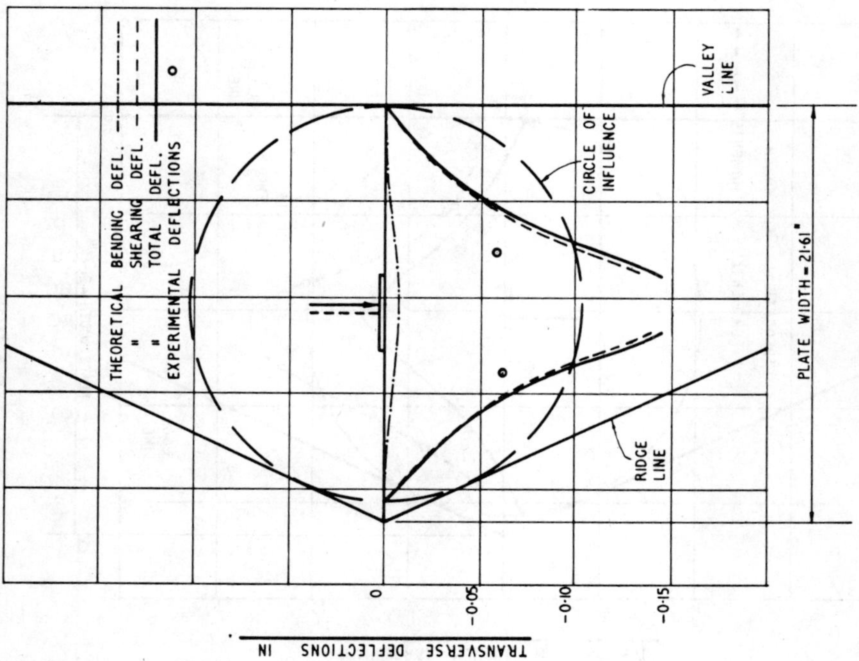

Fig. 204

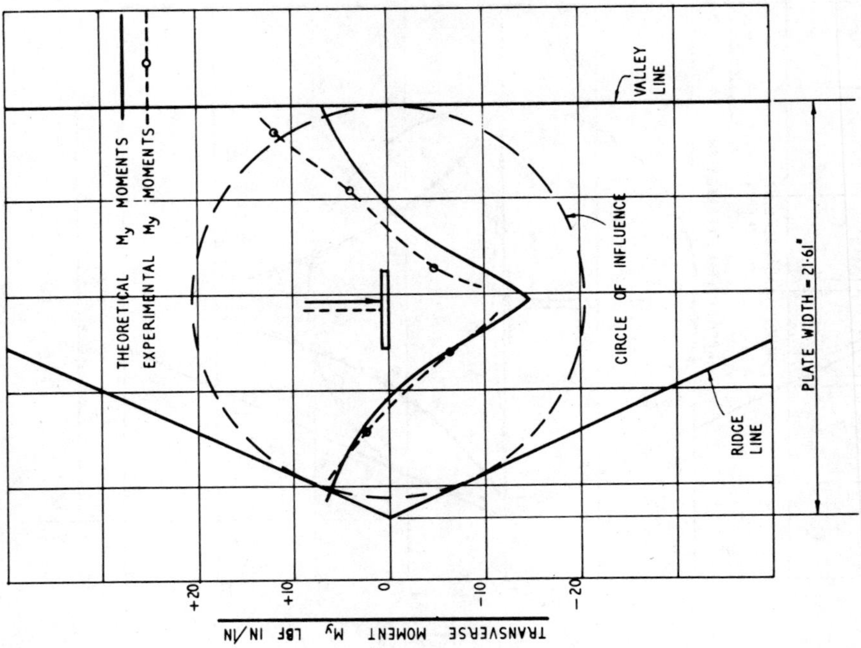

Fig. 207

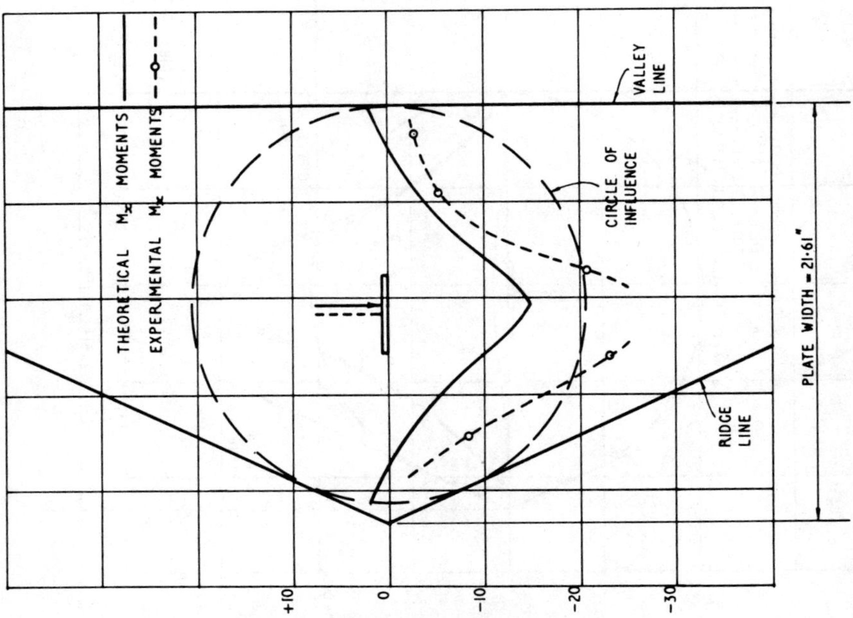

Fig. 206

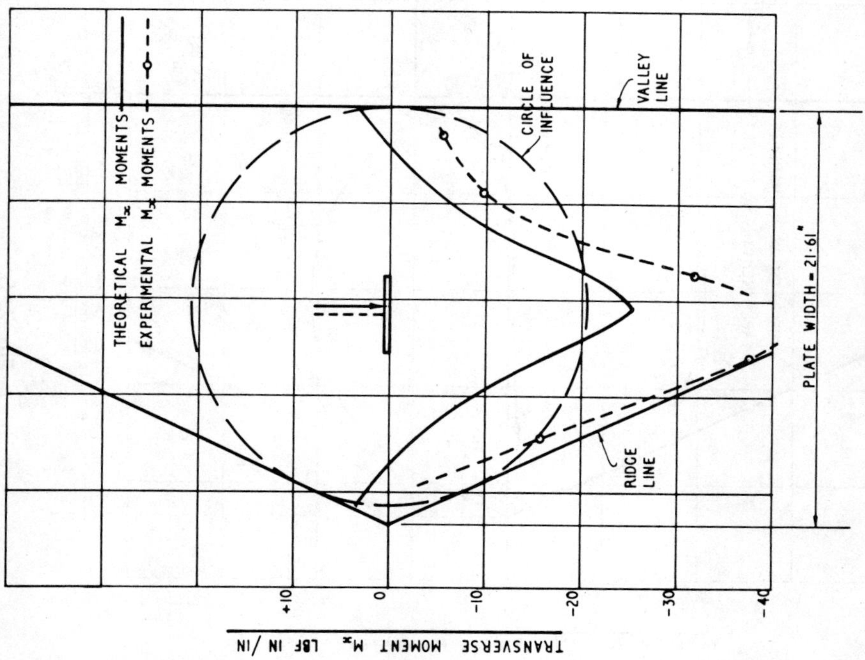

Fig. 208

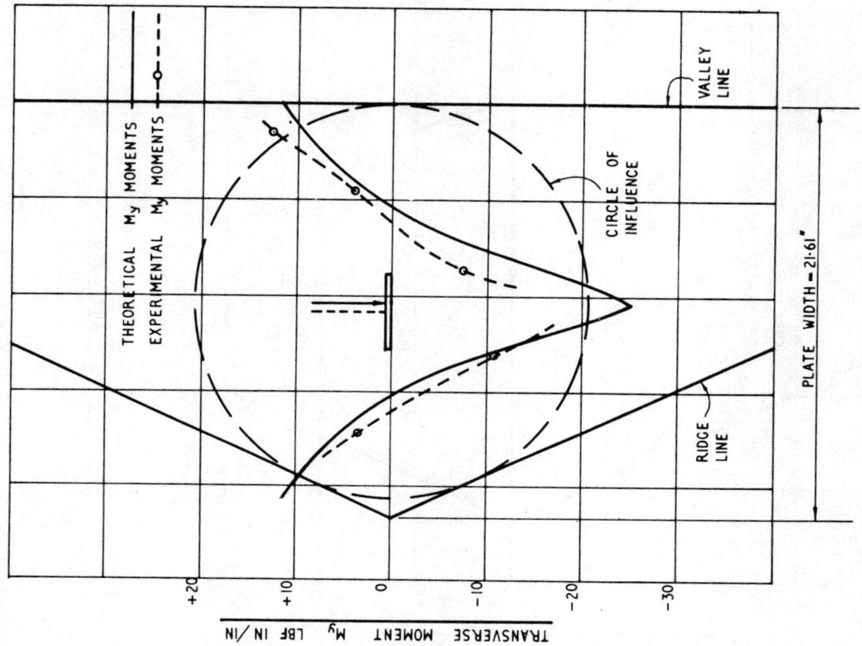

Fig. 209

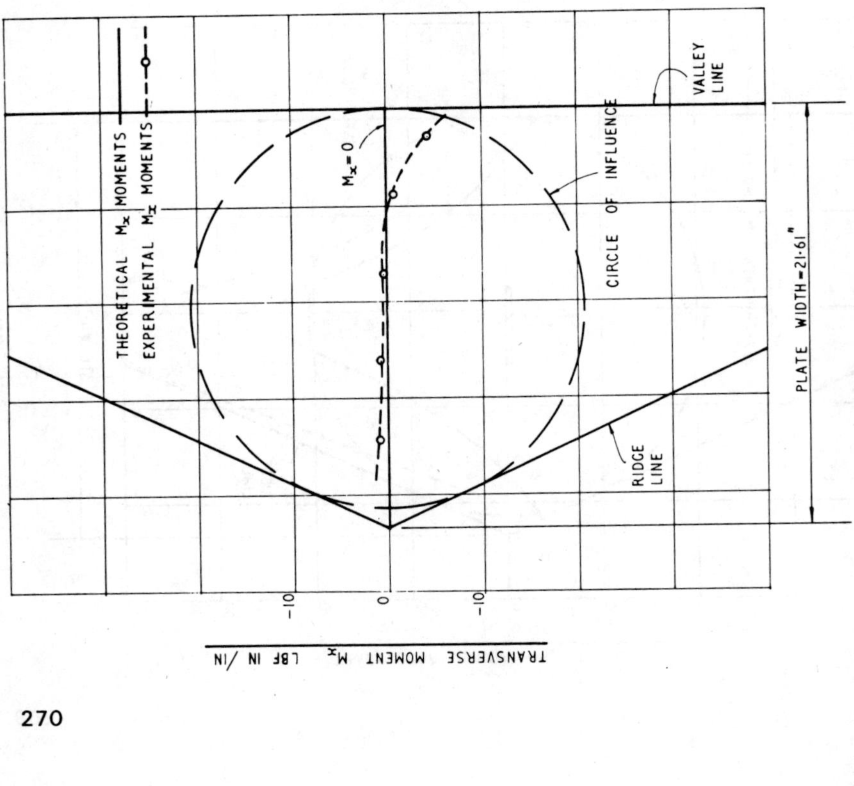

Fig. 210

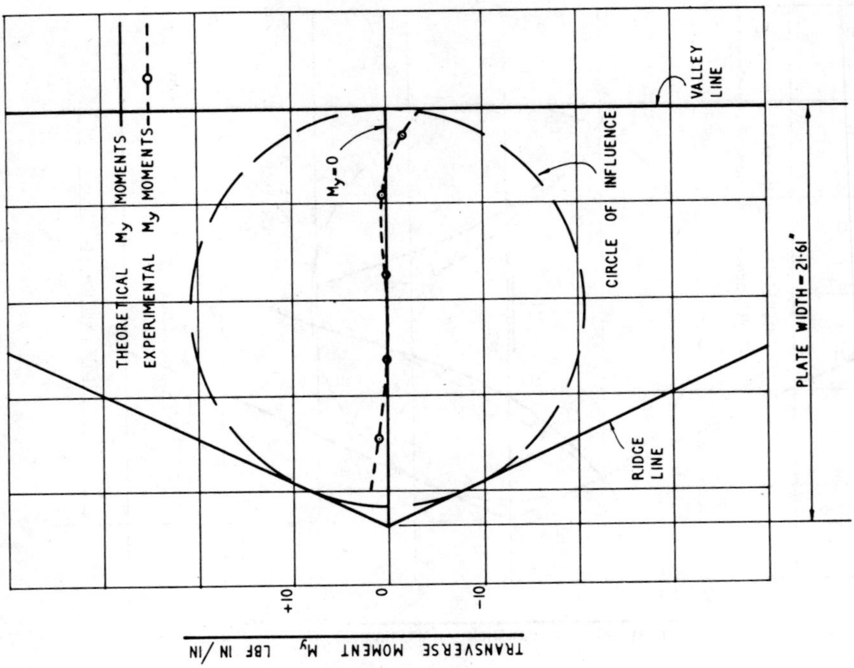

Fig. 211

shown in Figs. 212–215. The longitudinal stresses at $\Phi = +\pi/4$ and $\Phi = -\pi/4$ in loading Case 3 are shown in Fig. 216.

With the accuracy that can be expected from a full scale test of this nature, the correlation between the experimental and theoretical results in all the figures may be considered as being reasonably good. The experimental stresses and moments at any point are determined from the strain readings, in the longitudinal and lateral directions, at the top and the bottom of the sandwich. Each strain reading depends upon the thickness of the laminate under the strain gage and the modulus of elasticity and the Poisson's ratio of the laminate at that point. These latter constants E and ν themselves depend upon the resin/fiberglass content ratio. Hence, any stress value depends upon at least 10 quantities that vary all over the barrel vault, due to the variation in the

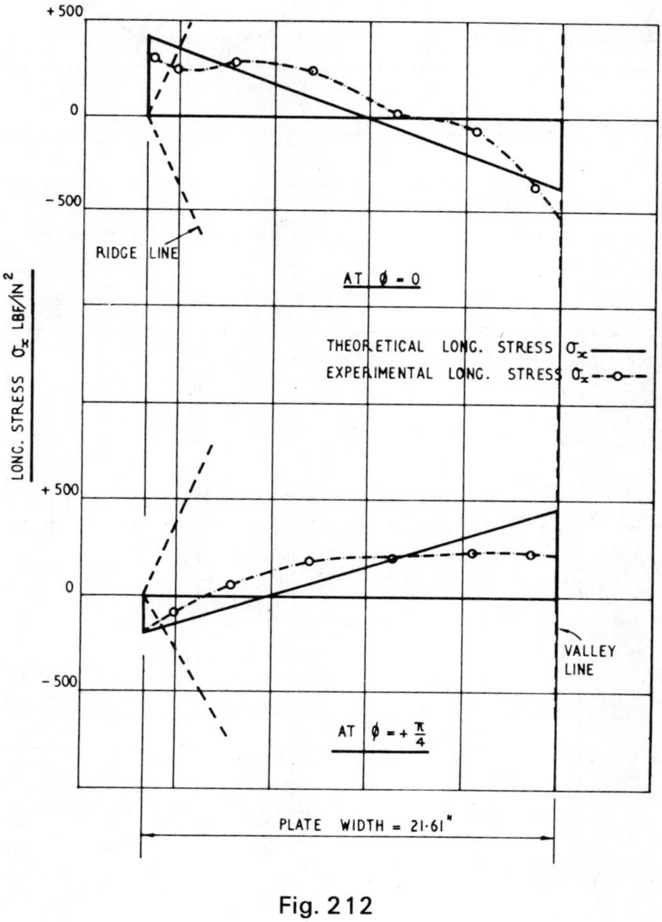

Fig. 212

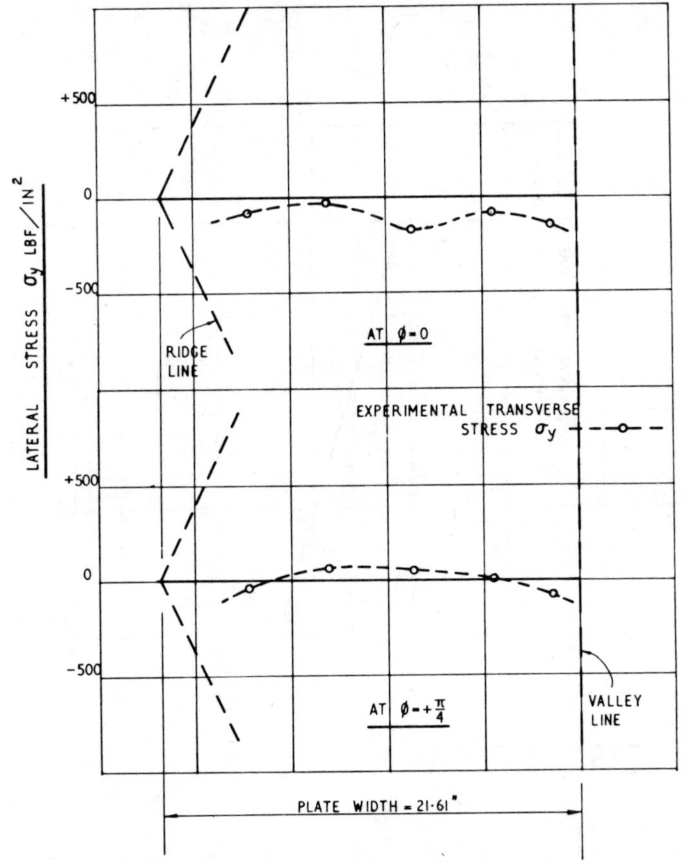

Fig. 213

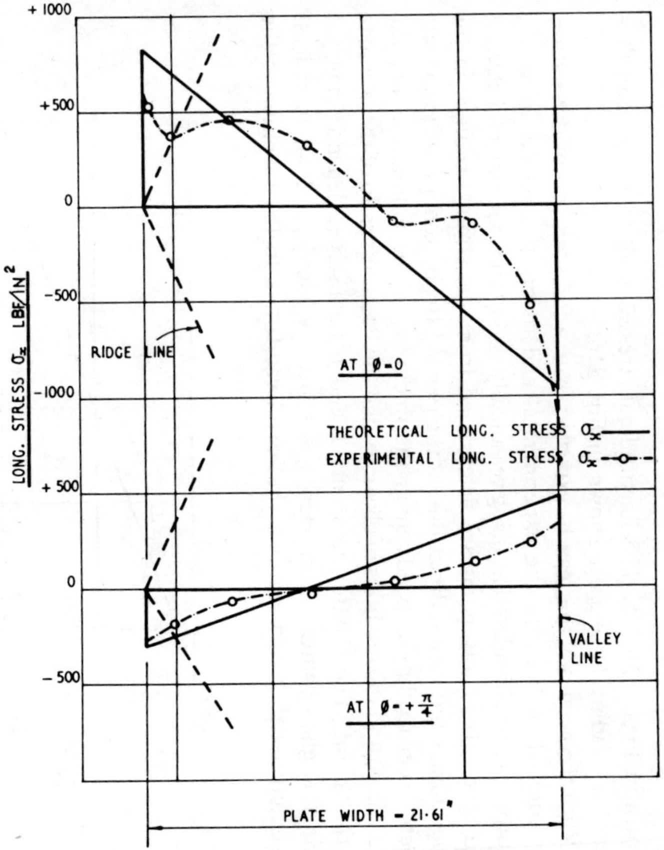

Fig. 214

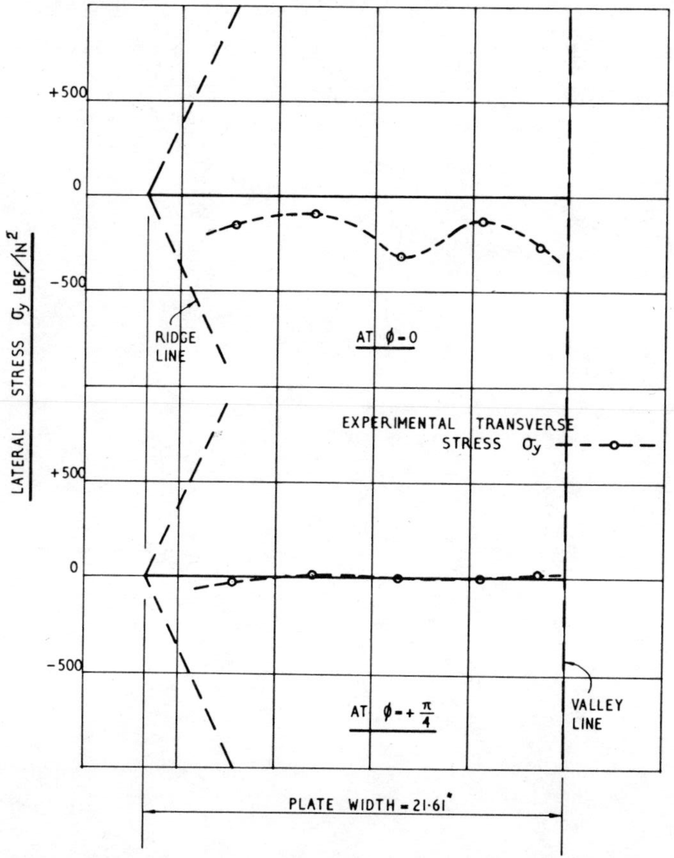

Fig. 215

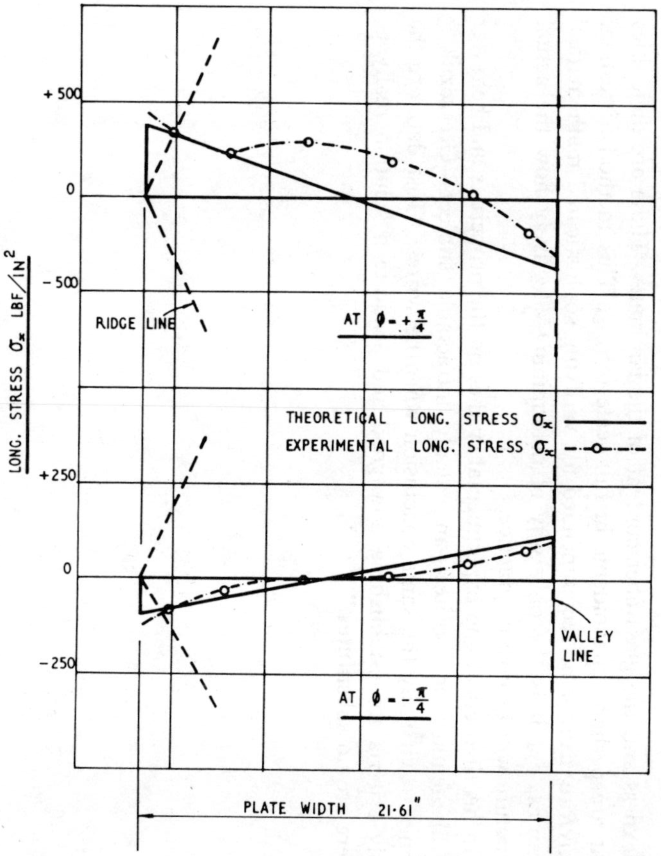

Fig. 216

thickness and the glass fiber content of the laminates. These are difficulties that are perhaps unavoidable in full scale testing. This method of testing, therefore, cannot be recommended for verifying sophisticated mathematical theories, but it is the only way of determining effectively how the actual structure will behave in practice.

In his theoretical and experimental studies on the properties and behavior of filamentary composites—an equally intractable subject—Greszczuk[165] comes to virtually the same conclusion when he writes: "Test data are the only criteria for establishing how good and accurate a solution really is; there are no substitutes."

CHAPTER 9

SHELL STRUCTURES

9.1 SHELL STRUCTURES IN PLASTICS

The analysis of various types of synclastic and anticlastic shells has been treated very exhaustively in the literature,[147,166-169] and there is no lack of information on the methods employed. However, the design of shell structures in plastics requires that the special material problems considered earlier to be taken into account. To illustrate this, the design of an FRP dome of single skin construction is considered in Section 9.2, the design of a long plastics barrel vault in Section 9.3, and the design of a hypar, also in single skin construction, is considered in Section 9.4.

9.2 DESIGN OF A FRP DOME OF SINGLE SKIN CONSTRUCTION

9.2.1 Design Data

Form	Hemispherical
Diameter	50 ft 0 in. (15.24 m)
Rise	25 ft 0 in. (7.62 m)
Special features	Dome surface to be divided into 24 segments connected to each other by bolts through FRP flanges. Full fixity at the connections may be assumed.
Specific gravity of FRP	1.5
Snow load	20 lb/ft^2 (97.6 kg/m^2) on plan area
Wind load	18 lb/ft^2 (87.9 kg/m^2) on a vertical surface
Incidental ladder load	336 lb (152.4 kg)

Ultimate short term flexural strength of FRP (at 0.1 hr)	20,000 psi (137.9 MPa)
Ultimate short term tensile strength of FRP (at 0.1 hr)	12,000 psi (82.7 MPa)
Ultimate short term compressive strength of FRP (at 0.1 hr)	15,000 psi (103.4 MPa)
Short term modulus of elasticity (at 0.1 hr)	1.2×10^6 psi (8.27 GPa)
Deflection restrictions	nil
Environmental conditions	Good. No harmful gases in the atmosphere. No special precautions necessary.
Design criteria	The dome is required to be designed on the assumption that self, snow and wind loads act simultaneously on it. Stability of the dome to be checked for dead and snow loads.

The dome is shown in Fig. 217.

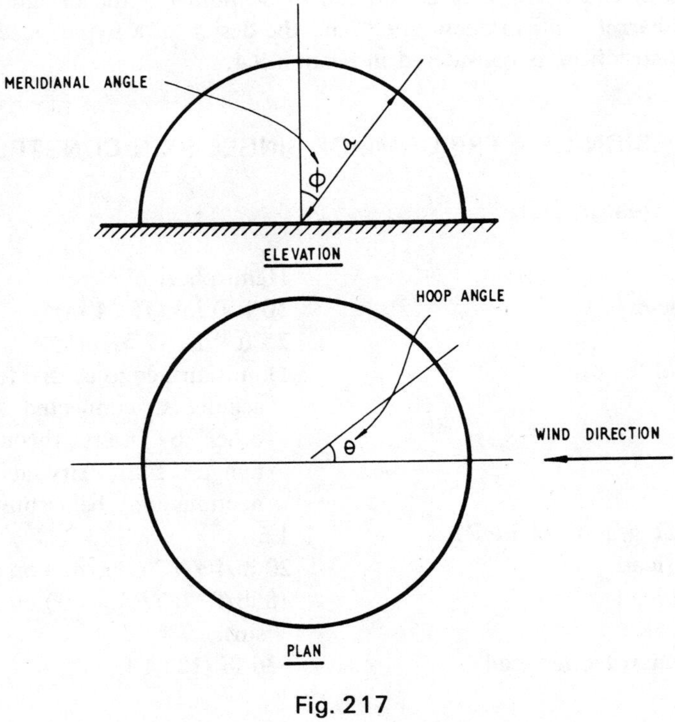

Fig. 217

9.2.2 Analysis for Self Load

The specific gravity of FRP is about 1.5, and therefore ⅛ in. (3 mm) thick FRP composite weighs 0.975 lb/ft² (4.76 kg/m²), say 1 lb/ft² (4.88 kg/m²). It is assumed that the dome has a minimum thickness of ¼ in. (6 mm). The average thickness of the dome for purposes of loading may be assumed to be 5/16 in. (8 mm) thick.

Self load of 5/16 in. (8 mm) thick laminate	= 2.5 lb/ft² (12.2 kg/m²)
Weight of flanges, nuts and bolts, incidental ladder load, fittings and fixtures, etc.	~ 1.5 lb/ft² (7.3 kg/m²)
Total	~ 4.0 lb/ft² (19.5 kg/m²)

The dome is first analyzed by the membrane theory. The membrane forces determined are fairly accurate if the dome is permitted to have an expansion at the base. Since the lower edge of the dome is fixed, however, bending moments are generated. The effect of the fixed edge is then determined by the application of a horizontal force H and a bending moment M at the lower edge, as shown in Fig. 218.

Membrane analysis:

Analysis of the structure by the membrane theory gives

$$N_\phi = -aq\left[\frac{1}{1+\cos\phi}\right] \tag{150}$$

$$N_\theta = aq\left[\frac{1}{1+\cos\phi} - \cos\phi\right] \tag{151}$$

where

a = radius of the dome in ft (m)
q = intensity of the self load in lb/ft² (kg/m²)

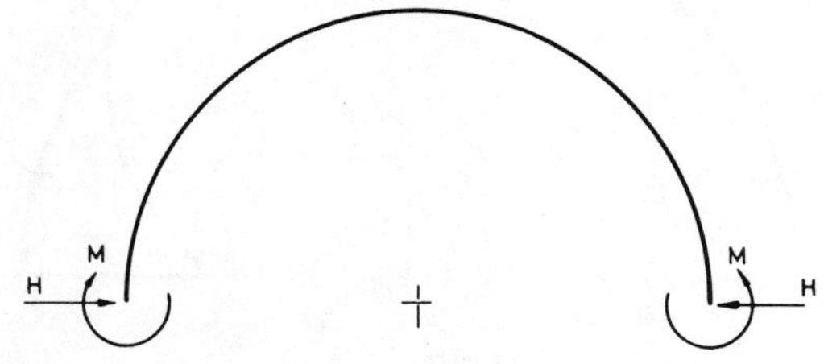

Fig. 218

Table 64

ϕ radians	N_ϕ lbf/ft. run	N_θ lbf/ft. run	vEt	wEt	$ET\delta_{vertical}$
.13090000	−50.214847	−48.929683	503.41659	4670.4507	4696.2033
.26180000	−50.866639	−45.725942	981.88796	4426.0957	4529.4113
.39270000	−51.978323	−40.409601	1411.0799	4027.0420	4260.4983
.52360000	−53.589861	−33.012621	1767.8604	3485.4070	3902.3810
.65450000	−55.761492	−23.573751	2030.8525	2817.7827	3471.8018
.78540000	−58.578691	−12.131859	2180.9264	2044.8747	2988.0928
.91630000	−62.159687	1.2837138	2201.6120	1191.0583	2471.7275
1.0472000	−66.666762	16.666974	2079.4055	283.86347	1942.7516
1.1781000	−72.323269	34.055180	1803.9440	−646.59174	1419.1905
1.3090000	−79.439724	53.558114	1368.0099	−1568.1981	915.52230
1.4399000	−88.454650	75.402363	767.31387	−2447.4528	441.30122
1.5708000	−100.00037	100.00074	−.02393592	−3250.0212	−.01196183

ϕ = meridianal angle in radians, as shown in Fig. 217
N_ϕ = meridianal force in lb/ft run (kg/m)
N_θ = hoop or tangential force in lb/ft run (kg/m).

Equations 150 and 151 can be programmed with great ease for a computer, and the meridianal and hoop forces for a loading intensity of 4 lb/ft² (19.5 kg/m²) are shown in Table 64. The forces are also shown plotted on the dome center line in Fig. 219. Compressive forces have been shown as negative. The vertical deflections of the dome are shown in Fig. 220.

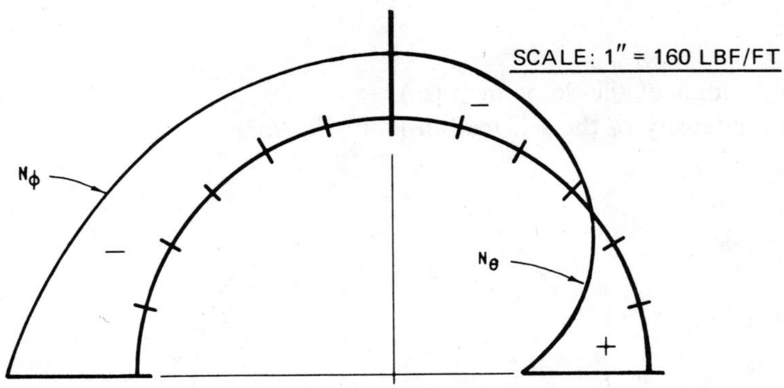

Fig. 219

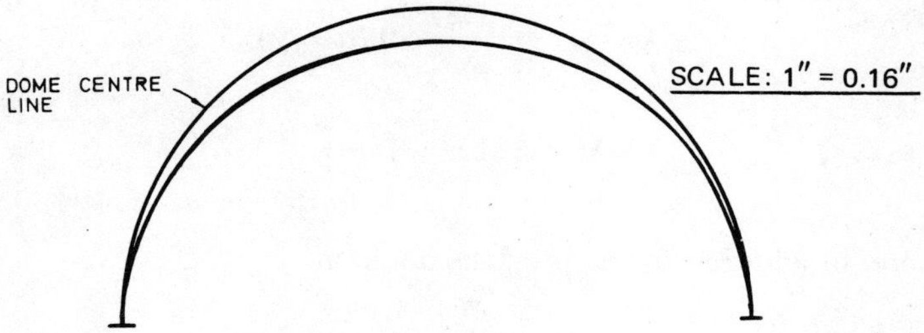

Fig. 220

Dome with a fixed edge:

As has been stated, the effects of edge fixity can be determined by the application of a horizontal force H and a bending moment M at the lower edge. This leads to the following solutions[161]:

$$Q_\phi = \frac{Ce^{-\lambda\chi}}{\sqrt{\sin(\alpha-\chi)}} \sin(\lambda\chi+\gamma) \tag{152}$$

$$V = -\frac{2\lambda^2 Ce^{-\lambda\chi}}{Eh\sqrt{\sin(\alpha-\chi)}} \cos(\lambda\chi+\gamma) \tag{153}$$

$$N_\phi = -\frac{Ce^{-\lambda\chi}}{\sqrt{\sin(\alpha-\chi)}} \sin(\lambda\chi+\gamma)\cot(\alpha-\chi) \tag{154}$$

$$N_\theta = \frac{C\lambda e^{-\lambda\chi}}{2\sqrt{\sin(\alpha-\chi)}} [2\cos(\lambda\chi+\gamma) - (k_1+k_2)\sin(\lambda\chi+\gamma)] \tag{155}$$

$$M_\phi = \frac{aCe^{-\lambda\chi}}{2\lambda\sqrt{\sin(\alpha-\chi)}} [k_1\cos(\lambda\chi+\gamma) + \sin(\lambda\chi+\gamma)] \tag{156}$$

$$M_\theta = \frac{aCe^{-\lambda\chi}}{4\nu\lambda\sqrt{\sin(\alpha-\chi)}}$$
$$\{[(1+\nu^2)(k_1+k_2) - 2k_2]\cos(\lambda\chi+\gamma) + 2\nu^2\sin(\lambda\chi+\gamma)\} \tag{157}$$

$$\delta = \frac{a\sin(\alpha-\chi)\, C\lambda e^{-\lambda\chi}}{Eh\sqrt{\sin(\alpha-\chi)}} [\cos(\lambda\chi+\gamma) - k_2\sin(\lambda\chi+\gamma)] \tag{158}$$

where

$$k_1 = 1 - \frac{1-2\nu}{2\lambda}\cot(\alpha-\chi) \tag{159}$$

$$k_2 = 1 - \frac{1+2\nu}{2\lambda} \cot(\alpha - \chi) \tag{160}$$

$$\lambda^4 = 3(1-\nu^2)\left(\frac{a}{h}\right)^2 \tag{161}$$

and where, in addition to the previous notation,

C, γ = constants
$\alpha = \pi/2$
$\chi = \alpha - \phi$
ν = Poissons's ratio = 0.3
h = thickness of the dome in ft (m)
E = modulus of elasticity of the material
Q_ϕ = shearing force in lb/ft run (kg/m)
V = angle of rotation in the meridianal direction in radians
M_ϕ = bending moment in the meridianal direction in lb ft/ft run (kgm/m)
M_θ = bending moment in the hoop direction in lb ft/ft run (kg m/m)
δ = horizontal component of the displacement in ft (m).

The constants C and γ are determined from the boundary conditions that the lower edge of the dome undergoes no rotation and suffers a horizontal displacement δ_m equal and opposite to that obtained by the membrane theory. This gives

$$-\frac{4\lambda^3 M}{Eahk_1} + \frac{2\lambda^2 \sin\alpha\, H}{Ehk_1} = 0$$

$$\frac{2\lambda^2 \sin\alpha\, M}{Eh\, k_1} - \frac{\lambda a \sin^2\alpha}{Eh}\left(k_2 + \frac{1}{k_1}\right) H = \delta_m \tag{162}$$

Equations 162 determine both M and H, which are now applied separately to the dome.

When M alone is applied,

$$(M_\phi)_{\phi=\alpha} = M, \ (N_\phi)_{\phi=\alpha} = 0$$

This gives the value of C and γ as

$$C = \frac{M 2\lambda \sqrt{\sin\alpha}}{ak_1}, \ \gamma = 0$$

Table 65

key	ϕ / M_ϕ	Q_ϕ / M_θ	V / δ	N_ϕ	N_θ
$\phi = 7°30'$	.13089970	.00000064	.00011799	−.00000488	−.00001041
	.00000032	−.00000009	−.00002919		
$\phi = 15°0'$	.26179940	.00000066	−.00057493	−.00000246	.00001511
	.00000224	.00000112	.00010257		
$\phi = 22°30'$	.39269910	−.00000714	−.00057745	.00001725	.00010566
	−.00000531	−.00000130	.00096134		
$\phi = 30°0'$	.52359880	−.00000401	.01074954	.00000694	−.00037007
	−.00003468	−.00001428	−.00465193		
$\phi = 37°30'$	.65449850	.00015460	.00037945	−.00020149	−.00190132
	.00014925	.00004467	−.02801634		
$\phi = 45°0'$	.78539820	−.00008039	−.24785496	.00008039	.01063425
	.00063986	.00024359	.18756237		
$\phi = 52°30'$	.91629790	−.00367467	.27206002	.00281967	.03524201
	−.00436451	−.00135284	.68220678		
$\phi = 60°0'$	1.0471976	.00626339	5.9705878	−.00361617	−.31094699
	−.01132144	−.00411458	−6.7087123		
$\phi = 67°30'$	1.1780973	.08877078	−14.053683	−.03677006	−.57601803
	.12741594	.03943754	−13.049499		
$\phi = 75°0'$	1.3089970	−.27209757	−143.50904	.07290831	9.0433484
	.15595255	.05479683	217.85192		
$\phi = 82°30'$	1.4398967	−2.1080933	554.83283	.27753560	5.3768206
	−3.6794324	−1.1190474	131.20682		
$\phi = 90°0'$	1.5707964	10.113528	3342.0582	.00000072	−260.00000
	−.00000005	−.00000011	−6499.9999		

If H alone is applied,

$$(M_\phi)_{\phi=\alpha} = 0, \ (N_\phi)_{\phi=\alpha} = H \cos \alpha$$

This gives the value of C and γ as,

$$C = -\frac{H}{\sin \gamma}, \ \gamma = \tan^{-1}(-k_1)$$

The effects of the application of H and M as determined by the computer are shown in Tables 65 and 66. The values of N_ϕ, N_θ, M_ϕ, and M_θ are shown plotted in Figs. 221 and 222. It can be seen that fixing the edge has only very localized effects which die out rapidly towards the crown.

Table 66

key	ϕ M_ϕ	Q_ϕ M_θ	V δ	N_ϕ	N_θ
$\phi = 7°30'$	.13089970 -.00000027	.00000018 -.00000035	.00017173 -.00002588	-.00000140	-.00000835
$\phi = 15°0'$	.26179940 .00000173	.00000122 .00000067	-.00019582 -.00002848	-.00000455	-.00000577
$\phi = 22°30'$	.39269910 .00000143	-.00000284 .00000117	-.00147915 .00084672	.00000684	.00009056
$\phi = 30°0'$	.52359880 -.00003173	-.00001832 -.00001127	.00485933 .00026328	.00003173	.00003058
$\phi = 37°30'$	.65449850 .00000196	.00007831 -.00000640	.02573318 -.02931115	-.00010206	-.00195657
$\phi = 45°0'$	.78539820 .00072807	.00033415 .00024741	-.13915160 .02448130	-.00033415	.00128463
$\phi = 52°30'$	.91629790 -.00084347	-.00227102 -.00017800	-.46944792 .91351501	.00174261	.04658120
$\phi = 60°0'$	1.0471976 -.01751446	-.00587359 -.00574112	4.0470713 -1.8324470	.00339112	-.08361976
$\phi = 67°30'$	1.1780973 .04191420	.06593612 .01191888	7.5947589 -25.894507	-.02731163	-1.1293140
$\phi = 75°0'$	1.3089970 .42120663	.08052000 .13289391	-117.01180 85.334651	-.02157526	3.5273246
$\phi = 82°30'$	1.4398967 -1.6369520	-1.8957065 -.48915667	-70.328179 666.83398	.24957436	26.978395
$\phi = 90°0'$	1.5707964 -9.8349475	.00000000 -2.9504844	3342.0582 -3249.9999	.00000000	-130.00000

9.2.3 Analysis for Snow Load

Membrane analysis:

The structure is once again analyzed by the membrane theory, which gives[170]

$$N_\phi = -\frac{pa}{2} \tag{163}$$

$$N_\theta = -\frac{pa\cos 2\phi}{2} \tag{164}$$

where

p = intensity of the snow load in lb/ft² (kg/m²) of plan area.

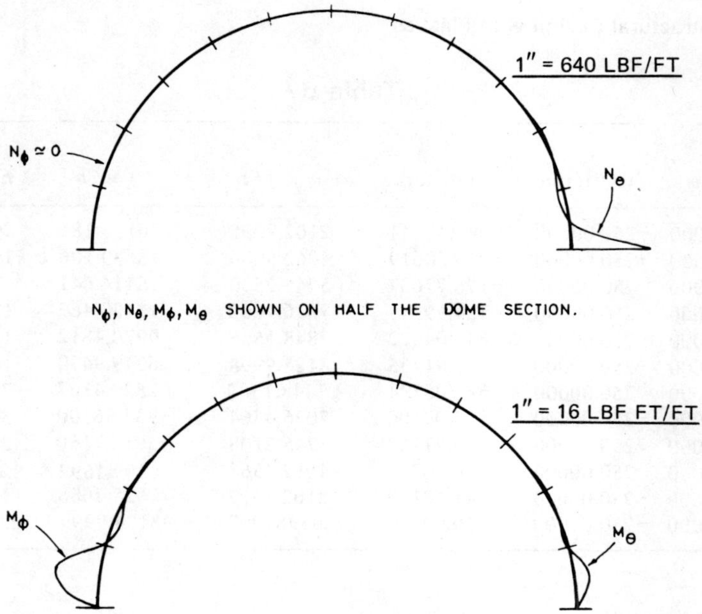

Fig. 221

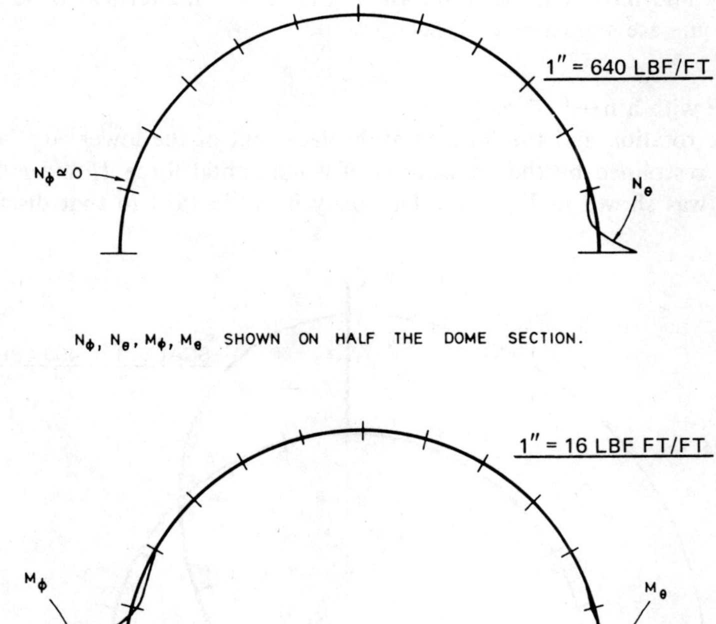

Fig. 222

Table 67

ϕ radians	N_ϕ lbf/ft. run	N_θ lbf/ft. run	vEt	wEt	$Et\delta_{vertical}$
.13090000	-250.00000	-241.48141	2102.9095	20135.181	20237.406
.26180000	-250.00000	-216.50619	4062.5086	18699.106	19113.403
.39270000	-250.00000	-176.77637	5745.2530	16414.641	17363.763
.52360000	-250.00000	-124.99947	7036.4662	13437.469	15155.422
.65450000	-250.00000	-64.704022	7848.1536	9970.4812	12687.768
.78540000	-250.00000	.00091735	8124.9998	6249.9470	10164.625
.91630000	-250.00000	64.705794	7848.1382	2529.4167	7766.1639
1.0472000	-250.00000	125.00106	7036.4364	-937.56100	5624.9627
1.1781000	-250.00000	176.77767	5745.2108	-3914.7160	3809.8018
1.3090000	-250.00000	216.50711	4062.4568	-6199.1592	2319.5930
1.4399000	-250.00000	241.48189	2102.8517	-7635.2088	1088.2933
1.5708000	-250.00000	250.00000	-.05987007	-8124.9999	-.02993503

The meridianal and hoop forces for a loading intensity of 20 lb/ft² (97.6 kg/m²) of plan area are shown in Table 67. The forces plotted on the dome center line have also been shown in Fig. 223. The vertical deflections of the dome are shown in Fig. 224.

Dome with a fixed edge:

The rotation and the horizontal displacement of the lower edge are once again restrained by the application of a horizontal force H and a moment M as was shown in Fig. 218. The analysis is identical to that discussed in

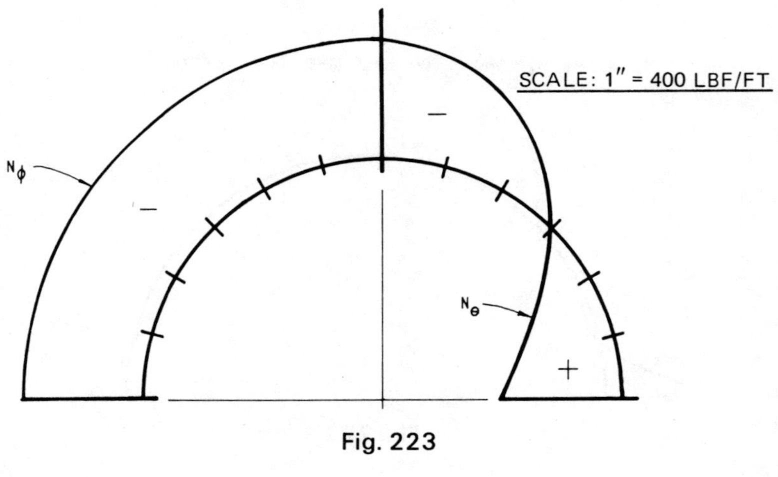

Fig. 223

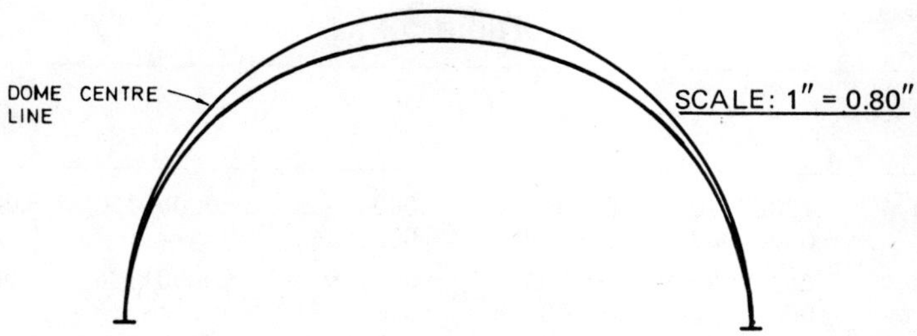

Fig. 224

Section 9.2.2, but a different value of the base expansion δ_m has to be used in Equations 162. The effects of H and M for a snow load intensity of 20 lb/ft² (97.6 kg/m²) of plan area, as determined by the computer, are shown in Tables 68 and 69. The values of N_ϕ, N_θ, M_ϕ and M_θ are shown plotted in Figs. 225 and 226.

Table 68

key	ϕ	Q_ϕ	V	N_ϕ	N_θ
	M_ϕ	M_θ	δ		
$\phi = 7°30'$	.13089970	.00000161	.00029497	−.00001220	−.00002602
	.00000080	−.00000023	−.00007298		
$\phi = 15°0'$	.26179940	.00000165	−.00143732	−.00000616	.00003778
	.00000559	.00000279	.00025643		
$\phi = 22°30'$	.39269910	−.00001786	−.00144362	.00004311	.00026414
	−.00001328	−.00000326	.00240335		
$\phi = 30°0'$	.52359880	−.00001002	.02687384	.00001736	−.00092518
	−.00008670	−.00003571	−.01162982		
$\phi = 37°30'$	.65449850	.00038651	.00094862	−.00050371	−.00475330
	.00037313	.00011168	−.07004084		
$\phi = 45°0'$	.78539820	−.00020097	−.61963740	.00020097	.02658562
	.00159965	.00060899	.46890591		
$\phi = 52°30'$	.91629790	−.00918668	.68015006	.00704919	.08810504
	−.01091127	−.00338211	1.7055170		
$\phi = 60°0'$	1.0471976	.01565847	14.926470	−.00904042	−.77736748
	−.02830360	−.01028645	−16.771781		
$\phi = 67°30'$	1.1780973	.22192695	−35.134208	−.09192514	−1.4400451
	.31853984	.09859384	−32.623746		
$\phi = 75°0'$	1.3089970	−.68024393	−358.77261	.18227077	22.608371
	.38988138	.13699209	544.62979		
$\phi = 82°30'$	1.4398967	−5.2702332	1387.0821	.69383901	13.442051
	−9.1985811	−2.7976186	328.01705		
$\phi = 90°0'$	1.5707964	25.283821	8355.1456	.00000179	−650.00001
	−.00000013	−.00000027	−16250.000		

Table 69

key	ϕ / M_ϕ	Q_ϕ / M_θ	V / δ	N_ϕ	N_θ
$\phi = 7°30'$	.13089970	.00000046	.00042933	−.00000350	−.00002088
	−.00000067	−.00000088	−.00006469		
$\phi = 15°0$	.26179940	.00000304	−.00048955	−.00001136	−.00001441
	.00000432	.00000168	−.00007121		
$\phi = 22°30'$	.39269910	−.00000709	−.00369787	.00001711	.00022639
	.00000358	.00000293	.00211681		
$\phi = 30°0'$	.52359880	−.00004580	.01214831	.00007932	.00007645
	−.00007932	−.00002818	.00065821		
$\phi = 37°30'$	.65449850	.00019578	.06433295	−.00025514	−.00489143
	.00000491	−.00001599	−.07327788		
$\phi = 45°0'$	.78539820	.00083538	−.34787899	−.00083538	.00321157
	.00182017	.00061853	.06120325		
$\phi = 52°30'$	.91629790	−.00567755	−1.1736198	.00435653	.11645300
	−.00210868	−.00044499	2.2837875		
$\phi = 60°0'$	1.0471976	−.01468397	10.117678	.00847779	−.20904940
	−.04378615	−.01435281	−4.5811174		
$\phi = 67°30'$	1.1780973	.16484029	18.986897	−.06827907	−2.8232849
	.10478551	.02979719	−64.736267		
$\phi = 75°0'$	1.3089970	.20129999	−292.52950	−.05393816	8.8183116
	1.0530166	.33223478	213.33663		
$\phi = 82°30'$	1.4398967	−4.7392662	−175.82045	.62393590	67.445987
	−4.0923800	−1.2228917	1667.0849		
$\phi = 90°0'$	1.5707964	.00000000	8355.1455	.00000000	−325.00000
	−24.587369	−7.3762109	−8124.9998		

9.2.4 Analysis for Wind Load

The intensity of the wind load p' has been given as 18 lb/ft² (87.9 kg/m²) on a vertical surface. It is assumed that the wind pressure distribution is given by the formula,

$$
\begin{aligned}
Z &= p' \sin\phi \cos\theta \\
X &= 0 \\
Y &= 0
\end{aligned}
\tag{165}
$$

where X and Y are the tangential components of the wind pressure and Z is the normal component of the wind pressure. Equation 165 gives an antimetrical load distribution with pressure on the windward side and suction on the leeward side, as shown in Fig. 227.

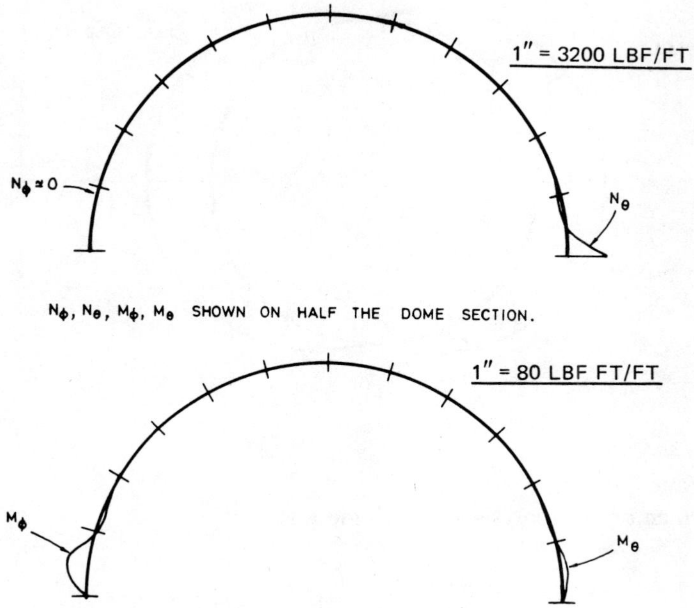

Fig. 225

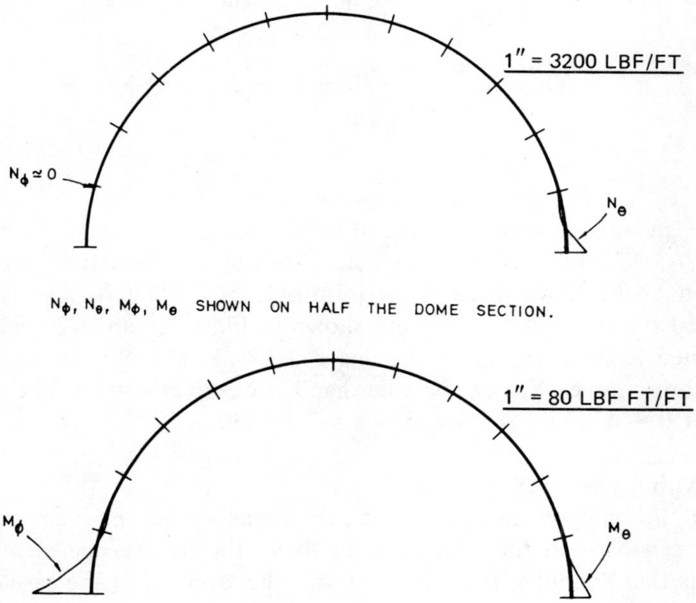

Fig. 226

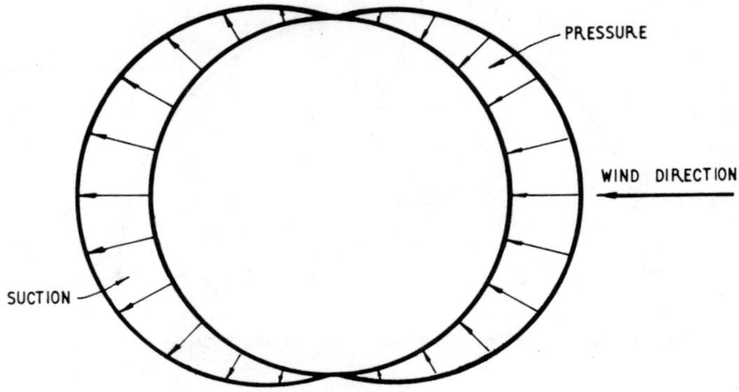

Fig. 227

Membrane analysis:

The membrane analysis of the dome gives

$$N_\phi = -\frac{p' a \cos\phi \cos\theta}{3 \sin^3\phi}[2 - 3\cos\phi + \cos^3\phi] \tag{166}$$

$$N_\theta = \frac{p' a \cos\theta}{3 \sin^3\phi}[2\cos\phi - 3\sin^2\phi - 2\cos^4\phi] \tag{167}$$

$$N_{\phi\theta} = -\frac{p' a \sin\theta}{3 \sin^3\phi}[2 - 3\cos\phi + \cos^3\phi] \tag{168}$$

where

$N_{\phi\theta}$ = membrane shearing force in lb/ft run (kg/m).

The N_ϕ, N_θ, and $N_{\phi\theta}$ forces as determined by the computer are far too numerous to be shown here in the form of a table. The N_ϕ and N_θ forces at section $\theta = 0°$ and $\theta = 45°$ are shown in Figs. 228 and 229. The wind is assumed to be acting along the line $\theta = 0°$. At $\theta = 90°$ $N_\phi = N_\theta = 0$. The shearing forces $N_{\phi\theta}$, on the other hand, are zero at $\theta = 0°$. The shearing forces at $\theta = 45°$ and 90° are shown in Fig. 230.

Dome with a fixed edge:

From the analyses carried out for the dome under self-load and snow load, it can be seen that the effect of fixing the lower edge creates edge moments that die out very rapidly towards the crown. For the case of wind loading, therefore, only the maximum moments and the horizontal force have been determined approximately as follows[171]:

$$H = \left[1 + \frac{1+\nu}{3\beta}\right]\frac{p'\,a}{\beta} \tag{169}$$

$$(M_\phi)_{\phi=\alpha} = -\left[\frac{1}{2} + \frac{1+\nu}{3\beta}\right]\frac{p'\,a^2}{\beta^2} \tag{170}$$

$$(M_\theta)_{\phi=\alpha} = \nu(M_\phi)_{\phi=\alpha} \tag{171}$$

where

$$\beta = \left(\frac{1-\nu^2}{4k}\right)^{1/4}$$

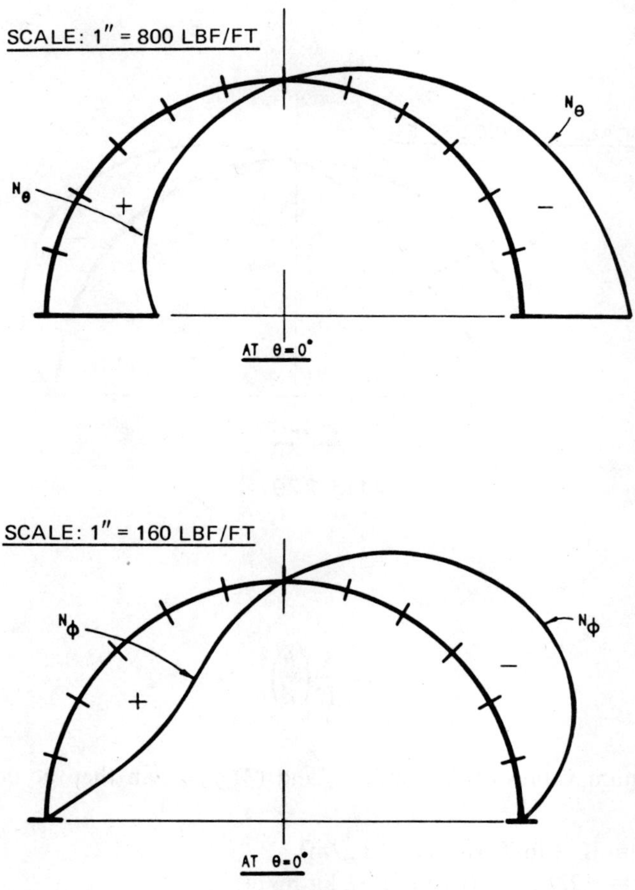

Fig. 228

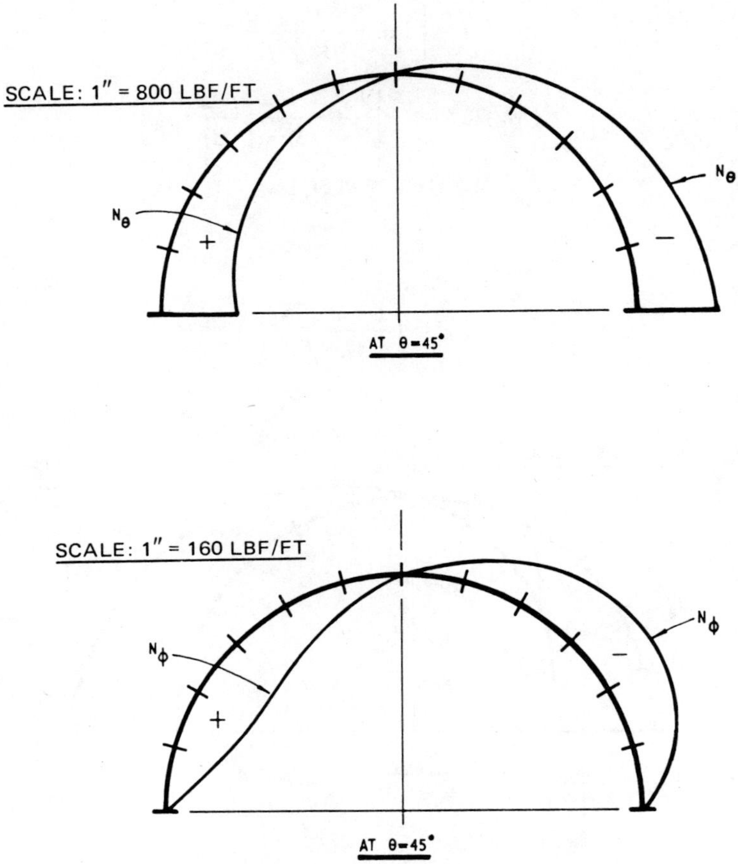

Fig. 229

and

$$k = \frac{1}{12}\left(\frac{h}{a}\right)^2$$

The maximum values of H, $(M_\phi)_{\phi=\alpha}$, and $(M_\theta)_{\phi=\alpha}$ can then be obtained as

$$H = 10.2 \text{ lb/ft run } (15.2 \text{ kg/m})$$
$$(M_\phi)_{\phi=\alpha} = -2.9 \text{ lb ft/ft run } (1.32 \text{ kg m/m})$$
$$(M_\theta)_{\phi=\alpha} = -0.9 \text{ lb ft/ft run } (0.41 \text{ kg m/m}).$$

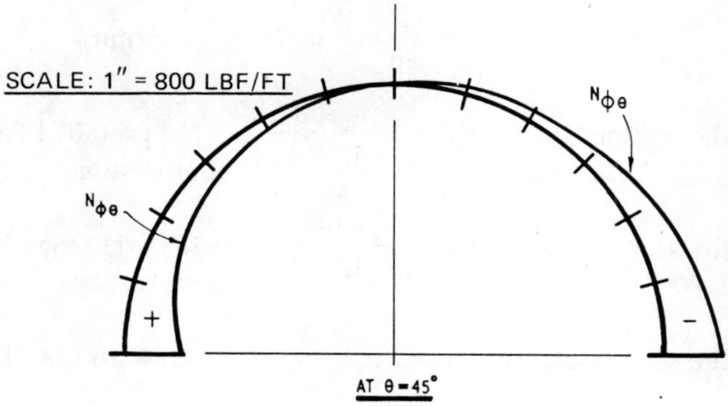

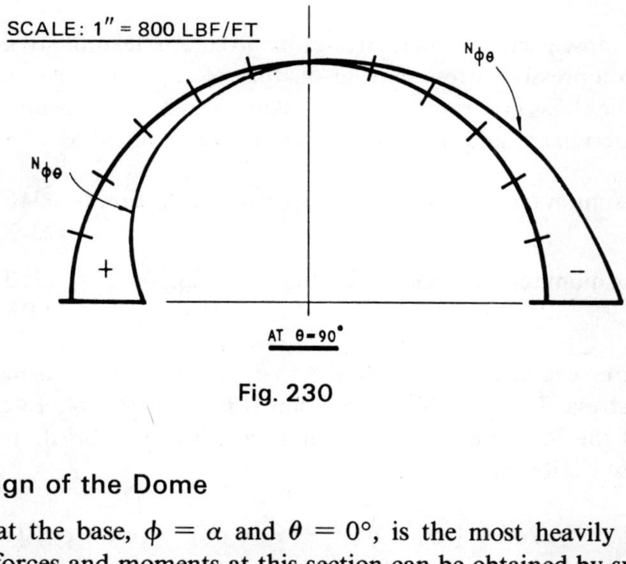

Fig. 230

9.2.5 Design of the Dome

The section at the base, $\phi = \alpha$ and $\theta = 0°$, is the most heavily stressed section. The forces and moments at this section can be obtained by summing up all the cases considered in Sections 9.2.2–9.2.4 as

$$N_\phi = -350 \text{ lb/ft run } (521 \text{ kg/m})$$
$$N_\theta = -1465 \text{ lb/ft run } (2180 \text{ kg/m})$$
$$M_\phi = -37.33 \text{ lb ft/ft run } (16.9 \text{ kg m/m})$$
$$M_\theta = -11.23 \text{ lb ft/ft run } (5.1 \text{ kg m/m}).$$

Assume that the thickness at the section is ¼ in. (6 mm); then

$$Z = \frac{1}{6} \times 1 \times \frac{1}{16} = \frac{1}{96} \text{ in.}^3\text{/in. (6.72 mm}^3\text{/mm).}$$

Stress in the section due to $N_\phi = -\dfrac{350}{12 \times 1/4} = -117$ psi (807 kPa) compression.

Stress in the section due to $N_\theta = -\dfrac{1465}{12 \times 1/4} = -488$ psi (3365 kPa) compression.

Stress in the section due to $M_\phi = \pm\dfrac{37.33}{1/96} = \pm 3584$ psi (24.71 MPa)

Stress in the section due to $M_\theta = \pm\dfrac{11.23}{1/96} = \pm 1078$ psi (7.43 MPa)

Shearing stress at the section $= 0$

For the most part, the main stresses involved are flexural stresses with small direct compressive stress. Erring on the safe side, however, and assuming the tensile stress criterion to apply to skin stresses, the maximum skin stresses in the meridianal and hoop directions can be obtained as

Maximum tensile stress in the meridianal direction = +3467 psi (23.90 MPa)

Maximum tensile stress in the hoop direction = +590 psi (4.07 MPa)

There are several failure criteria[172] that can be applied to materials under biaxial stress. The Von Mises criterion is more applicable to ductile failure, whereas the Rankine criterion is more applicable to brittle fracture. If the Von Mises criterion holds,

$$2\sigma^2 = [(\sigma_1 - \sigma_2)^2 + (\sigma_2 - \sigma_3)^2 + (\sigma_3 - \sigma_1)^2] \tag{172}$$

where $\sigma_3 = 0$. This gives

$$\sigma = +3210 \text{ psi (22.13 MPa)}$$

If the Rankine theory holds, then failure occurs when the numerically larger principal stress equals the failure stress, or

$$\sigma = +3467 \text{ psi (23.90 MPa)}$$

With the design life of the structure as 75 years, and referring to the discussion on the long term properties of FRP in Section 2.4, it can be assumed that at the end of 75 years, the ultimate strength of the material will be at least 40% of the ultimate short term strength. The long term tensile strength will be at least 4800 psi (33.09 MPa). This takes into account any wetting effects due to normal weathering. With the Von Mises criterion of failure, therefore,

$$\text{Factor of safety} = \frac{4800}{3210} = 1.50$$

With the Rankine theory of failure,

$$\text{Factor of safety} = \frac{4800}{3467} = 1.39$$

If there is any possibility of delamination, a similar analysis should be carried out for the compression surface as well. It may be noted that the factor of safety obtained is one that is valid at the end of the life of the structure. The factor of safety at the time of erection is considerably higher. Any excess stresses induced in handling and erection are, therefore, satisfactorily accounted for.

Stability of the dome:

The normal critical buckling load of a spherical shell can be expressed as,[147,173,174]

$$P_{cr} = CE\left(\frac{t}{R}\right)^2 \tag{173}$$

where

t = thickness of the shell
R = radius of the shell
E = modulus of elasticity
C = a constant

Various theoretical and experimental[147,175] values have been given for the constant C. Von Karman and Tsien suggested a theoretical value of 0.3652. In 1942, Tsien suggested that the value could be 0.312. Doing theoretical work on spherical domes, Mushtari and Surkin gave a value of 0.34. Feodosjew determined a theoretical value of C for spherical domes as 0.32 but recommended 0.20 in practice. Schmidt,[175] working with doubly curved aluminum shells, suggested a value of 0.32 to be used in practice. For domes of metal and plastics he recommended that this should be lowered to 0.20. Dealing with rigid plastics radomes, Davis,[176] on the basis of Sandorff's work, suggests that for stiffened shells such as radomes the value of the constant was 0.36.

It can be seen that the value of C ranges from about 0.20 to 0.36. For a single skin structure, without any stiffening, a value of 0.20 would be appropriate. However, the dome that is being analyzed here is to be divided into 24 segments that are to be connected by FRP flanges that afford full fixity. These flanges themselves form stiffening ribs in the meridianal direction. In addition, it would be appropriate, from an erection point of view, to have at least three lines of ribs in the hoop direction built into the structure. Forming the ribs in the meridianal and hoop directions during the actual manufacture of the units of the dome presents no difficulty, as seen in Table 7. In the circumstances, it is reasonable to take the value of the constant C to be equal to 0.32.

It can be seen from Equation 173 that the critical buckling load of the shell depends upon the value of E. The modulus of elasticity is time dependent and, if the buckling stability is to be examined at the end of the design life, it is essential to estimate the reduction in the modulus due to creep of the material. Referring to Fig. 31 in Section 2.12 it may be assumed that, if the stress level is low, the modulus of elasticity of the material does not reduce by more than 33%. Once again this takes into account normal ambient temperatures and natural weathering. The modulus of elasticity may, therefore, be assumed to be at least 0.8×10^6 psi (5.52 GPa) at the end of 75 years. The critical buckling load of the shell can, therefore, be determined as

$$P_{cr} = \frac{0.32 \times 0.8 \times 10^6 \times (1/4)^2}{(25 \times 12)^2} \times 144$$

$$= 25.6 \text{ lb/ft}^2 \text{ (125 kg/m}^2\text{)}$$

The factor of safety against buckling at the end of 75 years is, therefore

$$\frac{25.6}{24} = 1.07$$

This factor of safety is very low, and the thickness should be increased to ⅜ in. (10 mm) when the factor of safety is 2.40. The factor of safety at the time of erection is, of course, considerably higher. The composite thickness for purposes of design was assumed to be ¼ in. (6 mm). A ⅜ in. (10 mm) thickness, therefore, increases the loading on the dome by 1 lb/ft² (4.88 kg/m²). The factors of safety with a ⅜ in. (10 mm) thickness, however, are considerably higher than those obtained in Section 9.2.5, and so the calculations do not need to be revised.

The following design recommendations can be made:

1. The dome should have a guaranteed minimum thickness of ⅜ in. (10 mm).
2. The thickness should be increased to ½ in. (13 mm) at the connections to the base ring as well as to the flanges. The increase in thickness should be gradual and spread over at least 6 in. (152 mm).
3. The flanges must be capable of developing full fixity in order to validate the shell behavior that has been assumed in the design. A suitable flange connection has been shown in Fig. 198.
4. The segments of the dome should be stiffened by at least three lines of hoop stiffeners. These should be at least ½ in. (13 mm) thick and 6 in. (152 mm) deep.
5. Flanges and stiffeners should be formed integral with the shell during the manufacture of the segments.
6. The external surface of the shell should be suitably protected against weathering by a gel coat about 0.010–0.015 in. (0.25–0.38 mm) thick.
7. If the segments are made by a hand lay-up process, the internal surface should be finished smooth with surface tissue. If the segments are press molded, the internal surface will be finished fairly smooth.
8. The gel coat and laminating resins should be fire retardant.
9. The gel coat and laminating resins can be suitably pigmented, if desired.
10. The first segment manufactured should be cut up into specimens and tested. If the samples do not yield either the guaranteed minimum thickness or the guaranteed minimum strengths quoted for the material in Section 9.2.1, the design should be revised accordingly.
11. Other considerations, such as the type of gel coat and laminating polyester resins, type of glass and its finish, glass fiber content, resin fillers, process to be used for the manufacture of the segments, and the like, should be decided in consultation with the fabricator. Recommendations made in earlier chapters of this book, regarding properties of composites and their ability to meet fire and weathering requirements, should also be kept in mind.

9.3 THE BARREL SHELL

Cylindrical barrel shells[177] are shown in Fig. 231 and the close similarity between the prismatic folded plate structure of Fig. 153 and barrel shells

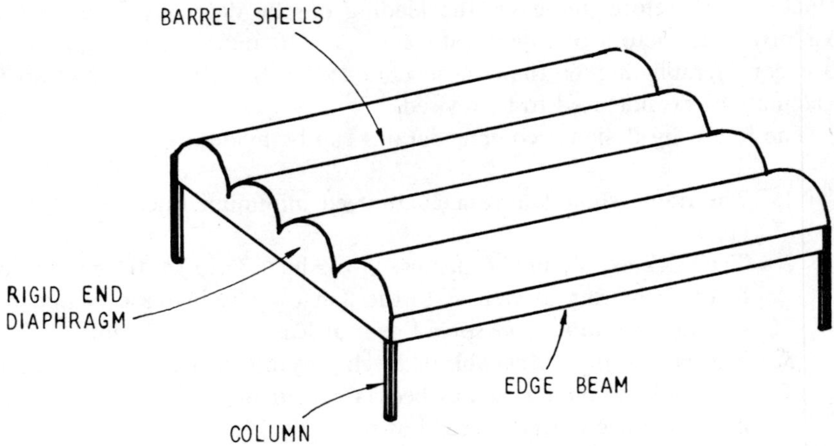

Fig. 231

can be easily seen. In the analysis of the Vee type folded plate structure, it was seen that the transverse action depends on the span of the transverse strip. As the number of plates in any one unit increases, the span becomes smaller and smaller, until finally, with an infinite number of plates, as in the barrel shell, the span is zero. The transverse plate or strip action is therefore eliminated for long barrel shells, leaving only the overall longitudinal action of the structure to be considered.

For the overall longitudinal action, it was seen for the Vee type folded plate unit in Section 7.2.2 that the internal unit in fact behaves like a beam of Vee cross section, spanning between the rigid end diaphragms. In a similar manner, the cylindrical barrel shell behaves like a beam of curved cross section, as shown in Fig. 232. The position of the neutral axis and the second moment of area, or moment of inertia, I_n can be determined as[178]

$$n = R\left(1 - \frac{\sin \phi_T}{\phi_T}\right) \tag{174}$$

$$I_n = 2tR^3\left(\frac{\phi_T}{2} + \frac{\sin 2\phi_T}{4} - \frac{\sin^2 \phi_T}{\phi_T}\right) \tag{175}$$

where

ϕ_T = central angle in radians as shown in Fig. 232
t = thickness of the shell
R = radius of the shell.

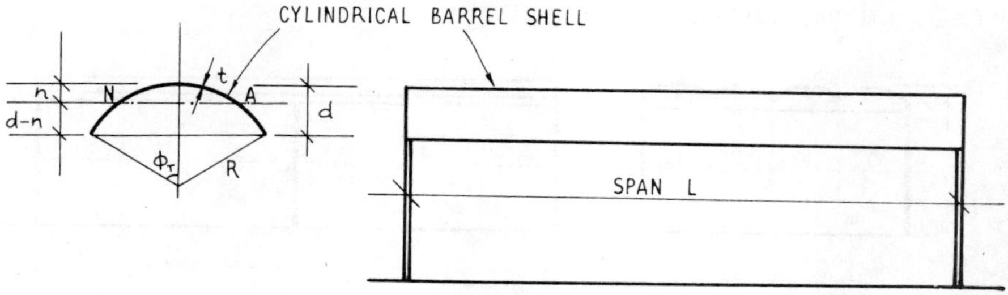

Fig. 232

Since the shell behaves as a simply supported beam on rigid end diaphragms, it follows that the maximum bending moment at the center is given by

$$BM_{\max} = M = WL/8 \tag{176}$$

The maximum compressive stress can then be obtained as

$$f_{bc} = Mn/I_n \tag{177}$$

For the simply supported barrel this is at the crown of the shell, at the center of the span.

The maximum tensile stress can be obtained as,

$$f_{bt} = M(d - n)/I_n \tag{178}$$

This stress occurs at the lower edges of the shell at the center of the span. The design of a sandwich barrel shell structure with FRP facings on a suitable core is considered in Section 9.3.1.

9.3.1 Design of a Two-span Continuous Barrel Shell

An area of 40 × 80 ft (12.2 × 24.4 m) is to be roofed by four long cylindrical barrel shells over two spans of 40 ft (12.2 m) each, as shown in Fig. 233. The chord width is 10 ft (3.05 m) and the rise of the barrel is 2 ft 6 in. (0.76 m). If the sandwich thickness is assumed to be 2 in. (51 mm), determine the maximum compressive stress and the maximum tensile stress in the facings of the sandwich, at positions of both maximum positive and maximum negative moments. The snow load is assumed to be 25 lbs/sq ft (122 kg/m^2). The self-weight of the barrel together with its finishing load may be assumed to be 4 lb/ft^2 (19.5 kg/m^2).

The radius of the cross-section of each barrel shell can be determined as

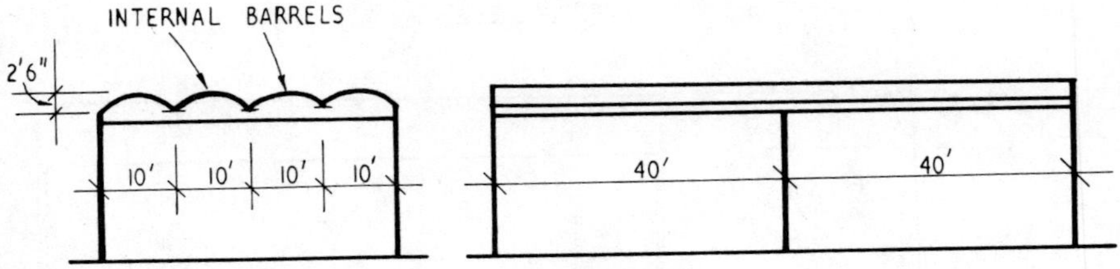

Fig. 233

$$R^2 = (R - 2.5)^2 + 5^2$$

$$R^2 = R^2 - 5R + 6.25 + 25$$

$$R = \frac{31.25}{5} = 6.25 \text{ ft } (1.91 \text{ m})$$

$$\sin \phi_T = \frac{5}{6.25} = 0.8000$$

$$\phi_T = 53°6'$$

$$\phi_T = \frac{\pi \times 53.1}{180} = 0.927$$

Using Equations 174 and 175 we have

$$n = 6.25\left(1 - \frac{0.8000}{0.927}\right) = 0.856 \text{ ft } (0.26 \text{ m})$$

$$I_n = 2 \times 2 \times (6.25 \times 12)^3 \left(\frac{0.927}{2} + \frac{0.9603}{4} - \frac{0.8000^2}{0.927}\right)$$

$$= 22275 \text{ in.}^4 \ (92{,}715 \times 10^5 \text{ mm}^4)$$

Also,

$$(d - n) = 2.500 - 0.856$$
$$= 1.644 \text{ ft } (0.50 \text{ m})$$

Therefore the section moduli for top and bottom fibers are

$$S = \frac{I_n}{n} = \frac{22275}{0.856 \times 12} = 2168.5 \text{ in.}^3 \ (3554 \times 10^4 \text{ mm}^3)$$

$$S = \frac{I_n}{(d-n)} = \frac{22275}{1.644 \times 12} = 1129.1 \text{ in.}^3 \ (1850 \times 10^4 \text{ mm}^3)$$

The total loading on the barrel can be calculated as

$$\text{Self-load} = 2 \times 0.927 \times 6.25 \times 1 \times 4 = 46 \text{ lb/ft (68 kg/m)}$$
$$\text{Snow load} = 25 \times 10 \times 1 = 250 \text{ lb/ft (372 kg/m)}$$

Therefore,

$$\text{Total load} = 296 \text{ lb/ft (440 kg/m)}$$
$$\text{Total load } W = 296 \times 40 = 11840 \text{ lb/span (5371 kg/span)}$$

Since the barrels are absolutely symmetrical and the loading is symmetrical as well, the structure must have a zero slope over the central diaphragm. Each internal barrel may therefore be regarded as two propped cantilevers. The maximum bending moments for a propped cantilever[177] are shown in Fig. 234. We have therefore,

$$BM_{max} \text{ positive} = +\frac{9WL}{128} = +\frac{9 \times 11840 \times 40}{128}$$

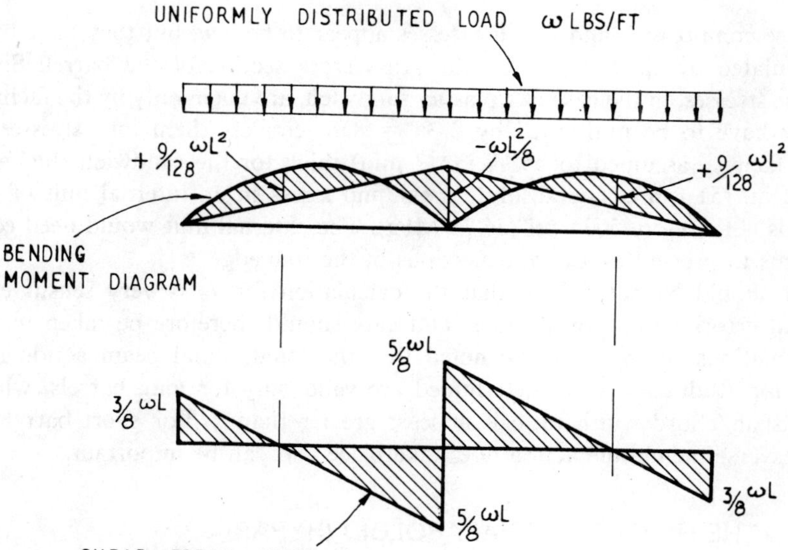

Fig. 234

$$= +33300 \text{ lb ft } (4604 \text{ kg m})$$

$$BM_{\text{max}} \text{ negative} = -\frac{WL}{8} = -\frac{11840 \times 40}{8}$$

$$= -59200 \text{ lb ft } (8185 \text{ kg m})$$

The maximum longitudinal compressive and tensile stresses in the middle of the barrel for the maximum positive moment are therefore

$$f_{bc} = \frac{33{,}300 \times 12}{2168.5} = 184.3 \text{ psi } (1271 \text{ kPa})$$

$$f_{bt} = \frac{33{,}300 \times 12}{1129.1} = 353.9 \text{ psi } (2440 \text{ kPa})$$

The maximum longitudinal compressive and tensile stresses in the barrel over the central diaphragm, for maximum negative moment are hence

$$f_{bc} = \frac{59{,}200 \times 12}{1129.1} = 629.2 \text{ psi } (4338 \text{ kPa})$$

$$f_{bt} = \frac{59{,}200 \times 12}{2168.5} = 327.6 \text{ psi } (2259 \text{ kPa})$$

These compressive and tensile stresses appear to be low, but they have been calculated by shell theory on the gross cross section of the barrel. Since these stresses, in the case of a plastics sandwich, are taken only by the facings, they have to be multiplied by 2/¼ = 8 to convert them into stresses in the facings, assumed to be ⅛ in. (3 mm) thick for the sandwich thickness of 2 in. (51 mm). The absolute maximum stress in an internal unit of the shells is therefore 5034 psi (34.71 MPa). The external unit would need edge beams to prevent excessive deflections of the free edge.

It should be noted here that the calculation for I_n is very sensitive to small errors or approximations; and care should therefore be taken in the calculations. It may also be noted that the longitudinal beam action and the longitudinal stresses determined are valid only for long barrels, where the span/chord width ratio is at least greater than 2. For short barrels, a transverse membrane action does take place and can be important.

9.4 THE HYPERBOLIC PARABOLOID (HYPAR)

The basic element of a hypar is shown in Fig. 235. As was stated, this is an anticlastic shell, where the principal curvatures are in opposite directions.

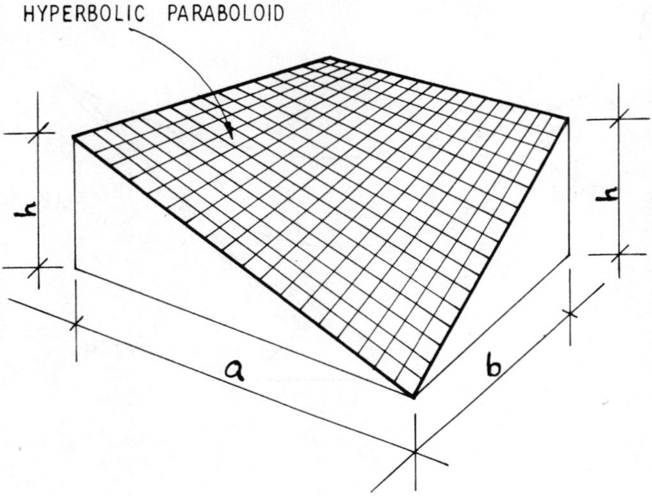

Fig. 235

The shell surface is derived by the movement of one parabola on another parabola of opposite curvature. It can also be generated by the movement of a line on two skew lines. This latter method of generating the surface implies that the formwork for a hyperbolic paraboloid can be constructed from straight elements. This could make the mold relatively simple.

The shell surface under the vertical uniformly distributed loading is extremely efficient in propagating the load to the supports. The shell behaves as a compression membrane with an arch action in one direction, and as a tension membrane with a cable action in the other. Further, the compression membrane is stabilized and prevented from buckling by the support it gets from the tension membrane. Since the parabolic arch, under uniformly distributed loading, is in pure compression, the compression membrane is under a pure compression. Since the parabolic cable under uniformly distributed load is in pure tension, the tension membrane is also under pure tension. The entire surface is therefore under pure compressive and tensile forces, with complete absence of moments—under a uniformly distributed vertical loading.

The propagation of the load by the surface as a membrane towards the supports is shown[177] in Fig. 236. The arch action of the compression membrane causes a thrust on the edge beams in one direction, as shown. The cable action of the tension membrane causes a pull on the edge beam at right angles. The resultant of these forces is an axial force that flows down the edge beam towards the foundations. The edge beams pick up all these

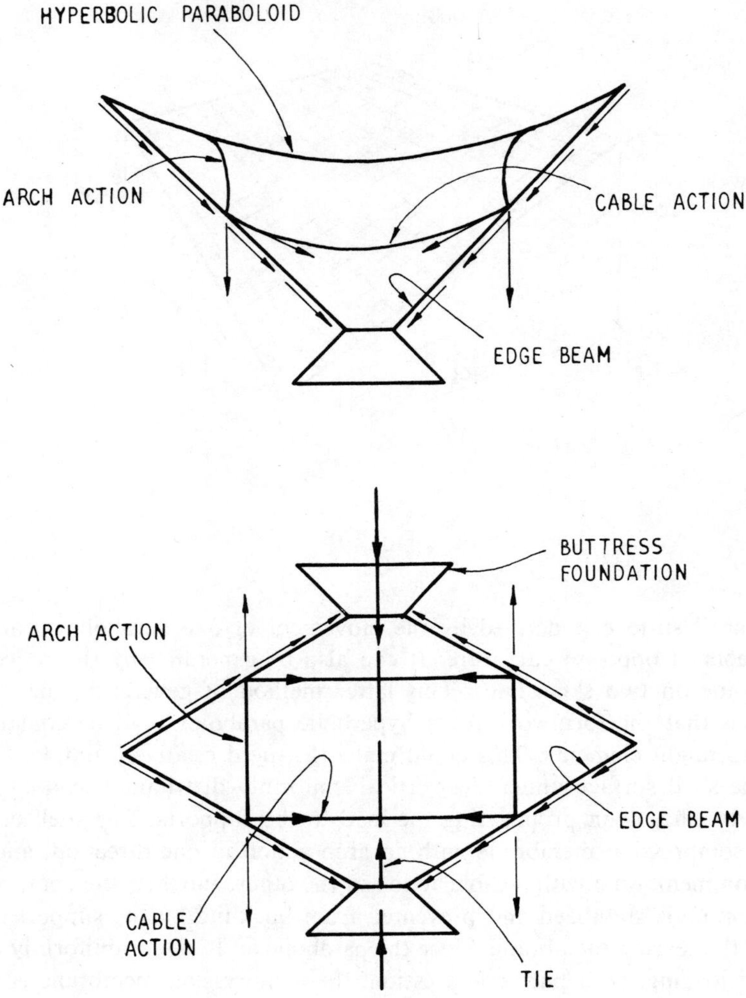

Fig. 236

axial forces from the continuous connection with the membrane and transmit them to the foundations. At the foundations, these axial thrusts have to be resisted, either by the provision of a tie, or by buttressing, as shown in Fig. 236.

The membrane compressive and tensile forces in a hypar can be obtained as[177,179]

$$C = -\frac{wab}{2h}$$
$$T = -\frac{wab}{2h} \tag{179}$$

where

a, b, and h are the dimensions of the hypar
w = loading in lb/ft^2 (kg/m^2).

The forces in the edge beams can be obtained as[177,179]

$$F_1 = \frac{wab}{2h}\sqrt{a^2 + h^2}$$
$$F_2 = \frac{wab}{2h}\sqrt{b^2 + h^2} \tag{180}$$

It should again be noted here that the shell is under tension and compression membrane forces only under a uniformly distributed vertical loading. Under any other loading, such as a point load, bending moments will arise. Since the edge beams must be able to withstand large compressive forces, they are larger than the shell membrane itself, imposing line loads along the edges. This leads to a thickening of the shell towards the edge.

The design of a simple hypar is shown in Section 9.4.1.

9.4.1 Design of a Hyperbolic Paraboloid Shell of Single Skin Construction

The design of the shell is carried out for the following dimensions:

a = 30 ft (9.14 m)
b = 40 ft (12.19 m)
h = 20 ft (6.10 m)
t = ½ in. (13 mm)
Snow load = 30 lb/ft^2 (146.5 kg/m^2)
Finishing load = 5 lb/ft^2 (24.4 kg/m^2).

The self-weight of ½ in. (13 mm) thick FRP laminate may be assumed to be about 4 lb/ft^2 (19.5 kg/m^2). Thus, total load on the shell is

Self-load + finishing load + snow load = 4 + 5 + 30 = 39 lb/ft² (190.4 kg/m²)

In a shell of this type, the actual surface area is not much greater than the plan area, and so a simple addition of the loads is acceptable.

The forces in the membrane can be obtained by the use of Equation 179 as

$$C = T = \mp \frac{39 \times 30 \times 40}{2 \times 20} = \mp 1170 \text{ lb/ft } (1741 \text{ kg/m})$$

The compressive and tensile stresses in the membrane are therefore

$$f_c = f_t = \frac{1170}{12 \times \frac{1}{2}} = 195 \text{ psi } (1344 \text{ kPa})$$

This is very low and quite acceptable. Other refinements, such as considering the long term strength of the laminate, have been considered at some length in the design of the spherical dome in Section 9.2 and are not repeated here.

The forces in the edge beams can be obtained by the use of Equation 180 as

$$F_1 = 1170 \sqrt{30^2 + 20^2} = 42{,}179 \text{ lb } (19{,}132 \text{ kg})$$

$$F_2 = 1170 \sqrt{40^2 + 20^2} = 52{,}322 \text{ lb } (23{,}733 \text{ kg})$$

The edge beams are compressive members, but they are prevented from buckling because they are restrained by the membrane along the entire length. However, the forces in the members are large, and so the design of the edge beams would be better carried out in a combination of FRP flanges with other nonplastics materials, such as timber. The foundations of the structure, of course, would be in reinforced concrete.

The buckling strength of the membrane should also be investigated. For hyperbolic paraboloids or saddle shells, the critical buckling load can be given by[179]

$$p_{cr} = 1.16 \left(\frac{th}{ab}\right)^2 E \qquad (181)$$

A rather larger factor of safety should always be allowed in such cases, because there are discrepancies between the theoretical and actual buckling loads of the structure. Using Equation 181 the critical buckling load of the hypar can be determined as,

$$p_{cr} = 1.16\left(\frac{\frac{1}{2} \times 20}{12 \times 30 \times 40}\right)^2 \times 1.2 \times 10^6 = 0.6713 \text{ psi}$$

$$p_{cr} = 0.6713 \times 144 = 96.7 \text{ lb/ft}^2 \text{ (472 kg/m}^2\text{)}$$

Since the actual load is 39 lb/ft² (190.4 kg/m²), we can compare the following:

$$\begin{aligned}\text{Factor of safey against buckling} &= 96.7/39 \\ &= 2.5\end{aligned}$$

For an increased factor of safety, the thickness *t* or the rise *h* of the hypar should be increased.

CHAPTER 10

STRESSED SKIN STRUCTURES

10.1 SKELETON ANALYSIS

The various practical forms of stressed skin structures are described in Chapter 5. It was seen that the analysis of the structure reduces primarily to the analysis of a skeleton space frame, as shown for the pyramid in Fig. 156. This is because such a structure would be of single skin construction. Plate buckling in the plates of the pyramids at low loads causes all the loads to be taken by the folds, therefore reducing the structure to a skeleton. The analysis of the skeleton is not particularly difficult. A matrix method of analysis would have to be used, the problem then being programmed for a computer. Commercially available programs can also be used, and the use of IBM's STRUDL program for the analysis of a single skin, folded plate, conical dome is considered in Section 10.8. There are several good books on the use of matrix methods of structural analysis[180-183] and the stiffness method of analysis is briefly outlined in Section 10.8. However, before the method, or a commercially available program, can be applied to an actual structure, it is necessary to know the areas of the plate that can be assumed to be effective at the folds. These edge areas and edge moments of inertia are then fed into the computer as data. This is where the real difficulty arises. The plates may not be prismatic, and the assumption that the effective plate width is constant along the edge can be erroneous. Experimental work carried out by the author and others, and certain theoretical work on prismatic plates, however, permits some conclusions to be readily reached.

10.2 FLAT PRISMATIC PLATES UNDER AXIAL COMPRESSION

Consider the flat plate supported along two edges, as shown in Fig. 237. Von Karman[184] has shown that the ultimate load P_u for the plate can be written as

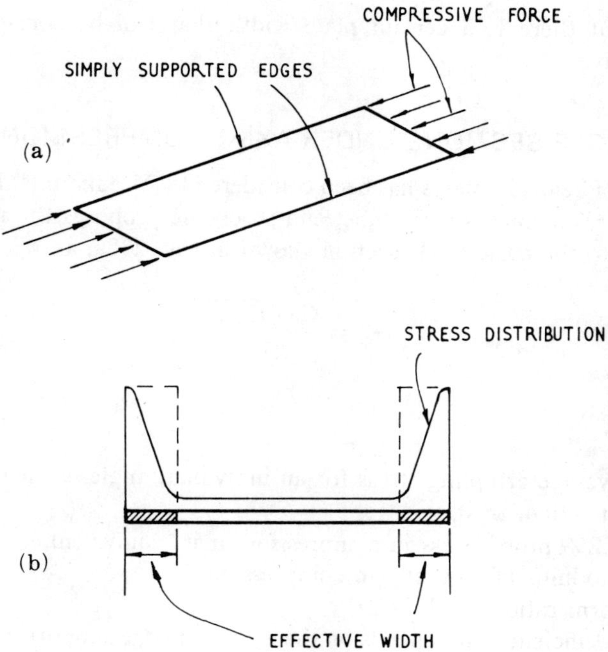

Fig. 237

$$P_u = \frac{\pi}{3(1-\nu^2)}\sqrt{(E'\,\sigma)_c}\,t^2 \qquad (182)$$

where

ν = Poisson's ratio for the material of the plate
t = thickness of the plate
$(E'\,\sigma)_c$ = product of the stress and tangent modulus in compression.

It can be seen that P_u varies as t^2 and that the ultimate load is independent of the dimensions of the plate. Equation 182 can be written as

$$P_u = C\sqrt{(E'\,\sigma)_c}\,t^2 \qquad (183)$$

where C is a constant that takes into account the various edge conditions of the plate, such as clamped edges, and the like.

The stress distribution across the plate is shown in Fig. 237b, and it can be seen that there is a certain plate width that can be considered to be truly effective.

10.3 ANGLE SECTIONS UNDER AXIAL COMPRESSION

The effect of bends in plates has been considered by Needham.[185] For sections such as Vee, channel, square tube, and so on, he proposes the relationship, applicable to the basic angle section shown in Fig. 238a, as

$$\sigma_{cc} = \frac{C_e\sqrt{E\,\sigma_{pc}}}{R_f^{0.75}} \tag{184}$$

where

σ_{cc} = average crippling stress for an individual angle element (σ_{cc} not greater than σ_{pc}),
σ_{pc} = 0.2% proof stress in compression or its equivalent,
E = modulus of elasticity in compression,
R_f = form ratio = $(a + b)/2t$
C_e = coefficient that depends on the degree of edge support from adjacent angle elements.

Then

$$P_u = \sigma_{cc} \times \text{total area } A, \tag{185}$$
$$P_u = \sigma_{pc} \times \text{ reduced area as shown in Fig. 240b.} \tag{186}$$

For purposes of computation, the angles in any section are separated midway between the junctions. Needham[185] has calculated the values of C_e for 90° angle shapes fabricated from a light aluminum alloy as

$C_e = 0.316$ for isolated angles
$C_e = 0.342$ for angles with only one edge free
$C_e = 0.366$ for angles with neither edge free.

The total area A can be obtained as

$$A = (a + b)\,t = 2\,R_f t^2$$

Equation 185 may be rewritten as

$$P_u = 2\,C_e\sqrt{E\sigma_{pc}}\,R_f^{0.25}\,t^2 \tag{187}$$

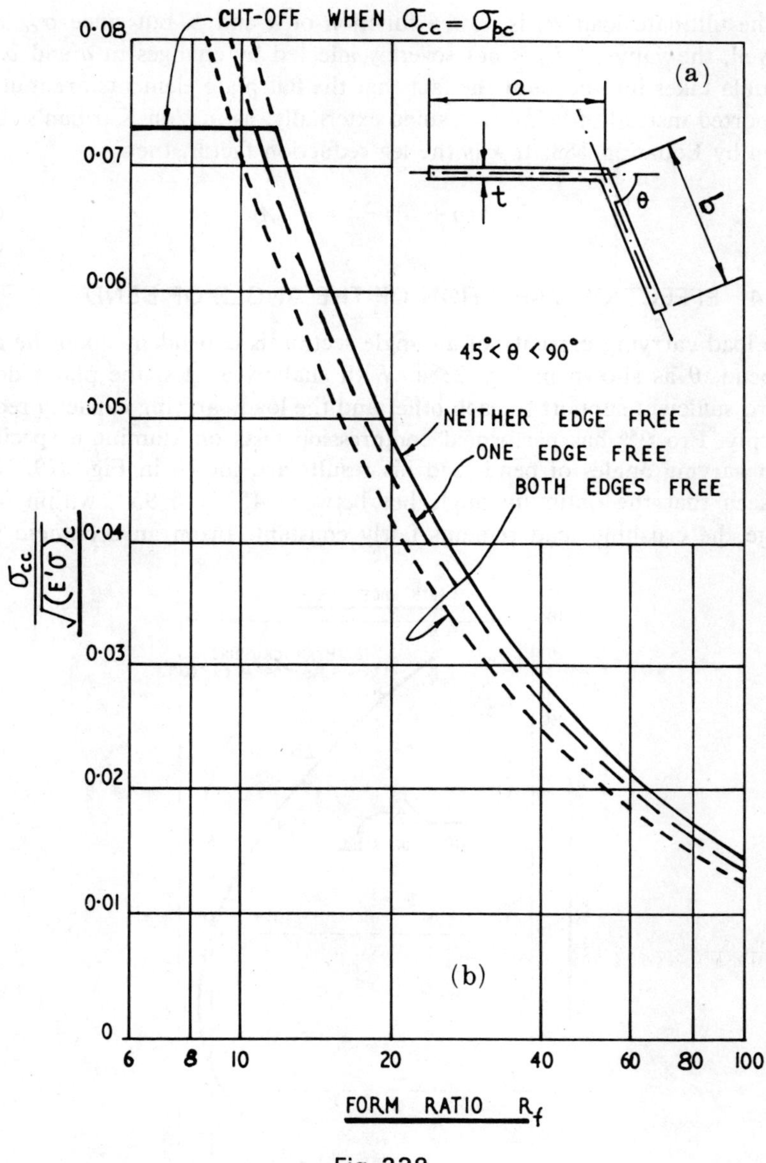
CUT-OFF WHEN $\sigma_{cc} = \sigma_{pc}$
(a)
a
t
θ
σ
45° < θ < 90°
NEITHER EDGE FREE
ONE EDGE FREE
BOTH EDGES FREE
$\frac{\sigma_{cc}}{\sqrt{(E'\sigma)_c}}$
0·08
0·07
0·06
0·05
0·04
0·03
0·02
0·01
0
6
8
10
20
40
60
80
100
(b)
FORM RATIO R_f

Fig. 238

The ultimate load P_u is now a function of a and b, but since σ_{cc} varies as well, the value of P_u is not severely affected by changes in a and b. The formula takes into account the fact that the flat plate elements are mutually supported instead of being restrained externally, as in Von Karman's theory given by Equation 183. If k is the leg reduction factor, then

$$k(a+b)t\sigma_{pc} = \sigma_{cc}A \tag{188}$$

10.4 EFFECT OF VARIATION OF THE ANGLE OF BEND

The load carrying capacity of an angle section is dependent upon the angle of bend, θ as shown in Fig. 238a. With shallow angles, the plates do not afford sufficient support to each other and the load carrying capacity reduces sharply. Frost[186] has performed compression tests on aluminum specimens with varying angles of bend, and his results are shown in Fig. 239. It can be seen that the optimum angle lies between 45° and 95°, within which range the crushing load remains fairly constant. From an economic point

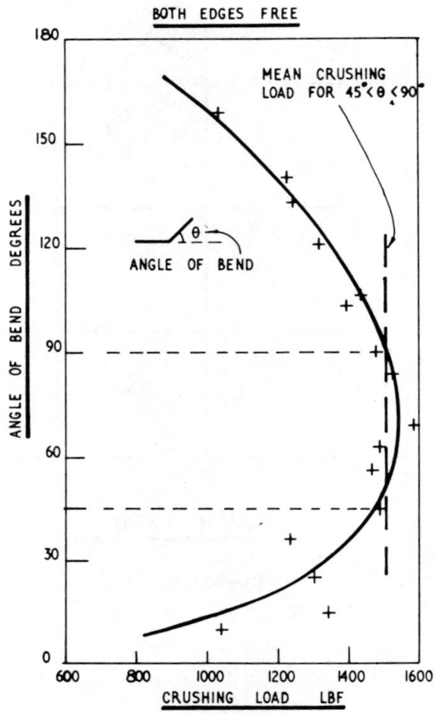

Fig. 239

of view therefore, the optimum angle is about 45°, since the shallower the angle the more area it can cover. All of the author's own work[7,145,146] has been carried out with an angle of bend of 50°.

10.5 EFFECT OF MATERIAL PROPERTIES

It was seen in Equation 182 that Von Karman used the term $(E'\sigma)_c$ in evaluating the ultimate load for the plate. For materials without a fairly well defined yield point, the value of $(E'\sigma)_c$ to be used is the maximum value of the product of the yield stress at any point and the tangent modulus at that point. In Equation 187, the value used is $E\sigma_{pc}$, which is again the same as $(E'\sigma)_c$ for materials with a sharp yield point. However, for materials without a well defined yield point, the difficulty arises that the maximum value of $(E'\sigma)_c$ can be quite different from the value $E\sigma_{pc}$. Frost[186] has found that it is possible to use $(E'\sigma)_c$ instead of $E\sigma_{pc}$ in Equation 187, provided Needham's coefficients are altered to

$C_e = 0.393$ for isolated angles
$C_e = 0.425$ for angles with only one edge free
$C_e = 0.455$ for angles with neither edge free.

The use of $(E'\sigma)_c$ instead of $E\sigma_{pc}$ permits a more general application of Equation 187 to any material such as FRP, whose compression characteristics have been previously determined. Then $\sigma_{pc} = \sigma_{ult}$. Equation 187 can be better represented by Frost's curve,[186] shown in Fig. 238. This gives the variation of $\sigma_{cc}/\sqrt{(E'\sigma)_c}$ against the form factor R_f for an angle section of any material. For a given angle section, therefore, the form factor can be determined, and the value of $\sigma_{cc}/\sqrt{(E'\sigma)_c}$ read off from the curve. Since the value of $(E'\sigma)_c$ is known for the material, σ_{cc} can be obtained, thus giving the ultimate load that the angle can withstand in crushing. The value of k can then be determined from Equation 188. The horizontal cut-off line at the top in Fig. 238, however, must be related to the particular material in question, since it is dictated by the fact that on no account can the value of σ_{cc} exceed the value of σ_{pc}. For FRP laminates made by a hand lay-up process, the maximum value of $(E'\sigma)_c \simeq 5.5 \times 10^9$ (psi)^2 (261.46×10^9 (kPa)^2) and the value of the ratio shown in Fig. 242 is approximately 0.42.

10.6 PRISMATIC FOLDED PLATES

The behavior of prismatic folded plates of single skin construction can be predicted by considering the section in compression as consisting of a number

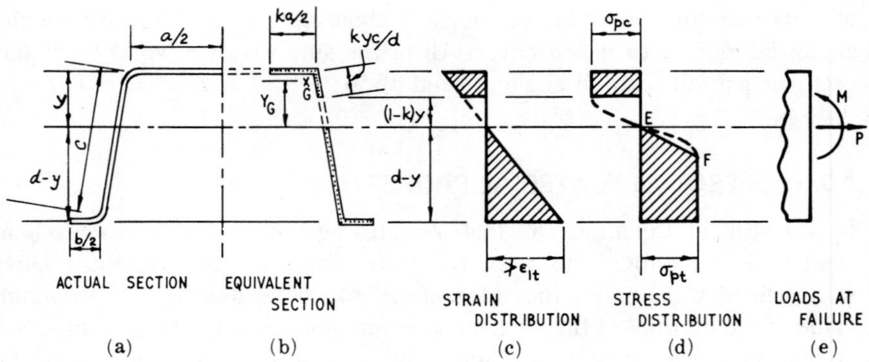

Fig. 240

of angle sections, and then applying the theory as set forth in Sections 10.3 and 10.5. The application to a trough shaped folded plate section, shown in Fig. 240, is considered here. The section has the linear dimensions *a, b, c,* and *d,* as shown in Fig. 240. Further, *r* is the radius of bend and *t* is the thickness. The section is put in bending together with an axial force, such that the portion above the assumed position of the neutral axis is in compression, whereas the part below the neutral axis is in tension. The position of the neutral axis is defined by $\bar{y}$. The portion in compression can, therefore, be idealized, as shown in Fig. 240b. The leg reduction factor is *k,* which is equal to

$$k = \frac{\sigma_{cc}}{\sigma_{pc}}\left[1 - \frac{3\left\{\tan\left(\frac{\beta}{2}\right) - \frac{\beta}{2}\right\}}{R_f}\right] \tag{189}$$

where

$$r = 3t,\ \beta = \sin^{-1}\left(\frac{d}{c}\right)$$

and where

$$R_f = \frac{\frac{a}{2} + \bar{y}\left(\frac{c}{d}\right)}{2t}$$

If the fold angles are sharp, then

$$k = \frac{\sigma_{cc}}{\sigma_{pc}}$$

The stress σ_{cc} can be obtained from Equation 184 as

$$\sigma_{cc} = \frac{0.455\sqrt{(E'\sigma)_c}}{R_f^{0.75}} \tag{190}$$

The assumption is made that the strain distribution across the section is linear, as shown in Fig. 240c. The limiting strains are ϵ_{pc} and ϵ_{lt}, where

ϵ_{pc} = strain at proof stress (or its equivalent) in compression
ϵ_{lt} = limiting strain in tension at failure.

The stress distribution across the section for the assumed position of the neutral axis is as shown in Fig. 240d. The portion of the curve EF in Fig. 240d can be assumed to be straight. From the known effective areas of the section and the stress distribution, the axial force and the bending moment on the section at failure can be determined as

$$\begin{aligned} F &= \Sigma\, \sigma \delta A \\ M &= \Sigma\, \sigma y \delta A \end{aligned} \tag{191}$$

An interaction diagram of M against F can be plotted, as shown by the typical plot in Fig. 241. This diagram then yields the various combinations of ultimate load that the section can withstand. The method is not applicable to very shallow angles (less than 20°), where the fold may be subject to snap-through buckling.

Frost[186] has determined design curves for Vee and top hat sections in bending, which are shown in Figs. 242 and 243. Figure 242 shows the ratio of the ultimate bending moment on the Vee section to the fully plastic moment of the section, plotted against the L/t ratio. It shows the plot for different values of $(E'\sigma)_c/E\sigma_{pc}$, which makes it very useful for application to a variety of materials. Figure 243 shows the same plot for top hat sections, which are rectangular, when the b/d ratio is equal to 1, and for different values of the a/d ratio. The curves have been determined for $(E'\sigma)_c/E\sigma_{pc} = 0.446$ and are, hence, strictly valid only for that value. However, they are safe for all values of $(E'\sigma)_c/E\sigma_{pc} > 0.446$.

The method suggested above can be used effectively for prismatic folded plates when it is necessary to determine the ultimate load carrying capacity of the section. However, situations arise where the prismatic folded plate

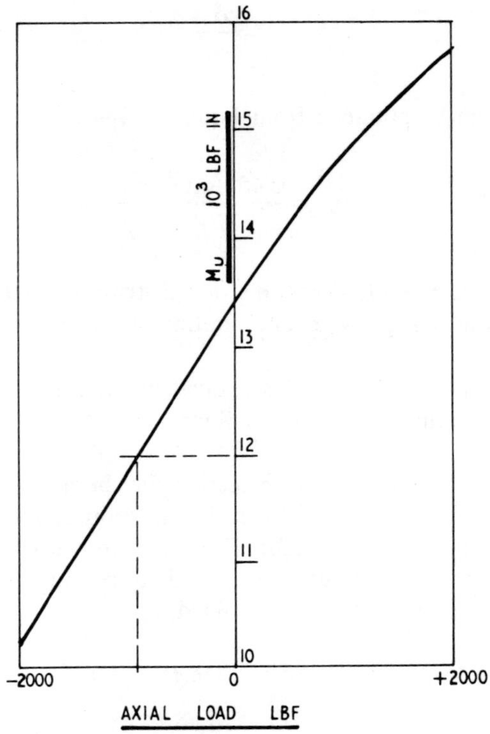

Fig. 241

forms part of a larger structure, such as a folded plate rigid frame. It may also be necessary to determine the behavior of the structure at working loads. In such cases, if the very reasonable assumption is made that the plates buckle at loads lower than the working loads, the idealized effective section (or widths) can easily be defined by determining the value of *k*, the leg reduction factor. The structure then reduces to discrete elements, and the cross-sectional areas and moments of inertia of the individual elements can easily be determined. These can then be used as data for a skeleton analysis as mentioned in Section 10.1. Such an analysis using IBM's STRUDL program is shown in Section 10.8.

10.7 NONPRISMATIC FOLDED PLATES

The behavior of nonprismatic folded plate structures in bending and axial compression, when susceptible to plate buckling, is very difficult to determine.

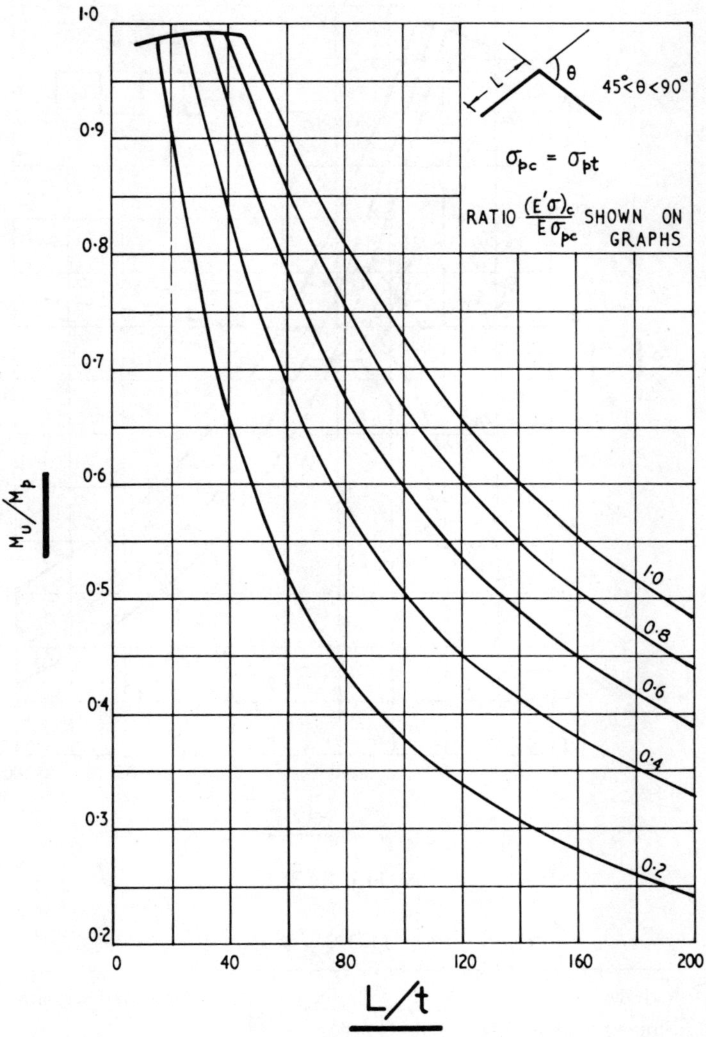

Fig. 242

The determination of the effective plate widths, under various loading conditions, is a very complex problem. The author has carried out some experimental work on the subject, and the results[7,138] are described here.

The structure tested was a model, folded plate, barrel vault of the type shown in Fig. 160. The model was formed from sheets of polymethylmethacrylate, and the details of the model are given in Table 70. The theoretical

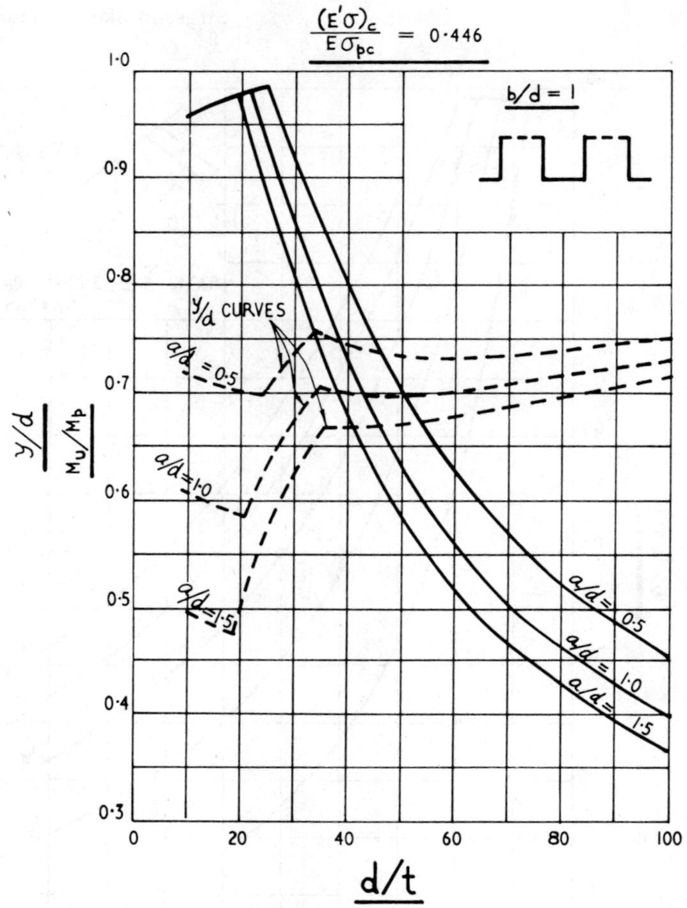

Fig. 243

Table 70

Material:	Polymethylmethacrylate
Number of arch ribs:	6
Shape:	Semi-circular on the neutral axis
Maximum span of the barrel:	94.1250 in.
Maximum height of the barrel:	47.0625 in.
Neutral axis span of the barrel:	90.4918 in.
Length of each unit:	36 in.
Plate thickness:	0.0625 in.
Fold angle, θ:	25°
Angle of unit, α:	11° 15′
m:	8
ν:	0.35

analysis of the model under various loading conditions was performed on the basis of two assumptions:

1. The overall section of the plates was effective.
2. An idealized section with a plate width of $16t$ (where t is the thickness of the plate) was effective on either side of the ridge and valley lines.

The theoretical results under the two assumptions, and the experimental results, are compared in Figs. 244–249. In all cases the model was tested at $T = 20°C$ (68°F) and $t = 1$ hour. The theoretical and experimental arch deflections for three loading cases are shown in Figs. 244–246. It can be seen that plate buckling causes the arch to deflect much more than is theoretically predicted by the use of assumption 1. The use of assumption 2, that a reduced section is effective, leads to better correlation, particularly in Figs. 245 and 246. The longitudinal stresses at various sections, such as $\phi = 0$, $\phi = \pi/4$, and so on, for the different loading cases, are shown in Figs. 247–249. Here again, the very simple assumption that only a small fixed width of plate is effective over the entire structure leads to a fairly accurate estimation of the behavior of the structure. It can also be seen that since the effective plate width is generally small, the nonprismatic nature of the plates is not very important. The effective width stays constant along the fold, irrespective of the shape of the fold, until it becomes less than the actual available width, at which point the average stress increases. This has also been found by Gilkie[148,149] in his tests on FRP pyramids of the type shown in Fig. 154c. He found that, except for small values of load, the effective strut area, assumed to be constant along the ridge, was fairly constant. Though the central part of the plate is not effective in taking load, it must be realized that it cannot actually be removed, since it serves to stabilize the fold and helps to prevent torsional buckling.

The author[145] has also carried out several tests on a folded plate conical dome under vertical loading at the apex. Some details of the model structure in acrylic are shown in Fig. 268. The stresses along the ridge and valley lines are shown in Figs. 250a and 250b. The ridge line of the structure was unsupported at the point B shown in Fig. 250. The experimental work shows clearly, once again, that the shape of the plate does not materially affect the effective width of the plate, unless the effective width is less than the actual available width. In Figs. 250a and 250b, it is quite clear that, neglecting edge effects at B, the effective width stays constant until about point C, after which the stresses increase very rapidly towards the crown.

10.8 STIFFNESS METHOD OF ANALYSIS

This method of analysis, also termed the "equilibrium" method[178,183,187,188] is a matrix method of structural analysis. Consider a linear skeletal structure

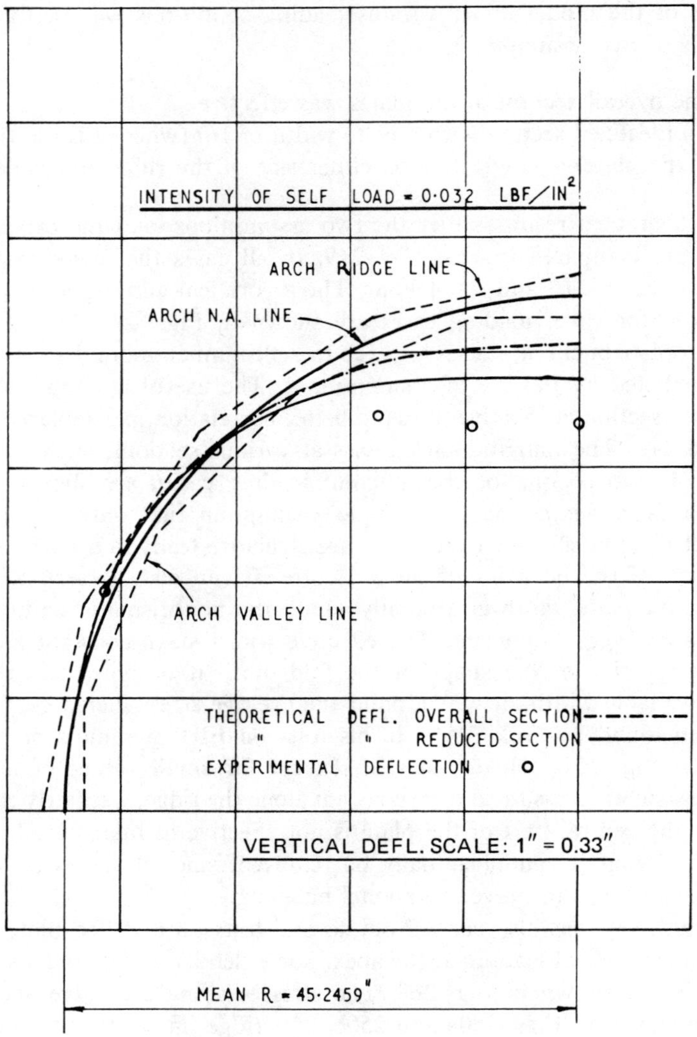

Fig. 244

under a general system of loading. The actual loading system is then replaced by an equivalent loading system, which is applied only at the joints and produces joint displacements identical to those of the actual loading system. The components of these external loads at any joint form the elements of a column vector. All the joints in the structure are numbered, thereby also identifying each member incidence by the numbers of its terminal joints.

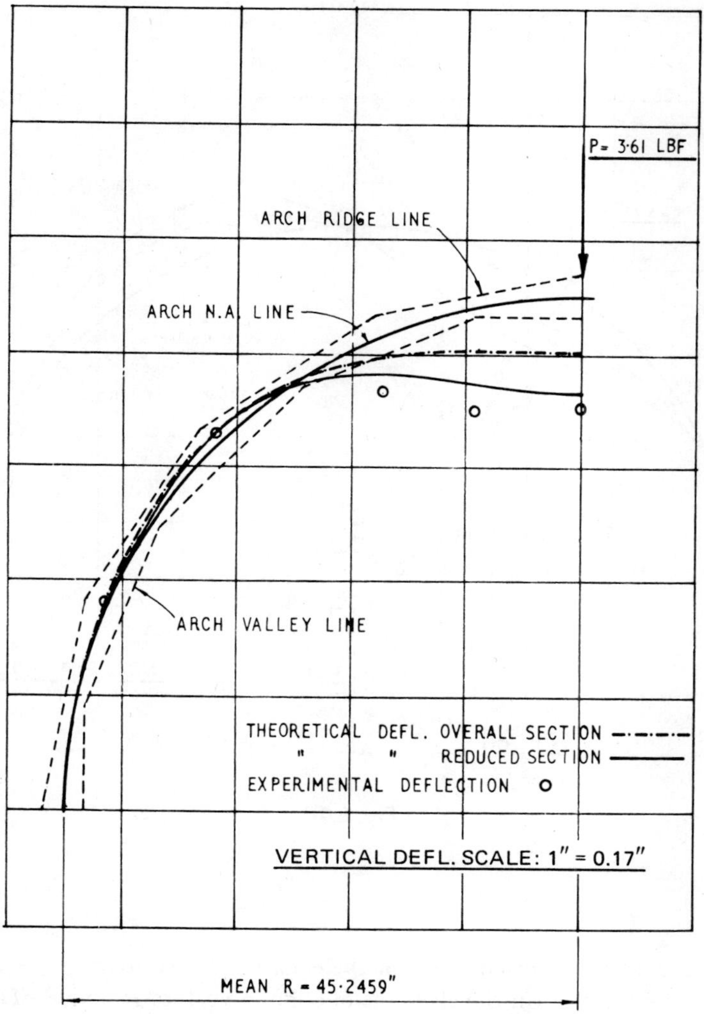

Fig. 245

With each member is now associated a cartesian coordinate system *x, y, z,* called the member coordinate system and shown in Fig. 251a. The axes of this coordinate system are chosen so that the *x* axis passes through the centroid of the cross section and the *y* and *z* axes lie along the principal axes of inertia of the cross section. With the structure as a whole is also associated a frame coordinate system of axes *X, Y, Z,* shown in Fig. 193a.

Consider a member *ij* lying between joints *i* and *j* of the structure. The

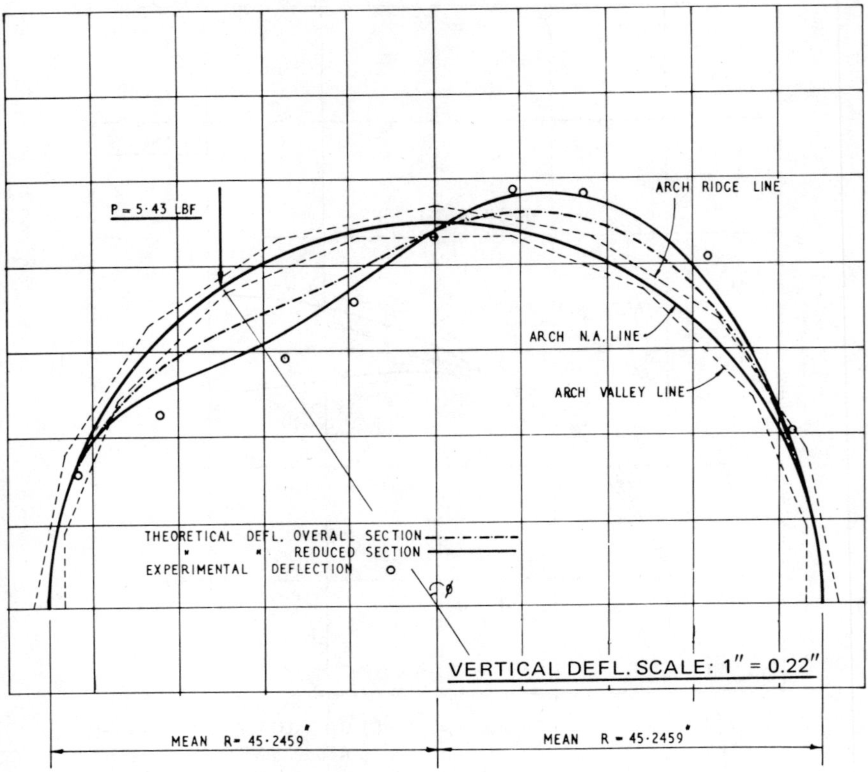

Fig. 246

components of the internal forces at the ends *i* and *j*, relative to the member coordinate system, form the elements of column vectors P_i and P_j. The term "force" is used in the general sense to denote both forces and moments. In the most general case, P would be equivalent to

$$P = \begin{bmatrix} P_x \\ P_y \\ P_z \\ M_x \\ M_y \\ M_z \end{bmatrix} \tag{192}$$

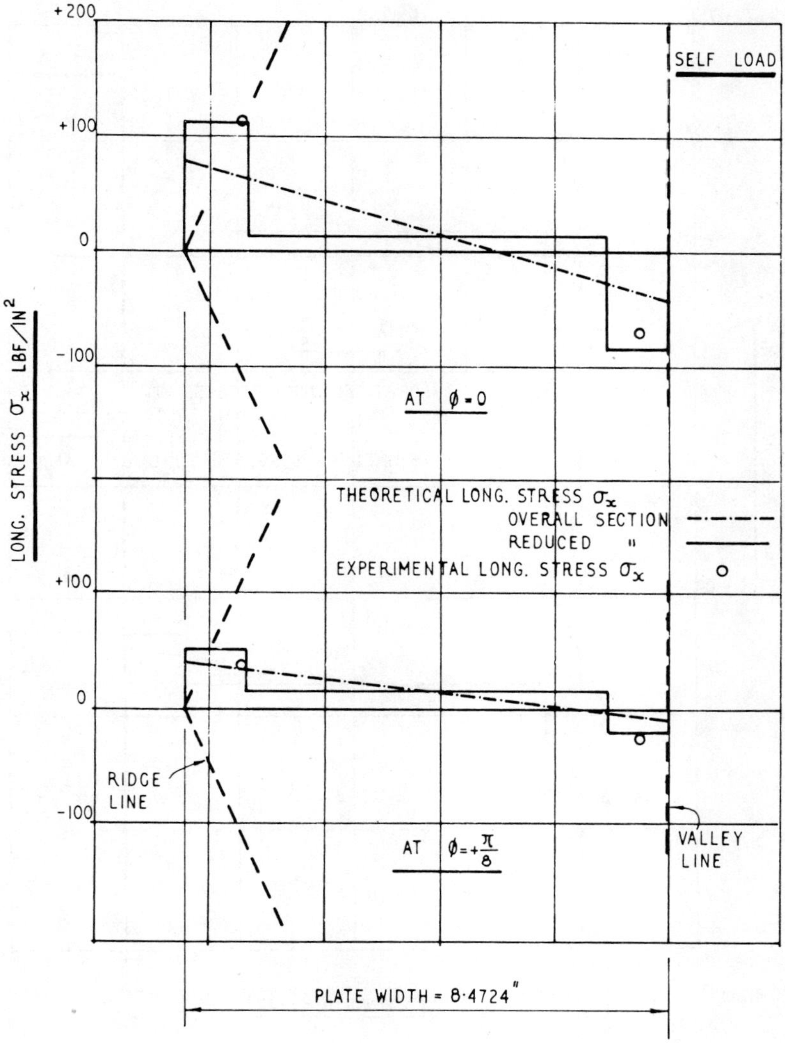

Fig. 247

The components of the displacements at the ends i and j, relative to the member coordinate system, also form the elements of column vectors. Here again, the term "displacement" is used in the general sense to denote both translation and rotation. In the most general case, the end displacement vector d would be equivalent to

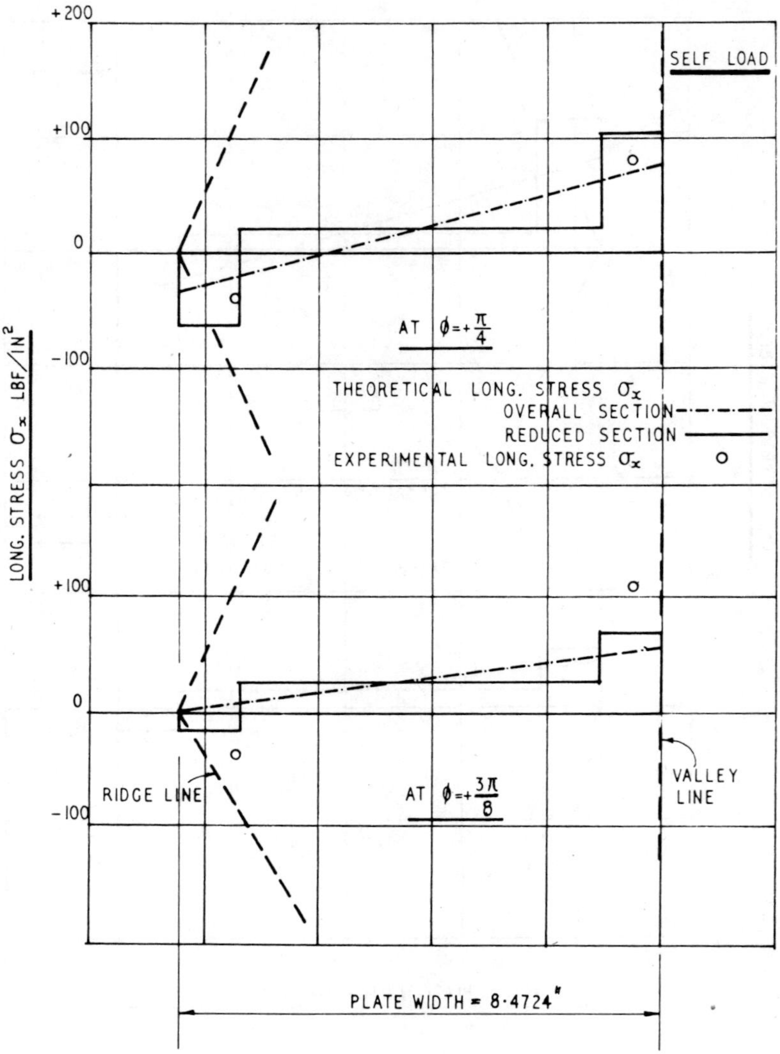
+200
+100
0
-100
+100
0
-100
SELF LOAD
LONG. STRESS σ_x LBF/IN^2
AT $\phi = +\frac{\pi}{4}$
THEORETICAL LONG. STRESS σ_x
OVERALL SECTION
REDUCED SECTION
EXPERIMENTAL LONG. STRESS σ_x
RIDGE LINE
AT $\phi = +\frac{3\pi}{8}$
VALLEY LINE
PLATE WIDTH = 8·4724"

Fig. 248

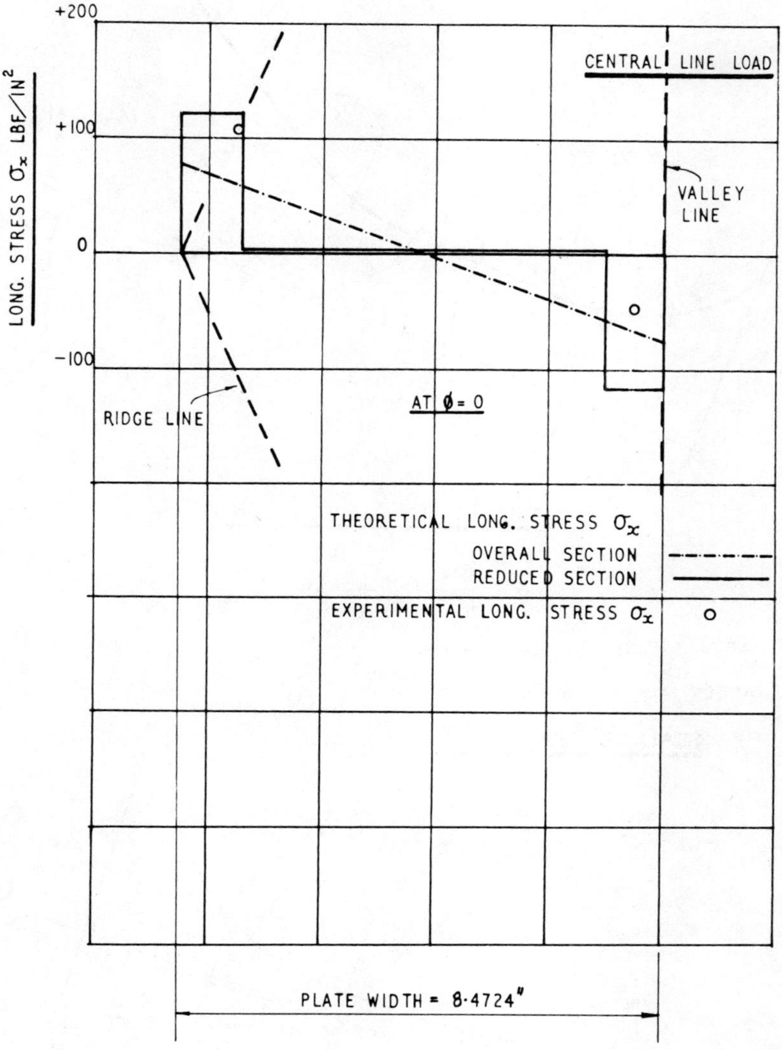

+200
+100
0
−100
LONG. STRESS σ_x LBF/IN^2
CENTRAL LINE LOAD
VALLEY LINE
RIDGE LINE
AT ϕ = 0
THEORETICAL LONG. STRESS σ_x
OVERALL SECTION
REDUCED SECTION
EXPERIMENTAL LONG. STRESS σ_x
PLATE WIDTH = 8·4724"

Fig. 249

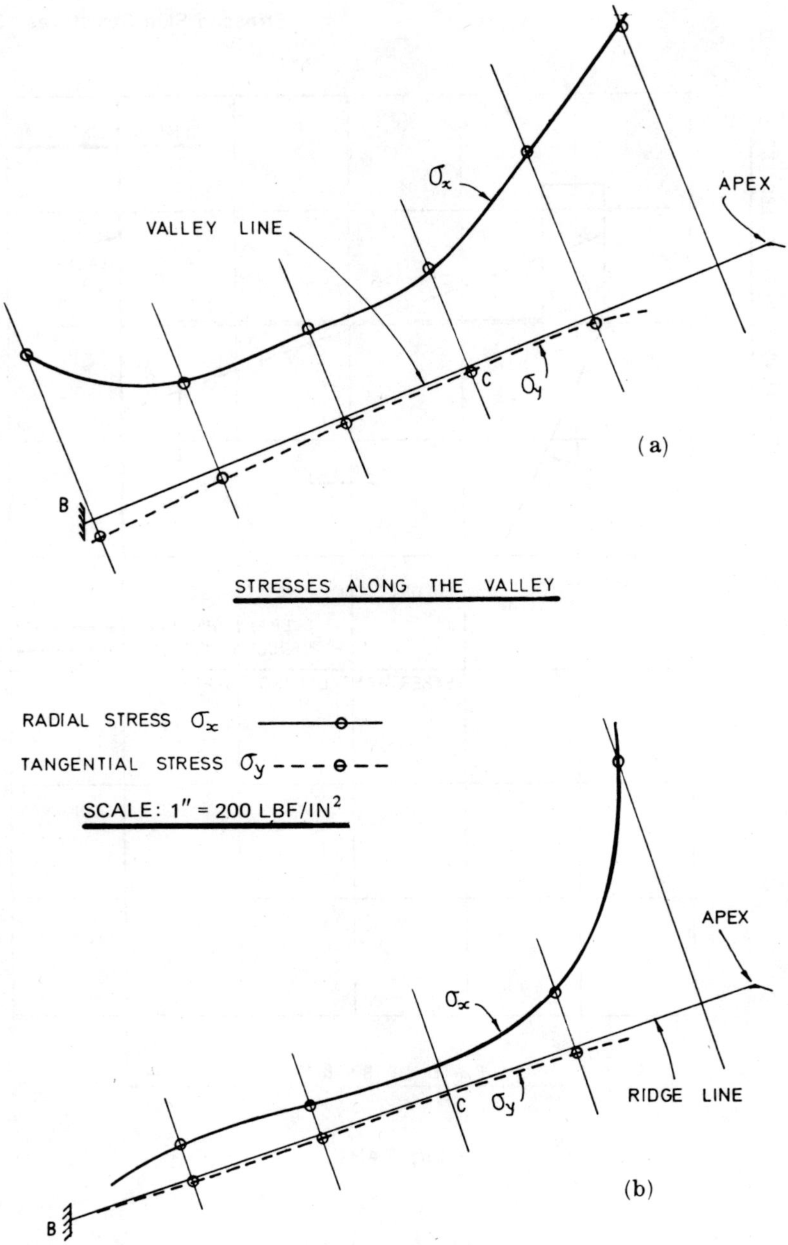

STRESSES UNDER A LOAD W = 200 LBF AT THE APEX

Fig. 250

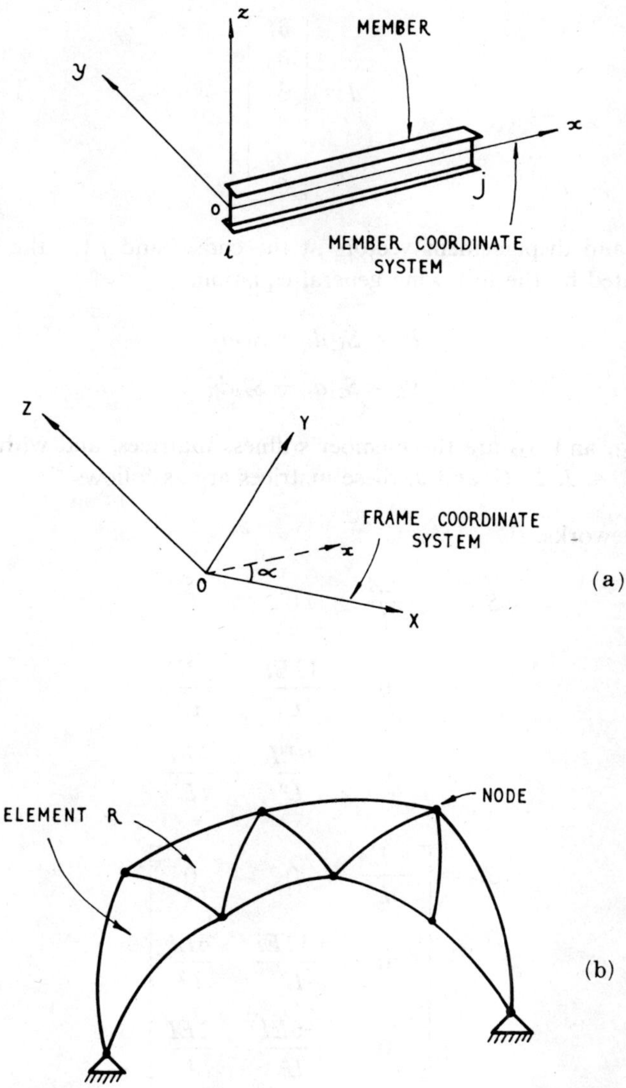

z
MEMBER
y
x
o
j
i
MEMBER COORDINATE
SYSTEM
Z
Y
FRAME COORDINATE
SYSTEM
x
α
O
X
(a)
ELEMENT R
NODE
(b)

Fig. 251

$$d = \begin{bmatrix} \delta_x \\ \delta_y \\ \delta_z \\ \theta_x \\ \theta_y \\ \theta_z \end{bmatrix} \tag{193}$$

The force and displacement vectors at the ends i and j for the member ij can be related by the following general equation:

$$\begin{aligned} P_{ij} &= S_{11} d_{ij} + S_{12} d_{ji} \\ P_{ji} &= S_{21} d_{ij} + S_{22} d_{ji} \end{aligned} \tag{194}$$

S_{11}, S_{12}, S_{21}, and S_{22} are the member stiffness matrices, and with the usual notation E, A, I, L, G, and J, these matrices are as follows[187]:

Plane frameworks:

$$S_{11} = \begin{matrix} \frac{EA}{L} & 0 & 0 \\ 0 & \frac{12EI}{L^3} & \frac{6EI}{L^2} \\ 0 & \frac{6EI}{L^2} & \frac{4EI}{L} \end{matrix} \tag{195}$$

$$S_{12} = \begin{bmatrix} \frac{-EA}{L} & 0 & 0 \\ 0 & \frac{-12EI}{L^3} & \frac{6EI}{L^2} \\ 0 & \frac{-6EI}{L^2} & \frac{2EI}{L} \end{bmatrix} \tag{196}$$

$$S_{21} = \begin{bmatrix} \frac{-EA}{L} & 0 & 0 \\ 0 & \frac{-12EI}{L^3} & \frac{-6EI}{L^2} \\ 0 & \frac{6EI}{L^2} & \frac{2EI}{L} \end{bmatrix} \tag{197}$$

$$S_{22} = \begin{bmatrix} \frac{EA}{L} & 0 & 0 \\ 0 & \frac{12EI}{L^3} & \frac{-6EI}{L^2} \\ 0 & \frac{-6EI}{L^2} & \frac{4EI}{L} \end{bmatrix} \tag{198}$$

Grid frameworks:

$$S_{11} = \begin{bmatrix} \frac{GJ}{L} & 0 & 0 \\ 0 & \frac{4EI}{L} & \frac{6EI}{L^2} \\ 0 & \frac{6EI}{L^2} & \frac{12EI}{L^3} \end{bmatrix} \tag{199}$$

$$S_{12} = \begin{bmatrix} \frac{-GJ}{L} & 0 & 0 \\ 0 & \frac{2EI}{L} & \frac{-6EI}{L^2} \\ 0 & \frac{6EI}{L^2} & \frac{-12EI}{L^3} \end{bmatrix} \tag{200}$$

$$S_{21} = \begin{bmatrix} \frac{-GJ}{L} & 0 & 0 \\ 0 & \frac{2EI}{L} & \frac{6EI}{L^2} \\ 0 & \frac{-6EI}{L^2} & \frac{-12EI}{L^3} \end{bmatrix} \tag{201}$$

$$S_{22} = \begin{bmatrix} \dfrac{GJ}{L} & 0 & 0 \\ 0 & \dfrac{4EI}{L} & \dfrac{-6EI}{L^2} \\ 0 & \dfrac{-6EI}{L^2} & \dfrac{12EI}{L^3} \end{bmatrix} \tag{202}$$

Space frameworks:

$$S_{11} = \begin{bmatrix} \dfrac{EA}{L} & 0 & 0 & 0 & 0 & 0 \\ 0 & \dfrac{12EI_z}{L^3} & 0 & 0 & 0 & \dfrac{6EI_z}{L^2} \\ 0 & 0 & \dfrac{12EI_y}{L^3} & 0 & \dfrac{-6EI_y}{L^2} & 0 \\ 0 & 0 & 0 & \dfrac{GJ}{L} & 0 & 0 \\ 0 & 0 & \dfrac{-EI_y}{L^2} & 0 & \dfrac{4EI_y}{L} & 0 \\ 0 & \dfrac{6EI_z}{L^2} & 0 & 0 & 0 & \dfrac{4EI_z}{L} \end{bmatrix} \tag{203}$$

$$S_{12} = \begin{bmatrix} \dfrac{-EA}{L} & 0 & 0 & 0 & 0 & 0 \\ 0 & \dfrac{-12EI_z}{L^3} & 0 & 0 & 0 & \dfrac{6EI_z}{L^2} \\ 0 & 0 & \dfrac{-12EI_y}{L^3} & 0 & \dfrac{-6EI_y}{L^2} & 0 \\ 0 & 0 & 0 & \dfrac{-GJ}{L} & 0 & 0 \\ 0 & 0 & \dfrac{6EI_y}{L^2} & 0 & \dfrac{2EI_y}{L} & 0 \\ 0 & \dfrac{-6EI_z}{L^2} & 0 & 0 & 0 & \dfrac{2EI_z}{L} \end{bmatrix} \tag{204}$$

$$S_{21} = \begin{bmatrix} \frac{-EA}{L} & 0 & 0 & 0 & 0 & 0 \\ 0 & \frac{-12EI_z}{L^3} & 0 & 0 & 0 & \frac{-6EI_z}{L^2} \\ 0 & 0 & \frac{-12EI_y}{L^3} & 0 & \frac{6EI_y}{L^2} & 0 \\ 0 & 0 & 0 & \frac{-GJ}{L} & 0 & 0 \\ 0 & 0 & \frac{-6EI_y}{L^2} & 0 & \frac{2EI_y}{L} & 0 \\ 0 & \frac{6EI_z}{L^2} & 0 & 0 & 0 & \frac{2EI_z}{L} \end{bmatrix} \tag{205}$$

$$S_{22} = \begin{bmatrix} \frac{EA}{L} & 0 & 0 & 0 & 0 & 0 \\ 0 & \frac{12EI_z}{L^3} & 0 & 0 & 0 & \frac{-6EI_z}{L^2} \\ 0 & 0 & \frac{12EI_y}{L^3} & 0 & \frac{6EI_y}{L^2} & 0 \\ 0 & 0 & 0 & \frac{GJ}{L} & 0 & 0 \\ 0 & 0 & \frac{6EI_y}{L^2} & 0 & \frac{4EI_y}{L} & 0 \\ 0 & \frac{-6EI_z}{L^2} & 0 & 0 & 0 & \frac{4EI_z}{L} \end{bmatrix} \tag{206}$$

Equations 194 for the member ij are related to the member coordinate system. However, to consider the structure as a whole, it is necessary to convert these equations to the frame coordinate system by the use of the transformation matrix T. Let the two systems of axes x, y, z and X, Y, Z be related by

$$\begin{bmatrix} x \\ y \\ z \end{bmatrix} = [T] \begin{bmatrix} X \\ Y \\ Z \end{bmatrix} \tag{207}$$

Then Equation 194 can be written as

$$
\begin{aligned}
P'_{ij} &= T^{-1}S_{11}Td'_{ij} + T^{-1}S_{12}Td'_{ji} \\
P'_{ji} &= T^{-1}S_{21}Td'_{ij} + T^{-1}S_{22}Td'_{ji}
\end{aligned}
\tag{208}
$$

where P'_{ij}, P'_{ji}, d'_{ij}, and d'_{ji} are column vectors representing P_{ij}, P_{ji}, d_{ij}, and d_{ji} relative to the frame coordinate system.

For two-dimensional frameworks, the transformation matrix T can be obtained as

$$
T = \begin{bmatrix} \cos\alpha & \sin\alpha & 0 \\ -\sin\alpha & \cos\alpha & 0 \\ 0 & 0 & 1 \end{bmatrix}
\tag{209}
$$

where α is the angle between the two sets of axes as in Fig. 251a.

For space frameworks, the member coordinate system and the frame coordinate system can be related by

$$
\begin{bmatrix} x \\ y \\ z \end{bmatrix} = [T_\gamma]\,[T_\alpha]\,[T_\beta] \begin{bmatrix} X \\ y \\ Z \end{bmatrix}
\tag{210}
$$

where

$$
T_\alpha = \begin{bmatrix} \cos\alpha & 0 & \sin\alpha \\ 0 & 1 & 0 \\ -\sin\alpha & 0 & \cos\alpha \end{bmatrix}
\tag{211}
$$

$$
T_\beta = \begin{bmatrix} \cos\beta & \sin\beta & 0 \\ -\sin\beta & \cos\beta & 0 \\ 0 & 0 & 1 \end{bmatrix}
\tag{212}
$$

$$
T_\gamma = \begin{bmatrix} 1 & 0 & 0 \\ 0 & \cos\gamma & \sin\gamma \\ 0 & -\sin\gamma & \cos\gamma \end{bmatrix}
\tag{213}
$$

The conditions of compatibility at joint i require that all the member ends connected to the joint have the same components of displacement. This also applies to joint j and Equation 208 can be written as

$$\begin{aligned} P'_{ij} &= T^{-1}S_{11}Td'_i + T^{-1}S_{12}Td'_j \\ P'_{ji} &= T^{-1}S_{21}Td'_i + T^{-1}S_{22}Td'_j \end{aligned} \tag{214}$$

where d'_i and d'_j are column vectors defining the displacements of joints i and j relative to the frame coordinate system.

Let the components of the external load at the joint i, relative to the frame coordinate system, form the elements of a column vector W'_i. Then conditions of equilibrium at the joint i require that

$$\Sigma P'_{ij} = W'_i \tag{215}$$

where the summation extends over all the members whose ends are connected to i. Equations similar to Equation 215 can be written for all the active joints of the structure, leading to the relationship

$$[K]\ [d] = [W] \tag{216}$$

where K is the stiffness matrix for the structure. The solution of Equation 216 gives the joint displacements for the structure, and the substitution of these displacements into the force-displacement relationships for the individual members determines the internal forces on the members.

The stiffness method of analysis can be applied to virtually any form of skeletal plane or space structure, with pin-connected or rigid joints. Several authors[189-191] have considered the application of the method to specific structures such as braced domes, barrel vaults, and the like. The application of the method to structures with partially constrained joints has been considered by Nooshin.[188]

10.9 FINITE ELEMENTS METHOD

The stiffness method as applied to skeletal structures can be extended to structures that consist of interconnected elements (instead of straight bars), as shown in Fig. 251b. In such cases, it is assumed that the elements of type R are interconnected only at the nodes and, if the nodal displacements are known, the behavior of the element can be uniquely determined. This approach can be extended still further to continuum structures such as folded plates, shells, and the like. This approach, termed the finite elements method,[182,192-194] reduces the continuum to a number of discrete and finite elements that idealize the structure. These elements are assumed to be interconnected only at nodal points and not along continuous boundaries. The structure can then be analyzed without difficulty.

Consider a continuum structure and imagine its surface to be divided by hypothetical lines (or surfaces) into a number of finite elements that are connected only at the nodes. The elements can theoretically be of any shape, but regular shapes such as triangles or rectangles are much more convenient. It is now necessary to define the behavior of the element in terms of its nodal displacements. This defines the complete state of stress and strain within the element itself, as well as the forces along the boundaries in terms of the nodal displacements. This is equivalent to defining relationships similar to Equations 194 in Section 10.8. The boundary forces are then replaced by equilibrating forces applied only at the nodes. The structure has now been reduced to a discrete structure, and can be analyzed by applying Equation 216.

It can be seen that some approximations have been introduced in the analysis. The accuracy of the method depends to a large extent on the net chosen. A fine mesh with a large number of elements would lead to better results than a coarse mesh. Further, it is not always possible to ensure that continuity between adjacent elements is rigidly maintained. Finally, since the forces along the boundaries are replaced by equilibrating forces at the nodes, it is possible to satisfy equilibrium only in a general overall sense. This means that local violations can occur, both within the element itself as well as on the boundaries.

The finite elements method is a very powerful tool for the analysis of complex continuum structures, even in three dimensions.

10.10 STRUDL CODING AND ANALYSIS FOR A FOLDED PLATE, CONICAL, FRP DOME STRUCTURE IN SINGLE SKIN CONSTRUCTION

The stiffness method of analysis was explained in Section 10.8. It is not, however, necessary to write computer programs for each structure to be analyzed. Several excellent, commercially available programs can be used to analyze the structure. The program that the author uses was developed by MIT for IBM and is called STRUDL.[195] STRUDL stands for *Stru*ctural *D*esign *L*anguage and comes in two parts, STRUDL 1 and STRUDL 2. STRUDL 1 is for the analysis of skeletal structures and STRUDL 2 is for the analysis of concrete shells, shear walls, arch dams, and the like.

In the case of plastics construction in single skin structures, the very low modulus of elasticity, and the thin skin, causes buckling to occur at very low loads, the load then being carried by the rigid and unbuckled corners or edges, thereby converting the structure to a skeletal system, as was described in Section 10.1. The analysis of the structure can then be carried out by the use of STRUDL 1, provided it is properly defined and then coded for the program. Further, the results obtained from the program must then be properly interpreted to yield the forces and moments over the structure.

The coding of the structure for STRUDL is detailed in the STRUDL Coding Manual.[195] This section is no substitute for the reading and study of that manual, as it could hardly be discussed in its entirety here. Nevertheless, a basic, elementary knowledge of STRUDL Coding can be easily appreciated and the simplicity of the approach easily made apparent.

The folded plate conical dome structure to be analyzed by STRUDL 1 is shown in Fig. 252. The structure was developed by the author for use as a roof in low-cost housing for underdeveloped countries, and this aspect of the structure is discussed in greater detail in Section 11.4. The structure consists of three interconnected units, as shown in Fig. 252. The first step

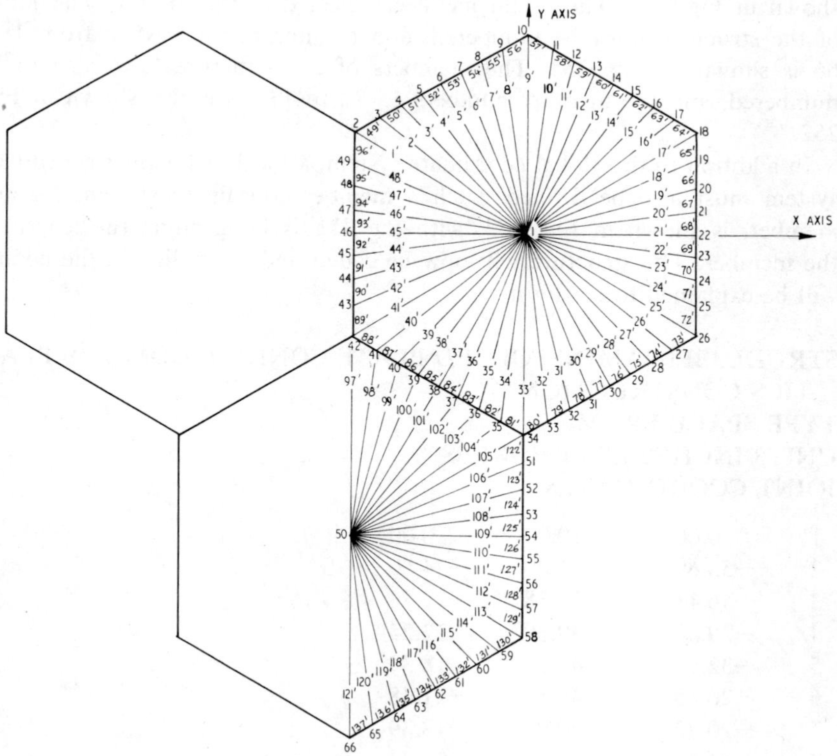

Fig. 252

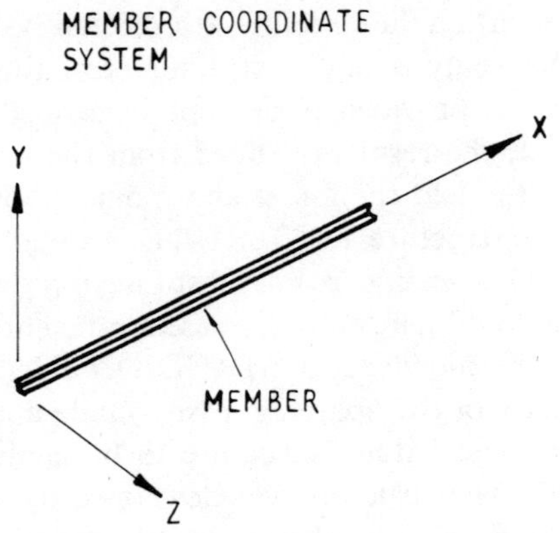

Fig. 253

in the Coding is to select an overall or global coordinate system of axes, as shown in Fig. 252. The origin has been selected at the joint 1. The joints of the structure must be numbered, and the numbering system from 1 to 66 is shown in Fig. 252. The members of the structure have also to be numbered, and the member numbers 1′, 2′, to 137′ are also shown in Fig. 252.

In addition to the global coordinate system, a local or member coordinate system must also be defined. Such a member coordinate system, for any member, is shown in Fig. 253, with the *X* axis lying along the length of the member. The structure will now be coded and every line in the coding will be explained to clarify it:

STRUDL 'BENJAMIN' FOLDED PLATE CONICAL DOME OF PLASTICS CONSTRUCTION
TYPE SPACE FRAME
UNITS INCHES LBS DEGREES
JOINT COORDINATES

1	0.00	0.00	0.00	
2	−53.69	31.00	−24.80	S
3	−46.11	35.38	−19.74	
4	−39.30	39.30	−22.24	
5	−32.97	42.97	−18.39	
6	−26.85	46.50	−21.48	
7	−20.72	50.03	−18.39	
8	−14.39	53.69	−22.24	
9	−7.59	57.61	−19.74	

10	0.00	62.00	−24.80	
11	7.59	57.61	−19.74	
12	14.39	53.69	−22.24	
13	20.72	50.03	−18.39	
14	26.85	46.50	−21.48	
15	32.97	42.97	−18.39	
16	39.30	39.30	−22.24	
17	46.11	35.38	−19.74	
18	53.69	31.00	−24.80	S
19	53.69	22.24	−19.74	
20	53.69	14.39	−22.24	
21	53.69	7.07	−18.39	
22	53.69	0.00	−21.48	
23	53.69	−7.07	−18.39	
24	53.69	−14.39	−22.24	
25	53.69	−22.24	−19.74	
26	53.69	−31.00	−24.80	
27	46.11	−35.38	−19.74	
28	39.30	−39.30	−22.24	
29	32.97	−42.97	−18.39	
30	26.85	−46.50	−21.48	
31	20.72	−50.03	−18.39	
32	14.39	−53.69	−22.24	
33	7.59	−57.61	−19.74	
34	0.00	−62.00	−24.80	S
35	−7.59	−57.61	−19.74	
36	−14.39	−53.69	−22.24	
37	−20.72	−50.03	−18.39	
38	−26.85	−46.50	−21.48	
39	−32.97	−42.97	−18.39	
40	−39.30	−39.30	−22.24	
41	−46.11	−35.38	−19.74	
42	−53.69	−31.00	−24.80	S
43	−53.69	−22.24	−19.74	S
44	−53.69	−14.39	−22.24	S
45	−53.69	−7.07	−18.39	S
46	−53.69	0.00	−21.48	S
47	−53.69	0.00	−21.48	S
48	−53.69	14.39	−22.24	S
49	−53.69	22.24	−19.74	S
50	−53.69	−93.00	0.00	S
51	0.00	−70.76	−19.74	

52	0.00	−78.61	−22.24	
53	0.00	−85.93	−18.39	
54	0.00	−93.00	−21.48	
55	0.00	−100.07	−18.39	
56	0.00	−107.39	−22.24	
57	0.00	−115.24	−19.74	
58	0.00	−124.00	−24.80	
59	−7.59	−128.38	−19.74	
60	−14.39	−132.30	−22.24	
61	−20.72	−135.97	−18.39	
62	−26.85	−139.50	−21.48	
63	−32.97	−143.03	−18.39	
64	−39.30	−146.69	−22.24	
65	−46.11	−150.61	−19.74	
66	−53.69	−155.00	−24.80	S

JOINT RELEASES FORCE *Z* MOMENT *X*
42 TO 50
43 TO 50 FORCE *Y*
MEMBER INCIDENCES

1	1	2
2	1	3
3	1	4
4	1	5
5	1	6
6	1	7
7	1	8
8	1	9
9	1	10
10	1	11
11	1	12
12	1	13
13	1	14
14	1	15
15	1	16
16	1	17
17	1	18
18	1	19
19	1	20
20	1	21
21	1	22

22	1	23
23	1	24
24	1	25
25	1	26
26	1	27
27	1	28
28	1	29
29	1	30
30	1	31
31	1	32
32	1	33
33	1	34
34	1	35
35	1	36
36	1	37
37	1	38
38	1	39
39	1	40
40	1	41
41	1	42
42	1	43
43	1	44
44	1	45
45	1	46
46	1	47
47	1	48
48	1	49
49	2	3
50	3	4
51	4	5
52	5	6
53	6	7
54	7	8
55	8	9
56	9	10
57	10	11
58	11	12
59	12	13
60	13	14
61	14	15
62	15	16
63	16	17

64	17	18
65	18	19
66	19	20
67	20	21
68	21	22
69	22	23
70	23	24
71	24	25
72	25	26
73	26	27
74	27	28
75	28	29
76	29	30
77	30	31
78	31	32
79	32	33
80	33	34
81	34	35
82	35	36
83	36	37
84	37	38
85	38	39
86	39	40
87	40	41
88	41	42
89	42	43
90	43	44
91	44	45
92	45	46
93	46	47
94	47	48
95	48	49
96	49	2
97	50	42
98	50	41
99	50	40
100	50	39
101	50	38
102	50	37
103	50	36
104	50	35
105	50	34

106	50	51
107	50	52
108	50	53
109	50	54
110	50	55
111	50	56
112	50	57
113	50	58
114	50	59
115	50	60
116	50	61
117	50	62
118	50	63
119	50	64
120	50	65
121	50	66
122	34	51
123	51	52
124	52	53
125	53	54
126	54	55
127	55	56
128	56	57
129	57	58
130	58	59
131	59	60
132	60	61
133	61	62
134	62	63
135	63	64
136	64	65
137	65	66

MEMBER PROPERTIES PRISMATIC
1 TO 48, 98 TO 120 AX 2.84E-1 IX 8.31E-3 IY 1.83E-1 IZ 7.49E-3
49 TO 80, 122 TO 137 AX 14.32 IX 2.03E-2 IY 89.3 IZ 13.15
97, 121 AX 1.42E-1 IX 4.15E-3 IY 0.92E-1 IZ 3.74E-3
81 TO 88 AX 14.32 IX 2.03E-2 IY 89.3 IZ 13.15
89 TO 96 AX 7.16 IX 1.01E-2 IY 44.7 IZ 6.58
CONSTANTS
E 500000.0 ALL
G 200000.0 ALL

```
BETA 0.0 ALL
LOADING 1 'INCIDENTAL LIVE LOAD'
JOINT LOADS
42 FORCE Z  -200.00
LOADING 2 'APEX LOADS'
JOINT LOADS
1  FORCE Z  -100.0
50 FORCE Z  -50.0
LOADING 3 'SNOW LOAD'
MEMBER LOADS
1 TO 48 FORCE Z GLOBAL LINEAR FR WA 0.0  WB -1.21
  LA 0.0  LB 1.0
97, 121 FORCE Z GLOBAL LINEAR FR WA 0.0  WB -0.60
  LA 0.0  LB 1.0
98 TO 120 FORCE Z GLOBAL LINEAR FR WA 0.0  WB -1.21
  LA 0.0  LB 1.0
LOADING LIST ALL
STIFFNESS ANALYSIS
LIST FORCES LOADS REACTIONS DISPLACEMENTS

DELETIONS
MEMBER PROPERTIES PRISMATIC
1 TO 48, 98 TO 120, 97, 121
ADDITIONS
MEMBER PROPERTIES VARIABLE
1 TO 48, 98 TO 120
SEG 1  AX 1.37E-1  IX 4.02E-3  IY 2.09E-2  IZ 8.40E-4  L 20.46
SEG 2  AX 4.10E-1  IX 1.21E-2  IY 5.64E-1  IZ 2.27E-2  L 20.46
SEG 3  AX 6.83E-1  IX 2.01E-2  IY 2.61  IZ 1.05E-1
97, 121
SEG 1  AX 0.68E-1  IX 2.01E-3  IY 1.04E-2  IZ 4.20E-4  L 20.46
SEG 2  AX 2.05E-1  IX 0.60E-2  IY 2.82E-1  IZ 1.13E-2  L 20.46
SEG 3  AX 3.41E-1  IX 1.00E-2  IY 1.30  IZ 0.52E-1
PRINT STRUCTURAL DATA
```

Each line in the coding will now be briefly explained.

Line 1: This gives the title of the project that the computer is to analyze.

Line 2: The type of structure is specified here. In this case it is a space frame.

Line 3: This specifies the units in which the program will be coded. The results will be in the same units.

Lines 4–70: The joint coordinates of each joint are now given to the computer. They are in inches. It is also specified that joints 2, 18, 34, 42–50, and 66 are supports.

Line 71: The program in the computer assumes that all joints are fixed, unless otherwise stated. This line specifies the releases.

Lines 72, 73: Line 72 specifies the joints that have the releases specified in Line 71. An additional release is specified in Line 73 for joints 43–50.

Lines 74–211: The joints have been previously specified but the description of the geometry of the space structure is not complete without complete knowledge of the number and arrangement of the members between the joints. This is termed "MEMBER INCIDENCES." For example, member 1 runs from Joint 1 to Joint 2, and so on. The member incidence of each member has to be supplied to the computer before the structure is completely described.

Lines 212–217: The properties of the members have now to be inputed. Line 212 specifies that the member properties are being assumed as prismatic. The remaining lines specify the assumed cross-sectional area *AX*, and the second moments of area or the moments of inertia *IX*, *IY*, and *IZ* about the *X*, *Y*, and *Z* axis.

Lines 218–220: These give the constants *E* (the modulus of elasticity) and *G* (the modulus of rigidity) for the material being used.

Line 221: This gives the angle of orientation of the member about its axis.

Lines 222–224: These specify the first loading case that is an incidental live load. The load is applied at the joint 42 in the negative *Z* direction and has a value of 200 lb.

Lines 225–228: These lines specify the apex loads applied at joints 1 and 50 with values of 100 and 50 lb, respectively.

Lines 229–233: This is snow load. The loads in this case are applied to the members and are referred to the global coordinate axes.

Lines 234–236: The final instructions to the computer ask for the loadings to be listed and the stiffness method of analysis to be used for the analysis of the folded plate conical dome. The last line, Line 236, specifies the results from the analysis that are required to be printed out. The results for the incidental live load of 200 lb are very briefly discussed at the end of this Section.

It can be seen that the assumption of member properties as prismatic could be in error, depending on the amount of plate buckling that had occurred in the plate, which would in turn depend on the value of the applied load. In STRUDL it is possible to delete some data and input additions, and this was carried out for the member properties, as is shown in Lines 237 to 249.

Lines 237–239: The member properties prismatic are deleted from the coding.

Lines 240–249: The member properties are assumed to be variable. The member is divided into three segments and the properties for each segment are separately given.

Line 250: The computer is now being asked, in view of such deletions and additions, to print out the structural data for the folded plate conical dome.

The results from such a computer analysis run into many thousands of lines, and it is neither possible nor necessary to reproduce them all here. A great deal of the information supplied by the computer is not essential for design. Some of the information is also in the nature of checks to ensure accuracy of the analysis that has been performed. The results for the incidental live load case, for member properties prismatic, for members 1 and 2, are shown in Table 71. The forces and moments at the ends of the members are given, and it can be seen that only the axial force and the bending moment about the Y axis are important. The results for the same loading case, but for member properties variable are given in Table 72.

The folded plate conical dome was tested experimentally in order to establish the accuracy of the analysis. Electrical resistance strain gauges were fixed to a unit of the structure at the positions shown in Figs. 254 and

Table 71

Member	Joint	Axial Force	Shear Force Y	Shear Force Z	Torsional Moment	Bending Moment Y	Bending Moment Z
1	1	25.7743	−0.0017	0.0789	−0.0208	−3.1399	−0.0587
1	2	−25.7743	0.0017	−0.0789	0.0208	−2.1296	−0.0572
2	1	21.7046	−0.0048	−0.0671	−0.0088	0.1495	−0.1283
2	3	−21.7046	0.0048	0.0671	0.0088	3.9697	−0.1688

Table 72

Member	Joint	Axial Force	Shear Force Y	Shear Force Z	Torsional Moment	Bending Moment Y	Bending Moment Z
1	1	28.6390	−0.0040	0.1812	−0.0629	−2.4263	−0.0429
1	2	−28.6390	0.0040	−0.1812	0.0629	−9.6739	−0.2227
2	1	19.8275	−0.0092	−0.1786	−0.0487	1.3564	−0.0977
2	3	−19.8275	0.0092	0.1786	0.0487	9.6057	−0.4699

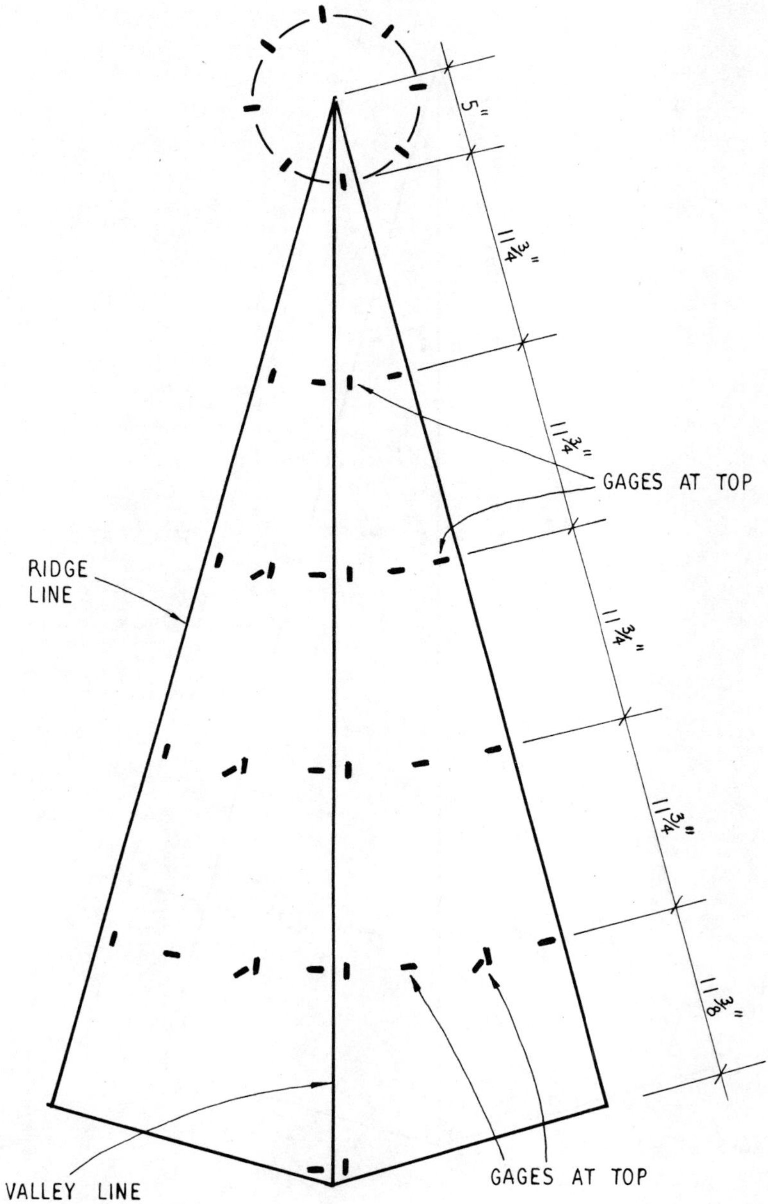
5"
11 3/4"
11 3/4"
GAGES AT TOP
11 3/4"
RIDGE LINE
11 3/4"
11 3/8"
VALLEY LINE
GAGES AT TOP

Fig. 254

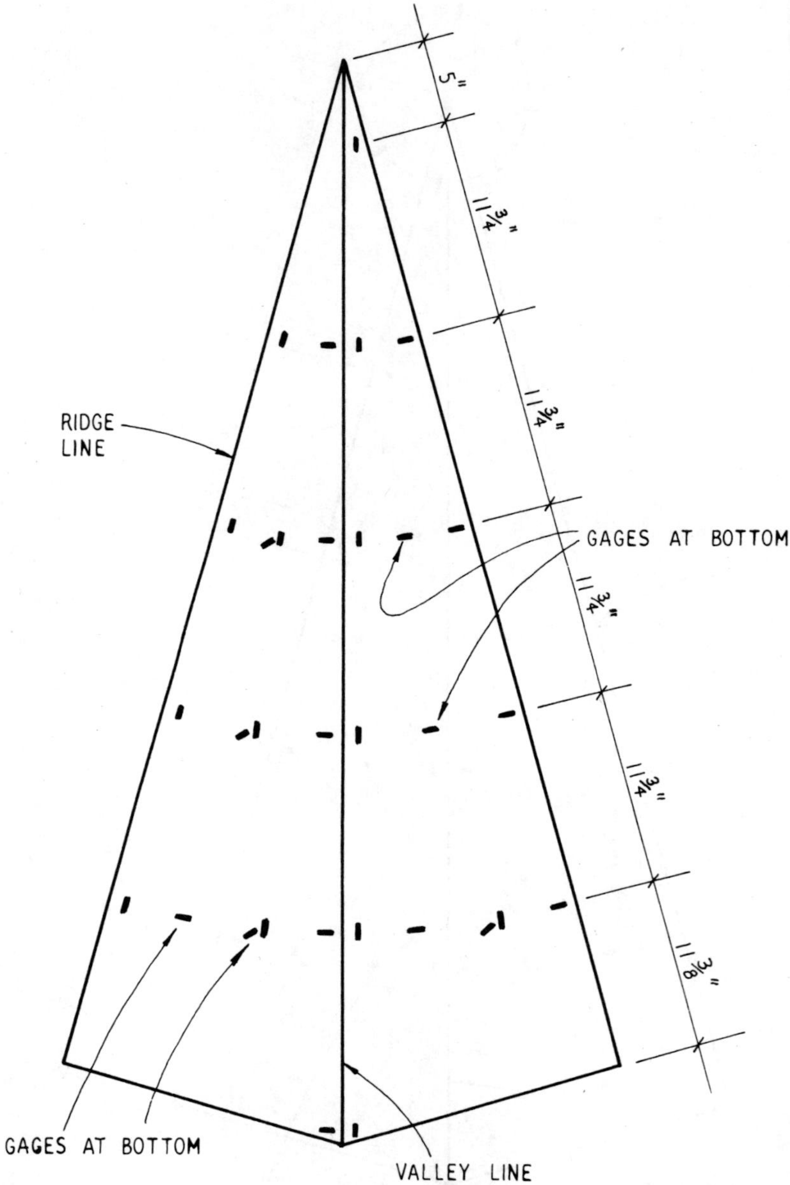
5"
11 3/4"
11 3/4"
11 3/4"
11 3/4"
11 3/8"
RIDGE LINE
GAGES AT BOTTOM
GAGES AT BOTTOM
VALLEY LINE

Fig. 255

255. The disadvantages of full scale testing have already been outlined in the last two paragraphs of Section 8.4. Within the limits of accuracy dictated by such testing, it can be concluded that the theoretical approach is entirely suitable for design purposes, but care must be taken in the assumptions made regarding the properties of the edge members that are fed into the computer program as data.

CHAPTER 11

MISCELLANEOUS STRUCTURES

11.1. INFLATABLES

These structures, which are also called pneumatic structures, depend for their stability on air pressure. In an inflatable structure the membrane carries all of its load in pure tension in much the same way as a soap bubble or a child's balloon. Inflatables are very popular for the roofing of very large areas such as greenhouses, athletic facilities, exhibition halls, tents, and the like. Because they can be deflated and reused again elsewhere they have also a special application as temporary structures. Inflatables are basically of two types—air supported structures and air inflated structures. A third type called the hybrid air structure is also sometimes referred to in the literature. This is a combination of an inflated roof and a more conventional supporting framework.

In all types of inflatables, anchorage is of some importance. The air pressure in an air supported structure causes an uplift that must be adequately resisted by the foundations. The structure has to be weighed down, or tied down, to ensure safety. This can be accomplished by earth or water berms, as is discussed in the following section. The air inflated structure has to be anchored by cables.

In all of these structures, however, the plastics component of the structure is the membrane, and this is discussed in greater detail.

11.1.1 Air Supported Structures

These are single skin fabric structures in which the membrane is supported by a small differential of air pressure between the inside and the outside of the structure. This air pressure differential is only of the order of 1/500 atmospheric pressure. This can be maintained by the continuous supply of air from a fan unit. It is not noticeable to the occupants of the building, but is sufficient to hold up the covering membrane. Doors and windows need to be air sealed, and an air lock has to be provided but, because the air pressure differential is very small, the actual loss of air through such openings is small and can be easily handled by the fans. Snow loads, wind loads, and incidental live loads can all be taken by a small increase in air pressure. The greatest danger lies in vandalism and rupture of the membrane. For this reason, the membrane used is usually a tough, light, plastics, cloth membrane made of neoprene/cotton, neoprene/nylon, terylene, polythene, PVC, fiberglass woven cloth covered with teflon, or fiberglass reinforced vinyl. The choice of the material would depend on the strength requirements and on the application. The long term life of such fabrics under constant stress has also to be considered, and the creep rupture of some fabrics is shown in Fig. 256.

An air supported membrane is shown[177] in Fig. 257. In this case, the anchorage has been provided by an earth berm around the structure. The disadvantage of a structure of this type is that it encloses large volumes

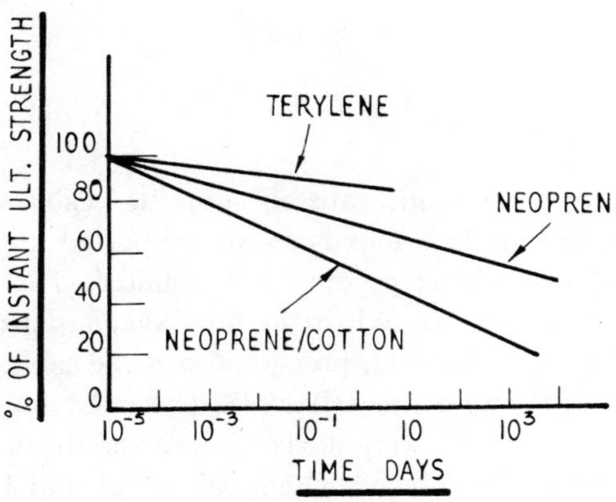

Fig. 256

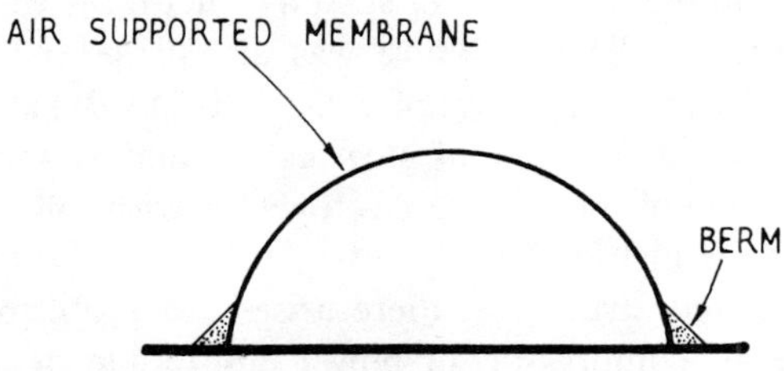

Fig. 257

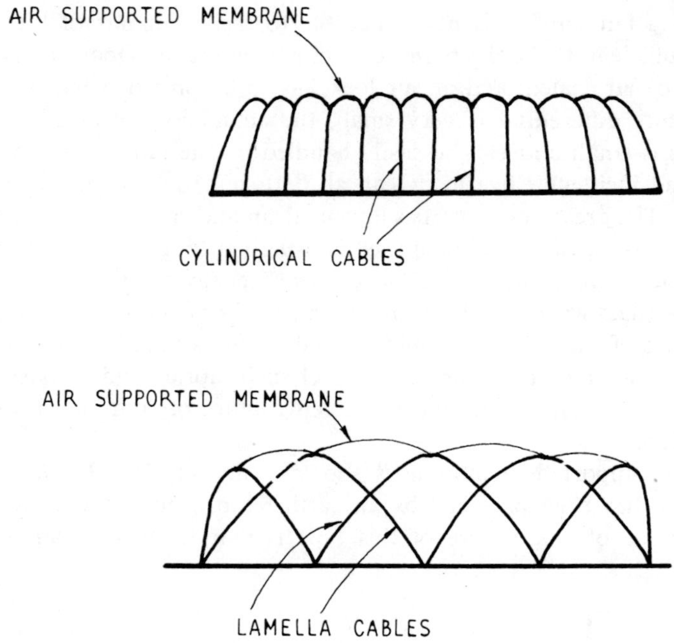

Fig. 258

with excessive headroom. To compensate, air supported cable structures can be used,[177] as shown in Fig. 258. This is an air supported net. The tension net determines the structure's shape once it is inflated. The membrane is inflated, lifting the cables with it. When the final shape, depending on the lengths of the cables is reached, the air pressure forces the cables into tension, thereby holding down the rise of the inflatable. The inflatable membrane is puffed out between the cables, allowing the headroom and the internal volume to be suitably controlled. The use of cylindrical cables and lamella cables is shown in Fig. 258. In all of the cases, the cables also serve to anchor the structure.

An excellent example of the use of such a structure is the U.S. Pavilion[196] at Osaka for Expo 70. The structure was an ellipse 262 × 462 ft (80 × 141 m) supported by an air pressure of 0.03 psi (0.21 kPa) above atmospheric. It was restrained by a network of steel cables and is shown in Fig. 259. The anchorage of one of the steel cables together with the concrete compression ring is shown[196] in Fig. 260.

In all air supported structures, there arises the problem associated with the failure of the air compressor, or power failure and decompression. This would cause the structure to deflate. The problem is not, however, as serious

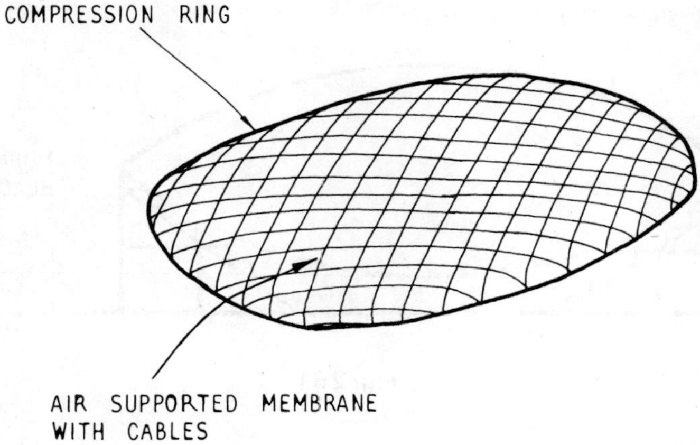

Fig. 259

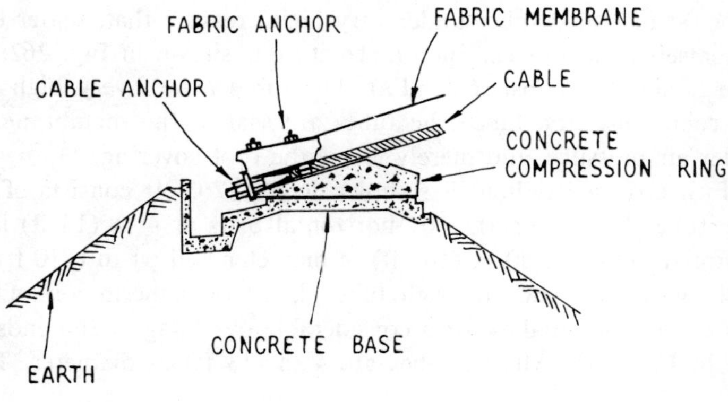

Fig. 260

as it seems, since the structure under such a small differential air pressure does not deflate instantly. In very important structures safety precautions can be taken. The hybrid air supported structure shown in Fig. 261 can be designed so that, even with a fully deflated membrane, a certain minimum headroom can always be maintained.[177] If the inflatable is to be set at ground level, the arena could of course be set in a pit for the same effect. It is also possible to provide in any of these structures, a secondary, independent, light structural system to hold up the sagging membrane, in the event it is accidentally deflated.

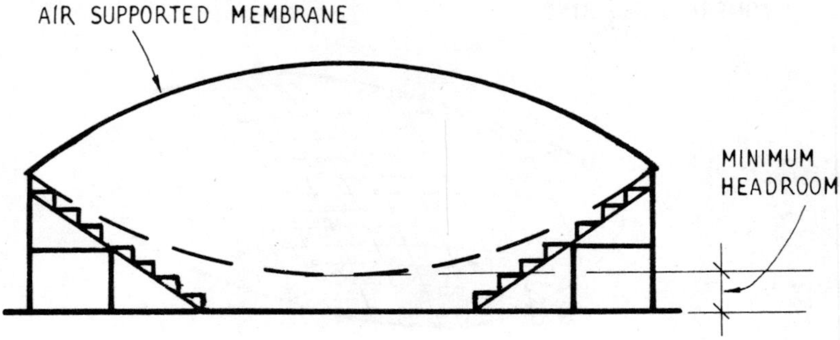

Fig. 261

11.1.2 Air Inflated Structures

Such structures work on the principle of the bicycle tube. When deflated the tube has no structural strength, but when inflated under high pressure, the tube springs into shape and has considerable strength. Air inflated structures consist of sealed tubes under very high pressure that, under inflation, erect themselves into place. Such a structure is shown in Fig. 262. The air pressure is about 100 psi (689 kPa). The tubes are of very high strength flexible rubber-plastics. Since the tubes are sealed, the membrane itself is not under air pressure, and merely forms the roof covering.

The Fuji Group Pavilion[197] is shown in Fig. 263. It consists of 16 high pressure tubes linked together by horizontal belts at 4 m (13 ft) intervals. The entire structure is 50 m (164 ft) in diameter and 64 m (210 ft) long at the roof, with both ends of each tube placed on a perimeter of a 50 m (164 ft) circle. This makes for a considerable overhang at the ends, as can be seen in Fig. 263. All the tubes are 4 m (13 ft) in diameter. They are

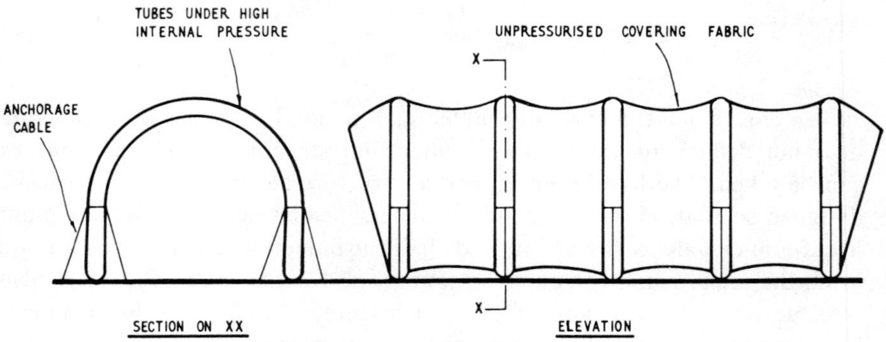

Fig. 262

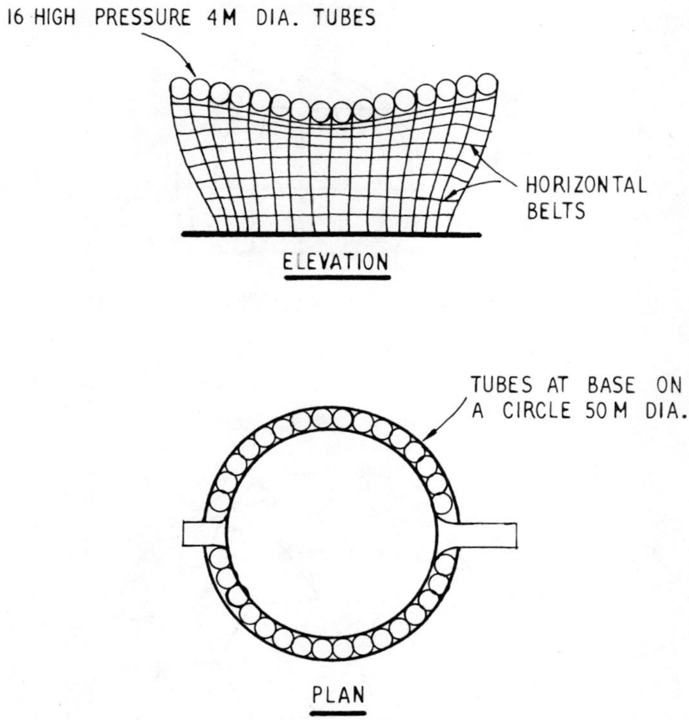

Fig. 263

made of a double-layered canvas of polyvinyl alcohol fiber bonded with neoprene to make a fabric that has a maximum rupture strength of 2844 psi (19,609 kPa). The outer surface is Hypalon. Normally the tube pressure is 800 mm (31.5 in.) of water, but during a typhoon it can go up to 2000–2500 mm (78.7–98.4 in.) of water pressure. If the air pressure exceeds 2500 mm (98.4 in.) of water, the air is bubbled out into a pond.

11.2 PLASTICS RADOMES

Ground radomes[135,174,176] used to protect radar installations are usually made of plastics, because plastics are transparent to radio waves, making them ideal materials for such an application. The material most commonly used is FRP. This can be used as a single skin laminate for small radomes. For larger radomes, FRP is generally used as the facing skins on a honeycomb core. In both cases the panels are connected by bolting through stiff flanges. Although the ground radome is generally spherical, there are several alternative arrangements to the panels forming the surface of the sphere. For small

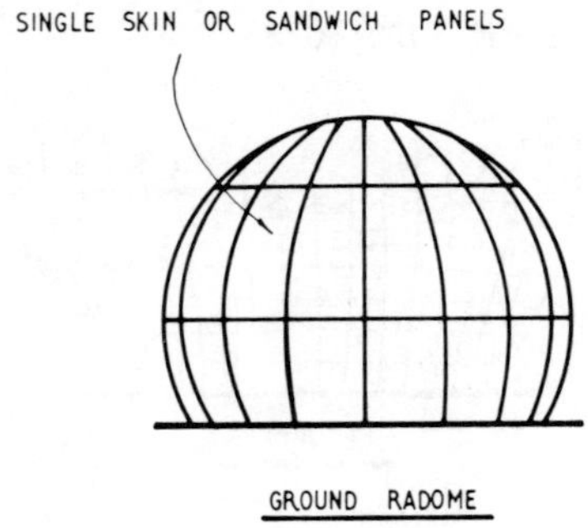

Fig. 264

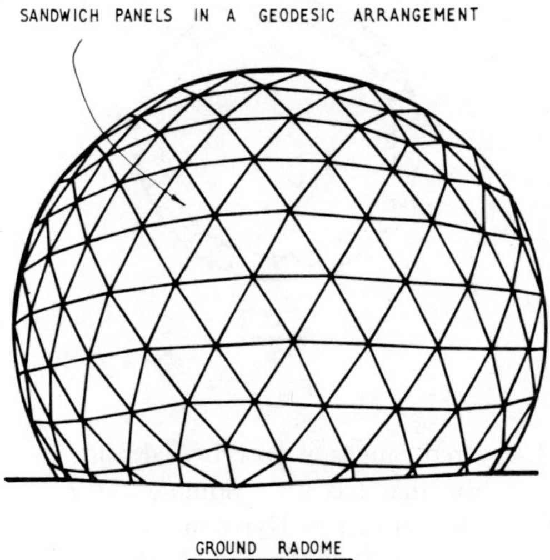

Fig. 265

radomes, the segmented form shown in Fig. 264 is quite economical. For larger radomes, the geodesic arrangement shown in Fig. 265 is more useful.

The analysis of plastics radomes can be carried out in the two ways outlined earlier in Chapters 9 and 10. If the panels are of sandwich construction, the entire structure can be considered to be a spherical shell subjected to dead, snow, and wind loads, and can then be analyzed in much the same way as the dome considered in Section 9.2. Alternatively, the finite elements method could be used. However, if the radome is of single skin construction, the thin skin would be subject to considerable compressive stresses. In this case it is advisable to consider the radome as a space frame structure with members concentrated along the flanges. This is particularly realistic, since the stiff flanges do, in fact, form a skeletal space structure. Matrix methods of analysis could then be easily applied to the analysis of the skeleton structure. The thin skin could be totally neglected in the analysis to yield a very safe design. Alternatively, a small fixed width of the skin could be assumed to be effective along each side of the flange line and the structure analyzed with member properties considered as prismatic, as in the STRUDL program for the folded plate conical dome of Section 10.10. In both cases, however, it may be safely assumed that the struts running along the flange lines are held in position by the thin membrane so that no buckling can occur. Yet another approximate method of analysis is the "shell analogy," which has been adequately covered by the author[198,199] and others[200] in the literature.

11.3 TENSIONED STRUCTURES

It has been seen that single skin laminates buckle at low loads when in compression. To obviate this disadvantage, the membrane can initially be placed in tension in such a way that the membrane, when under the external load, at no time carries compressive stresses. This principle was used in the single skin FRP roof[201] over one of the pavilions at the Swiss National Exhibition in 1964. The mushroom shaped units were formed of hyperbolic paraboloids connected in the manner shown in Fig. 266. After the units were made, they were post-tensioned so that the anticlastic surfaces were always in tension.

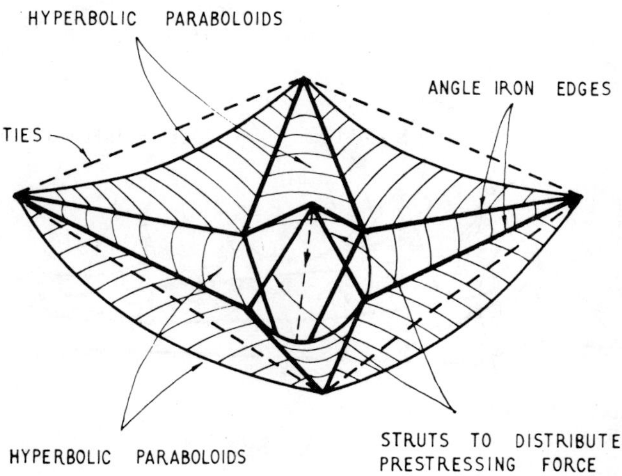

Fig. 266

When applied to plastics, however, this principle has some serious disadvantages. All plastics creep considerably with time, as has been seen, and this results in a loss of tensioning force. Hence, the complete creep behavior of the material over the life of the structure must be fully known. The loss of the tensioning force can then be allowed for by excess tensioning when the structure is erected. In the units forming the structure referred to above, hydraulic jacks were incorporated in each of the units to restore any loss of tension. This tends to make these structures expensive, particularly as periodic checks would be necessary over the entire life of the structure.

It should be noted here that the above type of structure should not be classified in the same category as tension structures of steel cables, which also use plastics as the covering membrane. Such types of structures are shown in Fig. 267. In these structures, the main structural system is the cable net, which loads onto steel pylons that form the main compression load carrying members. The plastics is used either as a fabric or as nonstructural panels, and both of these arrangements[202] have been shown in Fig. 267.

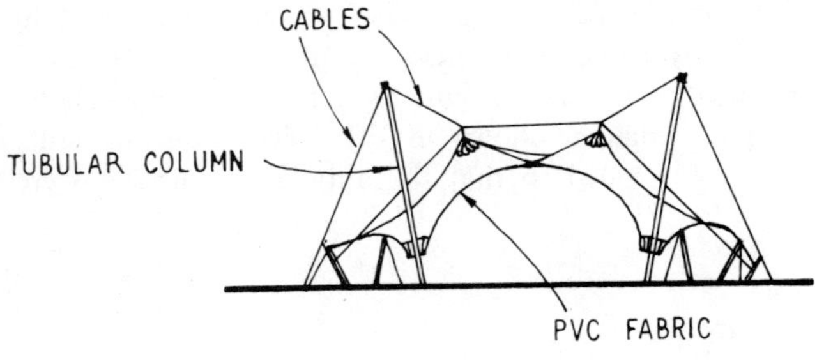

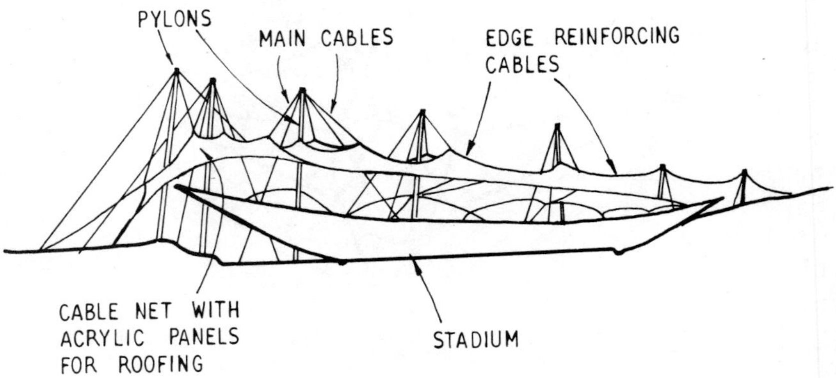

Fig. 267

11.4 PREFABRICATED LOW-COST HOUSING

In the first edition of this book, published more than ten years ago, the author made the following statement: "The age of instant housing is fast approaching, and it will soon be necessary to produce houses in precisely the same way as we produce cars—completely factory made, a few models from various companies and, perhaps, a new model on the market every year. This is particularly applicable to underdeveloped countries where millions may have to be adequately housed in brief spaces of time." In the ten years that have passed, it must be regretfully stated that that hope has not been realized. There are many reasons for this. Housing in underdeveloped countries has always been of low priority as compared to food or clothing. Funds for research and development have been all too scarce, and the little that was available through international agencies has been utilized in projects that were either too limited in their scope or failed to take into account the ethnic and cultural attitudes of the people for whom they were intended.

Nevertheless this hope was and is still shared by many,[203-205] and a few of the efforts made to realizing this hope are discussed here.

In underdeveloped countries, walls can generally be built very cheaply with local materials such as mud and cow dung plaster, but plastics could be very appropriately used to provide a low-cost roof of an adequate standard, and for a permanent dwelling. The author[206,207] developed a prefabricated FRP roof for use in a flexible system in which several units could be combined together to form rooms and hence houses. The unit, which was a folded plate conical dome, is hexagonal in plan and of single skin construction and is shown in Fig. 268. It is 10 ft (3 m) in diameter across the points of the hexagon and has 24 full folds with a fold angle of 25°. The large number of folds enable very thin skins to meet structural requirements. Including the thicknesss of the gel coat, the skin is only $\frac{1}{16}$ in. (1.6 mm) thick. In most underdeveloped countries thermal insulation is not important, but in extreme climates it can always be provided with local materials, or even by spraying rigid polyurethane foam on to one side of the roof.

Three or four units can be connected together to form a room as shown in Fig. 269. The three-unit room has an area of about 200 ft^2 (18.6 m^2), while the four unit room has an area of about 270 ft^2 (25.1 m^2). The units can be connected together by flanges, as shown in Fig. 270. The flanges, which are of sandwich construction, have a rigid polyurethane foam core of $\frac{5}{8}$ in. (16 mm) thickness, and a minimum depth of 5 in. (127 mm). The units are connected by simple bolting to the web of an aluminum channel as shown in Fig. 270. The whole assembly forms a stiff self-supporting grid system, thus eliminating the need for internal column supports. The sandwich flanges of two adjacent units also form a gutter for water drainage. Adequate

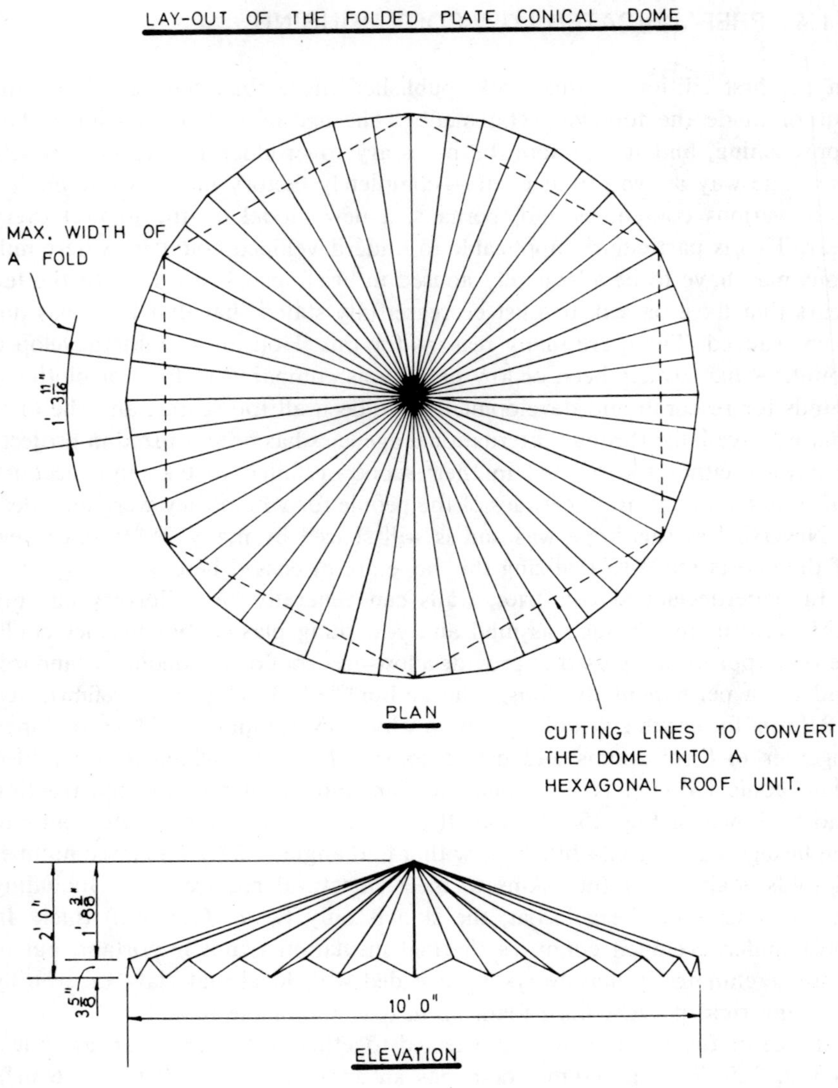

Fig. 268

slope can be obtained by inserting a suitable local filler material, shaped to give slope in desired directions and sealed with a mastic sealant. The entire roof of three units was subjected to both a theoretical analysis, as detailed in Section 10.10, and experimental testing for structural behavior and weathering characteristics. On the structural side, the roof was found to be excellent

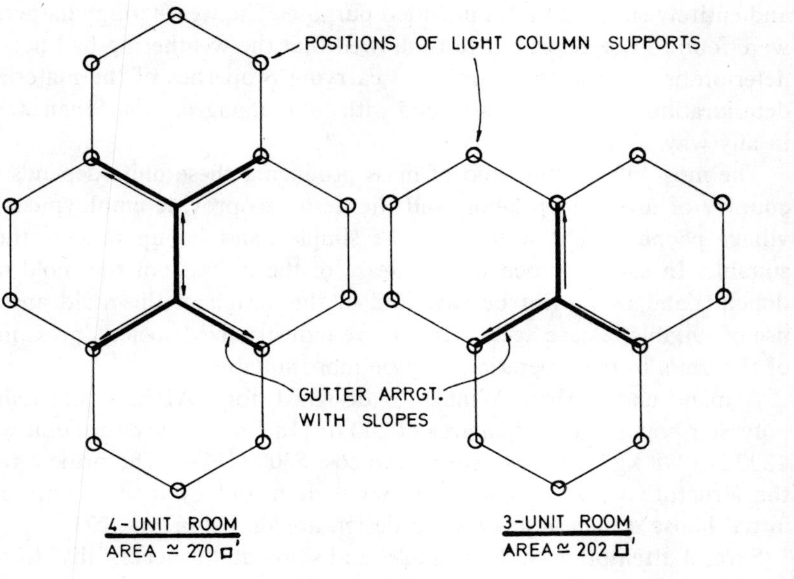

Fig. 269

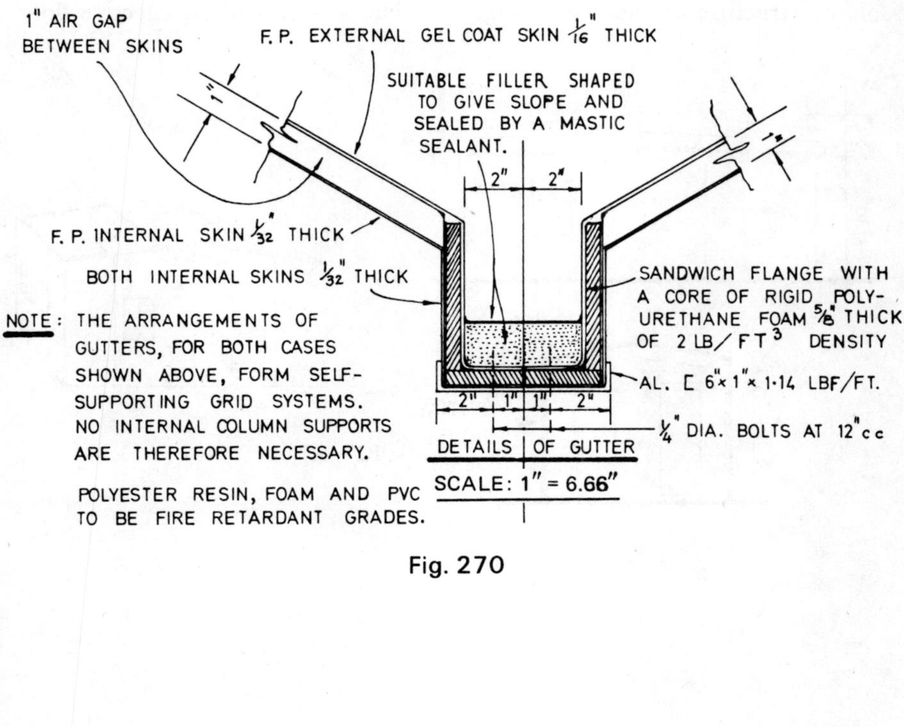

Fig. 270

and entirely suitable for its intended purpose. The weathering characteristics were found to be poor, but tests showed that the weathering had not caused deterioration in the structural load carrying properties of the material. The deterioration was visual, associated with color changes, rather than structural in any way.

The most efficient method of mass producing these units depends on the country of use. Cheap labor, and the desire to provide employment to the village people, could well make the simple hand lay-up process the most suitable. In such a process, the release of the units from the mold may be difficult, and care must be exercised in the design of the mold and in the use of suitable release agents. In a more industrialized society, press molding of the units in one operation may be more suitable.

Armand and Barbara Winfield[208] designed, for CARE, a jute reinforced polyester house that had an area of 200 ft^2 (18.6 m^2). The entire unit weighed 2200 lb (998 kg) and was expected to cost \$300 (1974). The project required the structure to be designed for Bangladesh and cyclone conditions. The initial house design and a revised design are shown in Fig. 271.

Several attempts have been made, and some quite successfully, to provide emergency housing or shelters for the victims of disasters. These range from simple tents, to domes that can be sprayed onto inflated membranes. The membrane is then deflated and reused again. Holes are then cut in the foam to form doors and windows.

In 1975, Avio-Fokker[209] designed and built an air transportable emergency shelter made entirely of flat panels for ease of transport. The plan and elevation of the structure are shown in Fig. 272. The shelter had an effective floor

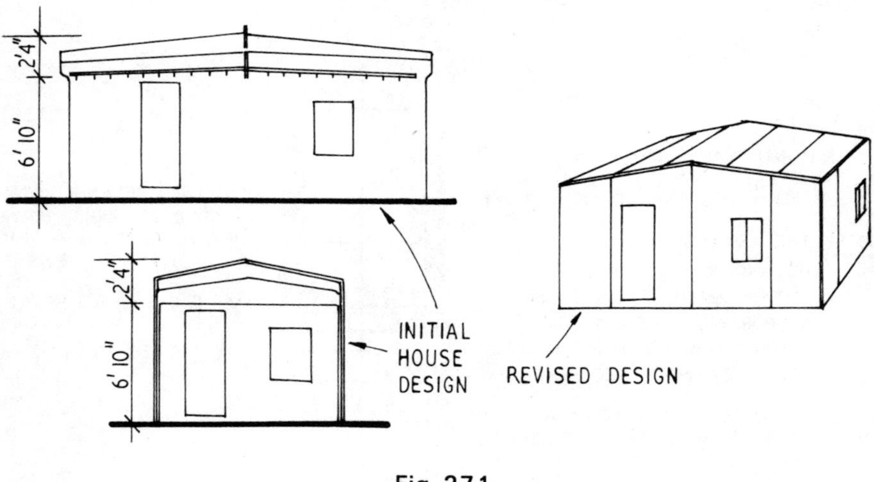

Fig. 271

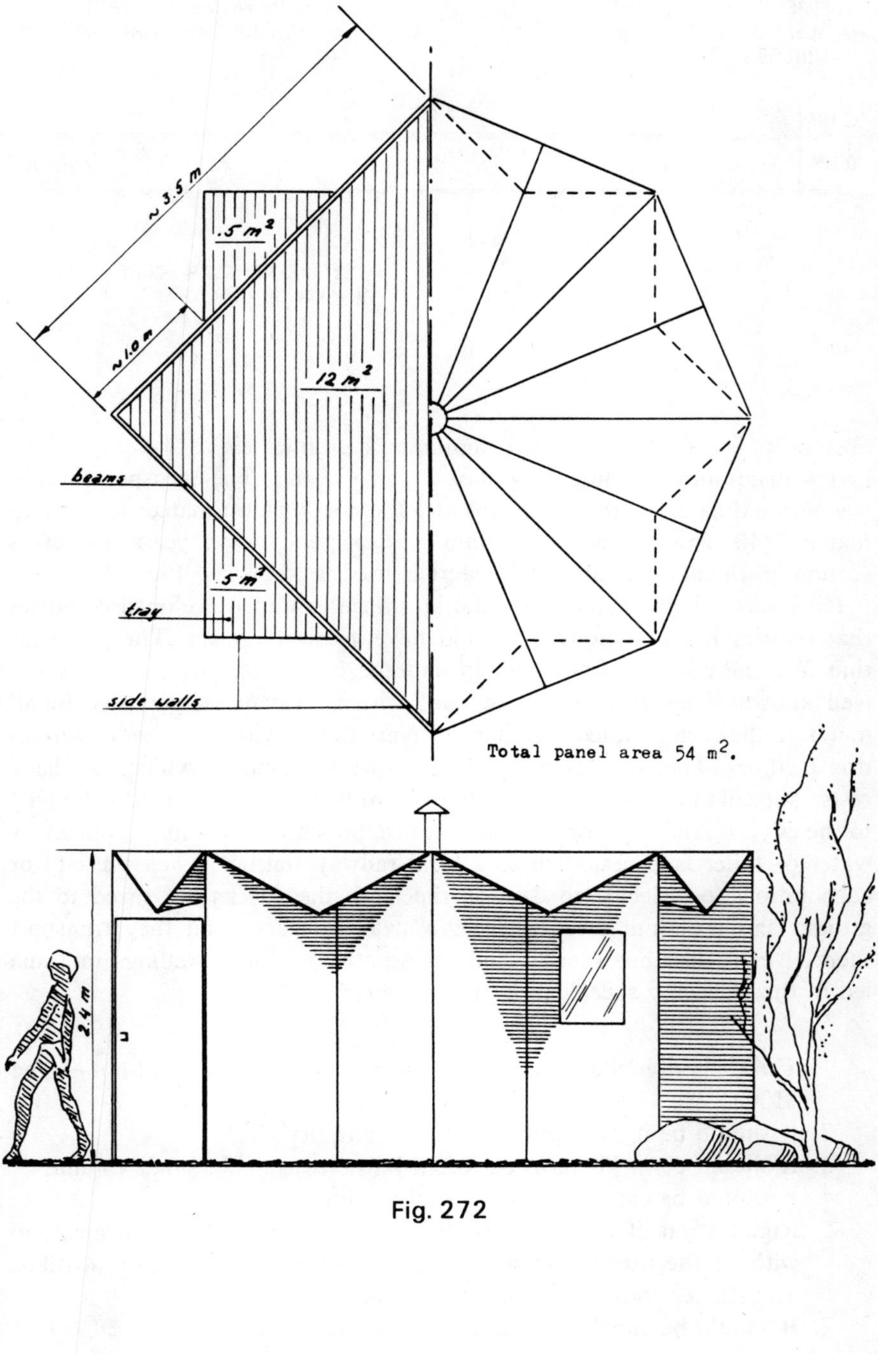
~ 3.5 m
~1.0 m
.5 m²
12 m²
.5 m²
beams
tray
side walls
Total panel area 54 m².
2.4 m

Fig. 272

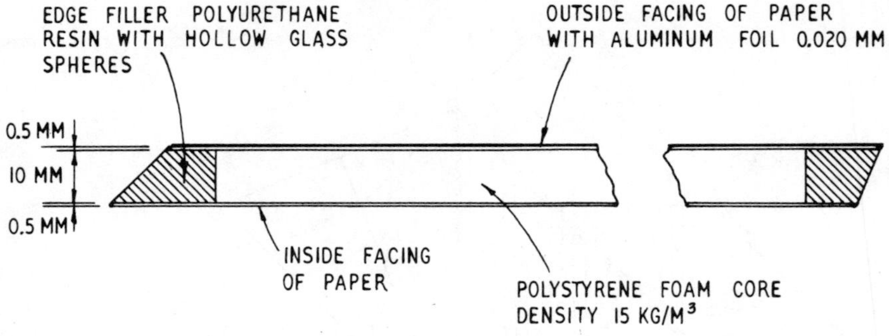

Fig. 273

area of 14 m^2 (151 ft^2) with a minimum standing height of 2 m (6.6 ft) and a maximum standing height of 2.4 m (7.9 ft). The transport weight was approximately 100 kg (220 lb) and the cost was estimated at approximately $440. The estimated minimum life expectancy was 2 years. The cross section of a panel, detailing the materials used, is shown in Fig. 273.

However, it is in the area of slum clearance in underdeveloped countries that plastics low-cost housing should have greatest impact. The problems that face the poorest of the poor in underdeveloped countries are only too well known. They live in shacks, built with any available scrap of found material, that are huddled together in shanty towns without water or sanitation facilities. The slum dweller does not own the land on which the shack rests. The colonies develop in a wild uncontrolled manner on land belonging to the city, the railways, or big corporations, possibly near some rudimentary water or toilet facilities, such as a local railway station. When the city or corporation does decide to develop the land, the shacks are razed to the ground and the slum dwellers forcibly evicted, whereupon they transport themselves to another vacant "lot." Research on slum dwellings in India led to the following specifications for a shelter[210-212]:

1. The shelter should be as low in cost as possible but should not exceed $100.
2. It should be lightweight and easily transportable.
3. It should be in kit form packaged into a reasonably small volume.
4. It should be capable of being easily erected.
5. Rigidization of the structure by mechanical or chemical means, and without the use of any sort of equipment, should be easily possible.
6. The shelter should be easily anchored.
7. It should be capable of being easily waterproofed.

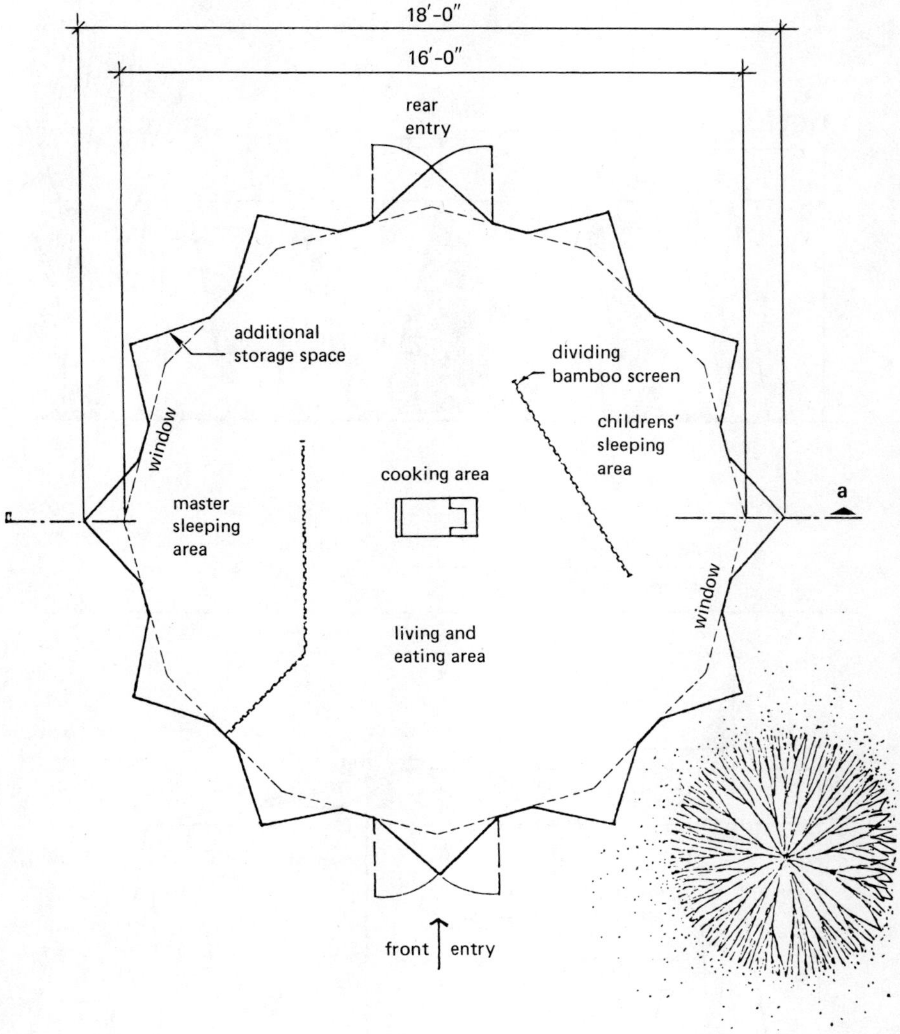

floor plan (floor area: 212 sft.)

scale: 1/4 ″ = 1′-0″

Fig. 274

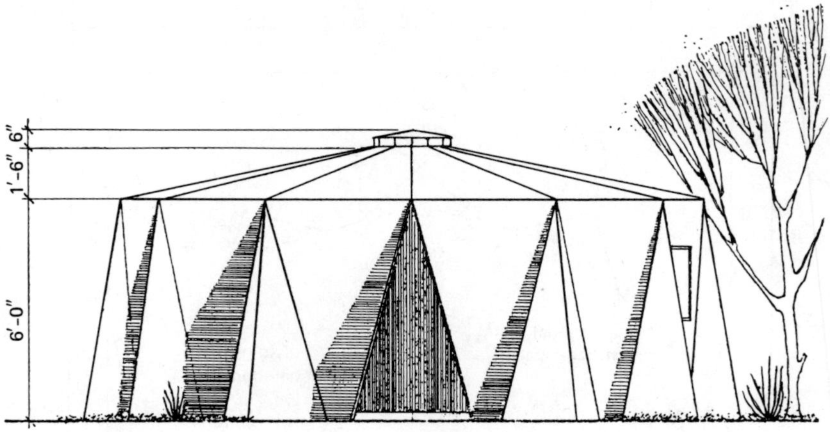

front elevation

scale: 1/4" = 1'-0"

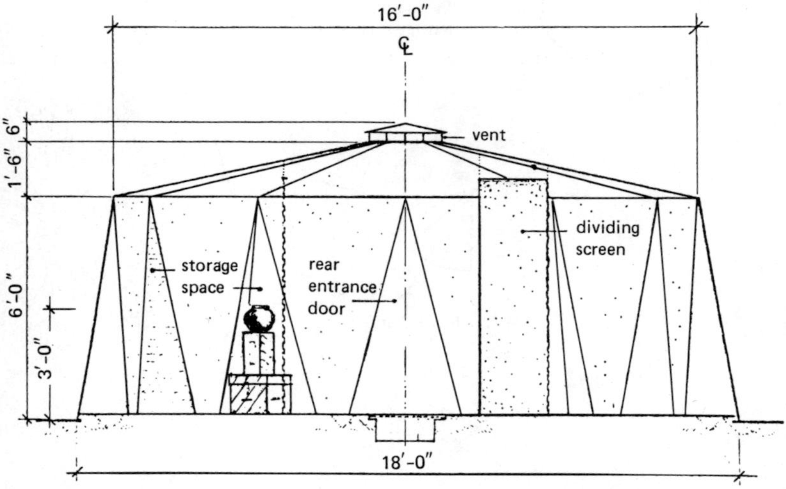

cross section "a"

scale: 1/4" = 1'-0"

Fig. 275

8. It should provide the minimum adequate standard space necessary for the slum dweller and his family.
9. It should be structurally capable of withstanding dead and wind loads.
10. It should have a reasonable degree of fire resistance.
11. It should have reasonable thermal insulation against heat and cold.

The shelter that is proposed here does meet all of these specifications in some reasonable measure. It should be emphasized, however, that this research is strictly applicable only to Indian conditions, and modifications would be necessary for applications to other underdeveloped countries.

The shelter is a folded plate dome structure with a floor area of about 212 ft^2 (19.7 m^2). A plan and elevation of the structure are shown in Figs. 274 and 275. The first stage is vertical. The minimum headroom is 6 ft (1.83 m) and the maximum headroom about 7 ft 6 in. (2.29 m). The cooking area occupies the center of the shelter, directly beneath a vent at the top of the dome. In dry weather, the whole cooking activity could be transferred outdoors. The sleeping areas may be shielded for privacy with bamboo or straw mat curtains, but this is not part of the shelter. A prototype, full scale structure[212] is shown in Fig. 276. The entire structure is constructed

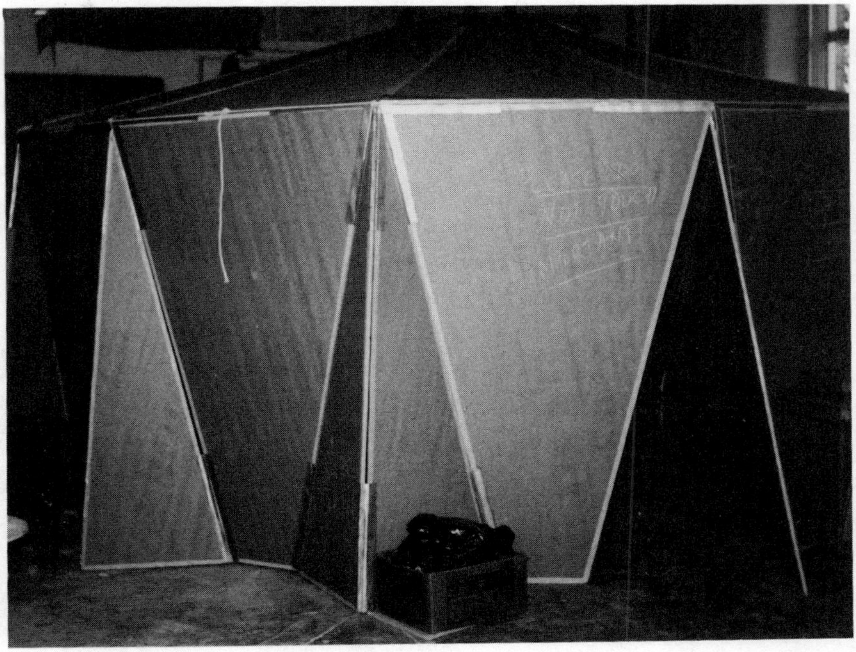

Fig. 276

from three types of panels—the trapezoidal panels of the roof and the walls, and the triangular panels of the folds. The architectural shape has great intrinsic strength and stability by virtue of its form and can withstand dead and wind loads effectively. Structurally some interesting facts come to light. The dome does not need an apex ring at the top for its stability. It does, however, need a horizontal ring tie at shoulder level, and this can be seen in Fig. 276. For the top to fall in, the shoulder has to move out, and this is effectively prevented by the ring at shoulder level. This leads to the important conclusion that the roof panels need only one set of joints at the apex and further that these joints can be flexible and need not be rigid. This conclusion is important from the viewpoint of the packaging characteristics of the dome. The entire structure can be packaged into a single kit 7 ft 3 in. (2.2 m) × 5 ft 3 in. (1.6 m) × 1 ft 3 in. (0.38 m). The entire package weighs only 140 lb (63.5 kg) and can be very easily handled and transported. The packaged dome is shown[212] in Fig. 277. The material of the dome panels was a sandwich with a low-density polyurethane foam core sandwiched between kraft paper and available as a board. However, other materials can be equally suitable depending on the budget. Joints were taped to provide freedom of rotation in both directions for the two panels meeting at the

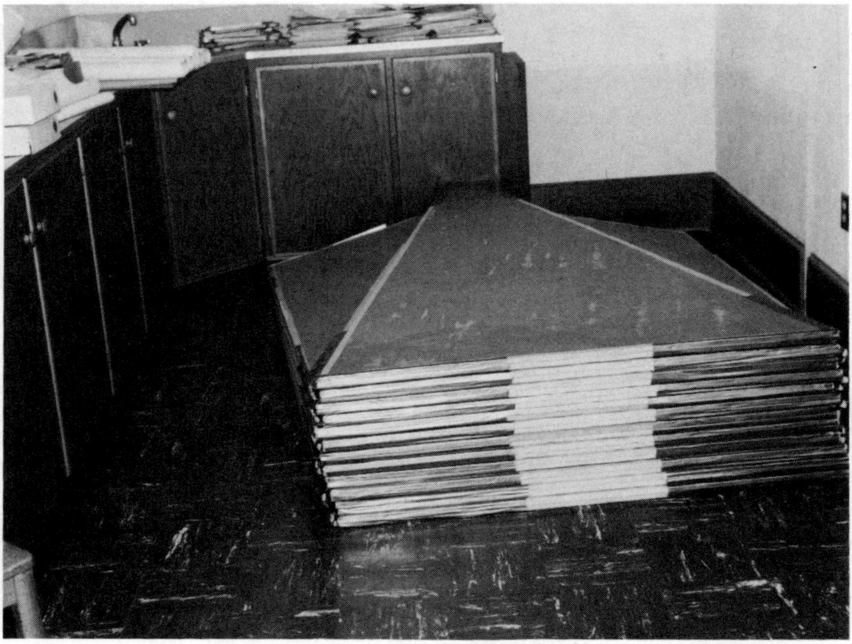

Fig. 277

joint. Since the entire structure folds into a single package, all the joints are factory joints except the joints at the apex and one joint at the shoulder, where the ends come together.

The waterproofing of the structure is important, particularly for a country like India, where the monsoon rains are very heavy. From the point of view of costs, it is totally unrealistic to waterproof all the joints in the structure. This is particularly difficult if the folding and packaging characteristics of the shelter are kept in mind. The author has therefore proposed a novel solution to waterproofing the shelter by the provision of a light, cheap, polythene raincoat. The raincoat could be tailored to an exact fit, or could be loosely drawn over the structure for the duration of the monsoon. The vent at the top would protrude through the raincoat. Suitable arrangements could be made for the door and windows. After the monsoon season, the raincoat could be stored for future use. In case of damage it could be replaced.

There are no major architectural or technological problems[210-212] involved in the final development of the shelter. Further development work on the shelter should be performed under local conditions, and this could include development of the vent, door and window openings, anchorage, precise nature of the taped joints, further examination of a variety of materials that could be used for the shelter panels, and, finally, an in-depth study of the economics, designed to make the shelter attractive to slum dwellers. The provision of water and sanitation facilities would go a long way to convincing the slum dweller that the small investment would be very worthwhile. He has the added assurance of knowing that his shelter is like a shell house on the back of a snail, and that if he is forced to move, it will move with him.

REFERENCES

1. Benjamin, B. S., "Plastics as a structural material," *The Consulting Engineer* (Nov., 1965).
2. Ronkka, Y., "The design and performance of acrylic domelights," in *International Conference on Space Structures,* London 1966, Blackwell Scientific Publications Ltd., 1967.
3. Building Research Station, Garston, England, "Cellular plastics for building," Digest, 2nd Series, No. 39.
4. Gallagher, R. B., Kamath, V. R., Kamens, E. R., and Harpell, G. A., "Lucel Azo foaming agents—A practical route to cellular plastics for unsaturated polyester and other resins," in *31st Annual Technical Conference, Reinforced Plastics/Composites Institute,* The Society of the Plastics Industry, Inc., 1976.
5. Smarook, Walter H., "Polysulfone reinforced 'dough-foam' moldings," in *32nd Annual Technical Conference, Reinforced Plastics/Composites Institute,* The Society of the Plastics Industry, Inc., 1977.
6. General Electric, "Structural foam resins," Data sheets and information, 1973.
7. Benjamin, B. S., and Makowski, Z. S., "The analysis of folded-plate structures in plastics," in *Conference on Plastics in Building Structures,* London, 1965. Pergamon Press, Oxford 1966.
8. Piano, R., "Experiments and projects with industrialized structures in plastic materials," in *International Conference on Space Structures,* London, 1966. Proc. published by Blackwell Scientific Publications Ltd., 1967.
9. Architectural Research Laboratory, "The potential use of foam plastics for housing in underdeveloped areas," University of Michigan, Ann Arbor, Michigan, March, 1963.
10. Davenport, A. J., "The application of reinforced plastics," *Applied Plastics* (Aug. and Sept., 1962).
11. Riddell, Malcolm N., and O'Toole, John L., "Significant properties of plastics for design," in *Structural Plastics—Properties and Possibilities,* American Society of Civil Engineers, Kentucky, April, 1969.
12. Imperial Chemical Industries, Ltd., "Perspex Acrylic Materials—Properties."
13. Morgan, P., *Glass Reinforced Plastics,* 3rd ed. Iliffe Books Ltd., London, 1961.
14. Crowder, J. R., "The weathering behavior of reinforced plastics sheeting," in *Fourth International Reinforced Plastics Conference,* London, 1964. British Plastics Federation, London, 1964.
15. Gumerman, C., and McKay, G. R., "Evaluation of Polyvinylfluoride film as a weatherable finish on reinforced polyesters," *Proceedings of 18th Technical and Management Conference Reinforced Plastics Division,* SPI, Feb., 1963.
16. Sleddon, G. J., "Building Regulations 1965," in *Fifth International Reinforced Plastics Conference,* Nov., 1966. British Plastics Federation, London, 1966.
17. Powell, D., "Private communication to the author," London, 1965.

18. Passaglia, E., and Knox, J. R., "Viscoelastic behavior and time-temperature relationships," in *Engineering Design for Plastics,* edited by E. Baer, Reinhold Publishing Corporation, New York, 1964, Chap. 3.
19. Hilton, H. H., "Viscoelastic analysis," in *Engineering Design for Plastics,* edited by E. Baer, Reinhold Publishing Corporation, New York, 1964, Chap. 4.
20. McCammond, D., "The substitution of time dependant data in Hookean formulae," *Trans. J. Plastics Institute* (April, 1967).
21. Turner, S., "The foundations of possible engineering design methods for plastics," *Trans. J. Plastics Institute* (June, 1966).
22. Baer, E., Knox, J. R., Linton, T. J., and Maier, R. E., *S. P. E. Journal,* 16, (1960).
23. Ratcliffe, W. F., and Turner, S., "Engineering Design: Data required for plastics materials," *Trans. J. Plastics Institute* (June, 1966).
24. Benjamin, B. S., *Discussion, Conference on Plastics in Building Construction,* London, 1964, Proc. edited by R. M. Davies, Blackie and Son, Ltd., London, 1965.
25. Illing, Gerhard, "Epoxy-coated glass spheres for reinforcement of nylon and polyester resin," *29th Annual Technical Conference, Reinforced Plastics/Composites Institute,* The Society of the Plastics Industry, Inc., 1974.
26. Sundstrom, G. A., Collister, J., and Hays, C. W., "Low density sheet molding (SMC) and bulk molding (BMC) compounds," in *30th Anniversary Technical Conference, Reinforced Plastics/Composites Institute,* The Society of the Plastics Industry, Inc., 1975.
27. Scott Bader and Co. Ltd., "Polyester Handbook," 1963.
28. Association of jute spinners and manufacturers, "Low-cost lightweight rigidity with jute fabrics," Information sheet, 1966.
29. Wells, H., Bowen, D. H., Macphail, I., and Pal, P. K., "The potential of jute and similar bast fibers as high specific stiffness reinforcements," in *35th Annual Technical Conference, Reinforced Plastics/Composites Institute,* The Society of the Plastics Industry, Inc., 1980.
30. Paramasivam, T., and Abdul Kalam, A. P. J., "On the study of indigenous natural-fiber composites," in *29th Annual Technical Conference, Reinforced Plastics/Composites Institute,* The Society of the Plastics Industry, Inc., 1974.
31. Cape Insulation, Ltd., " 'Noramite' Data sheets," Oct., 1966.
32. Ewchuk, William, "Asbestos/ebonite composites," in *28th Annual Technical Conference, Reinforced Plastics/Composites Institute,* The Society of the Plastics Industry, Inc., 1973.
33. Cameron, A. B., Heron, G. F., and Wicker, G. L., "Asbestos reinforced thermoplastics," in *28th Annual Technical Conference, Reinforced Plastics/Composites Institute,* The Society of the Plastics Industry, Inc., 1973.
34. Reis, James F., "Safe handling of chrysotile asbestos reinforced plastics," in *31st Annual Technical Conference, Reinforced Plastics/Composites Institute,* The Society of the Plastics Industry, Inc., 1976.
35. Hill, James E., and Pearson, Everett A., "Graphite fiber tow reinforcement in boat hull construction," in *28th Annual Technical Conference, Reinforced Plastics/Composites Institute,* The Society of the Plastics Industry, Inc., 1973.
36. Browning, C. E., "The effects of moisture on the properties of high performance structural resins and composites," in *28th Annual Technical Conference, Reinforced Plastics/Composites Institute,* The Society of the Plastics Industry, Inc., 1973.
37. Faughnan, Kevin A., "Longitudinal residual stresses in boron filaments," in *29th Annual Technical Conference, Reinforced Plastics/Composites Institute,* The Society of the Plastics Industry, Inc., 1974.
38. Riewald, P. G., and Zweben, C., "Kevlar 49 hybrid composites for commercial and aerospace applications," in *30th Annual Technical Conference, Reinforced Plastics/Composites Institute,* The Society of the Plastics Industry, Inc., 1975.

39. Milewski, John V., "The field is wide open for reinforcements," *Plastics Engineering,* Vol 34, No. 11 (Nov., 1978).
40. Woodhams, R. T., and Xanthos, M., "Polyester resins reinforced with HAR Mica," in *29th Annual Technical Conference, Reinforced Plastics/Composites Institute,* The Society of the Plastics Industry, Inc., 1974.
41. Xanthos, M., Hawley, G. C., and Antonacci, J., "Novel developments and applications for Suzorite Mica in reinforced plastics," in *33rd Annual Technical Conference, Reinforced Plastics/Composites Institute,* The Society of the Plastics Industry, Inc., 1978.
42. Schrock, B. Jay, and Rangel, G., "Reinforced plastic mortar pipe breakthrough in sewer and irrigation applications," in *30th Anniversary Technical Conference, Reinforced Plastics/ Composites Institute,* The Society of the Plastics Industry, Inc., 1975.
43. Bacon, C. E., "Towards a competitive design material," in *Third International Reinforced Plastics Conference,* London, Nov., 1962. British Plastics Federation, London.
44. Rosato, D. V., "Cost comparison of plastic parts for the transportation industry based on different manufacturing processes," in *28th Annual Technical Conference, Reinforced Plastics/Composites Institute,* The Society of the Plastics Industry, Inc., 1973.
45. Trampenau, Richard H., and Wilson, Thomas B., Jr., "The use of fire retardant polyester in Willard boat works' 120 foot motor vessel," in *30th Anniversary Technical Conference, Reinforced Plastics/Composites Institute,* The Society of the Plastics Industry, Inc., 1975.
46. Scott, R. J., and Sommella, J. H., "Feasibility study of glass reinforced plastic cargo ship," in *28th Annual Technical Conference, Reinforced Plastics/Composites Institute,* The Society of the Plastics Industry, Inc., 1973.
47. Desai, Ramesh R., "Corrosion control fiber reinforced plastic enclosures," *28th Annual Technical Conference, Reinforced Plastics/Composites Institute,* The Society of the Plastics Industry, Inc., 1973.
48. Permali Ltd., "Data sheets," 1966.
49. Kabelka, J., "The behavior of glass-reinforced polyester under long-term stress," in *Fifth International Reinforced Plastics Conference,* London, Nov., 1966, British Plastics Federation, London.
50. Abbott, B. W., and Broutman, L. J., "Determination of the modulus of elasticity of filament reinforced plastics using stress wave techniques," in *Proc. 21st Technical and Management Conference, Reinforced Plastics Division,* SPI, 1966.
51. Green, Douglas H., and Guevin, Paul R., "Polyurethanes—the new FRP finish," in *28th Annual Technical Conference, Reinforced Plastics/Composites Institute,* The Society of the Plastics Industry, Inc., 1973.
52. Pritchard, Geoffrey, and Taneja, Neelam, "Nature and extent of water damage in glass laminates," *29th Annual Technical Conference, Reinforced Plastics/Composites Institute,* The Society of the Plastics Industry, Inc., 1974.
53. Dixon, B. G., Longenecker, D. M., and Ferrarini, J., "Curing of unsaturated polyester resins for corrosion resistant structures," *31st Annual Technical Conference, Reinforced Plastics/Composites Institute,* The Society of the Plastics Industry, Inc., 1976.
54. Brink, N. O., "Mechanical behavior of reinforced plastics at cryogenic temperatures," *S.P.E. Journal* (Oct., 1964).
55. Hertz, J., "The effect of cryogenic temperatures on the mechanical properties of reinforced plastic laminates," *S.P.E. Journal* (Feb., 1965).
56. Dietz, A. G. H., *Engineering Laminates,* John Wiley and Sons, New York, 1949.
57. Sonneborn, R. H., "*Fiberglas Reinforced Plastics,*" Reinhold Publishing Corporation, New York, 1964.
58. Shaffer, B. W., "The influence of filament orientation on the material properties of reinforced

plastics," in *Proc. 19th Technical and Management Conference, Reinforced Plastics Division,* SPI, Feb., 1964.

59. Thompson, A. W., "Further work on the fatigue and creep properties of plastics laminates," *First International Reinforced Plastics Conference,* London, Oct., 1958, British Plastics Federation, London.
60. Boller, K. H., "The effect of long-term loading on glass-reinforced plastic laminates," in *Proc. 14th Technical and Management Conference, Reinforced Plastics Division,* SPI, 1959.
61. Ainsworth, L., "Properties of glass for plastics reinforcement," *Second International Reinforced Plastics Conference,* London, Nov., 1960, British Plastics Federation, London.
62. Bax, J., "Deformation behavior and failure of glass fiber reinforced resin material," Text of lecture, Plastics Institute, Cranfield, England, Jan., 1969.
63. Goldfein, S., "Time-temperature relationship for rupture stresses in reinforced plastics," *Proc. A.S.T.M.,* 54, 1954.
64. Findley, W. N., "The effect of temperature and combined stresses on creep of plastics," in Second International Reinforced Plastics Conference, London, Nov., 1960, British Plastics Federation, London.
65. Findley, W. N., "Prediction of performance of plastics under long-term static loads," *Trans. J. Plastics Institute* (June 1962).
66. McCorsley, C. C., "ENCRON Reinforcement of polyester laminates," in *28th Annual Technical Conference, Reinforced Plastics/Composites Institute,* The Society of the Plastics Industry, Inc., 1973.
67. Haddad, G. N., "Recent innovations in PVC-FRP composite pipe," in *32nd Annual Technical Conference, Reinforced Plastics/Composites Institute,* The Society of the Plastics Industry, Inc., 1977.
68. Montella, R., "Designing for rigidized acrylic," in *32nd Annual Technical Conference, Reinforced Plastics/Composites Institute,* The Society of the Plastics Industry, Inc., 1977.
69. Rolston, J. Albert, and Gilbu, Agnar, "Design, production and testing of large diameter reinforced plastic underground effluent pipe," in *28th Annual Technical Conference, Reinforced Plastics/Composites Institute,* The Society of the Plastics Industry, Inc., 1973.
70. Benjamin, B. S., "The weathering of an RP/C roof structure," in *29th Annual Technical Conference, Reinforced Plastics/Composites Institute,* The Society of the Plastics Industry, Inc., 1974.
71. Philip, A. R., "Plastics reinforcement with jute cloth," *Engineering Materials and Design,* (July, 1965).
72. Lant, T. P. R., "The testing of reinforced-plastics laminates for use as structural materials," *Trans. J. Plastics Institute* (Dec., 1961).
73. Hart, Gerald L., and Pritchard, Geoffrey, "Role of the interface in the corrosion of graphite fiber composites," in *29th Annual Technical Conference, Reinforced Plastics/Composites Institute,* The Society of the Plastics Industry, Inc., 1974.
74. Price, Howard L., and Lucy, Melvin H., "Effect of volatile removal during molding on the properties of two phenolic-fiber composites," in *29th Annual Technical Conference, Reinforced Plastics/Composites Institute,* The Society of the Plastics Industry, Inc., 1974.
75. Thomas, R. E., and Enos, J. H., "Very high strength SMC in automotive structures," in *33rd Annual Technical Conference, Reinforced Plastics/Composites Institute,* The Society of the Plastics Industry, Inc., 1978.
76. Breitigam, W. V., and Brass, L. L., "Engineering properties of epoxy-based vinyl ester resin laminates," in *35th Annual Technical Conference, Reinforced Plastics/Composites Institute,* The Society of the Plastics Industry, Inc., 1980.
77. Szymanski W. A., and Vaccarella, P. W., "New chemical resistant vinyl ester resins," in *33rd Annual Technical Conference, Reinforced Plastics/Composites Institute,* The Society of the Plastics Industry, Inc., 1978.

78. Jaspers, Hans, and Wynes, Tom, "DIACRYL, A new high performance, styrene free vinyl ester resin," in *35th Annual Technical Conference, Reinforced Plastics/Composites Institute,* The Society of the Plastics Industry, Inc., 1980.
79. Penn, Lynn and Chiao, T. T., "Fiber composites for energy storage flywheels," in *31st Annual Technical Conference, Reinforced Plastics/Composites Institute,* The Society of the Plastics Industry, Inc., 1976.
80. Anthony, F. M., and Hanley, D. P., "Design procedure for a graphite fiber composite structure," in *Symposium on graphite fiber composites,* ASME, Winter Annual Meeting, Pittsburgh, Nov., 1967.
81. Hanson, M. P., and Chamis, C. C., "Experimental and theoretical investigation of HT-S/PMR-PI composites for application to advanced aircraft engines," in *29th Annual Technical Conference, Reinforced Plastics/Composites Institute,* The Society of the Plastics Industry, Inc., 1974.
82. Liber, T., and Daniel, I. M., "Edgeless composite laminate specimen for static and fatigue testing," in *33rd Annual Technical Conference, Reinforced Plastics/Composites Institute,* The Society of the Plastics Industry, Inc., 1978.
83. Liber, T. M., Daniel, I. M., and Chamis, C. C., "Static behavior and the effects of thermal cycling in hybrid laminates," in *32nd Annual Technical Conference, Reinforced Plastics/Composites Institute,* The Society of the Plastics Industry, Inc., 1977.
84. Chamis, C. C., "Prediction of fiber composite mechanical behavior made simple," in *35th Annual Technical Conference, Reinforced Plastics/Composites Institute,* The Society of the Plastics Industry, Inc., 1980.
85. Wolkowitz, William, and Poveromo, Leonard M., "The effect of long-time thermal exposure on the mechanical properties of graphite/polyimide composites," in *28th Annual Technical Conference, Reinforced Plastics/Composites Institute,* The Society of the Plastics Industry, Inc., 1973.
86. Serafini, T. T., and Vannucci, R. D., "Tailor making high performance graphite fiber reinforced PMR polyimides," in *30th Anniversary Technical Conference, Reinforced Plastics/Composites Institute,* The Society of the Plastics Industry, Inc., 1975.
87. Hensel, Joseph D., "When, where, and how to use polyimides," *Plastics Engineering,* Vol. 33, No. 10 (Oct., 1977).
88. Alvarez, Robert T., and Darmory, Franklin P., "KINEL polyimide molding compounds," in *29th Annual Technical Conference, Reinforced Plastics/Composites Institute,* The Society of the Plastics Industry, Inc., 1974.
89. Sturgeon, J. B., "Fatigue and creep testing of unidirectional carbon fiber reinforced plastics," in *28th Annual Technical Conference, Reinforced Plastics/Composites Institute,* The Society of the Plastics Industry, Inc., 1973.
90. Jutte, Ralph B., "Structural SMC: Material, process and performance review," in *33rd Annual Technical Conference, Reinforced Plastics/Composites Institute,* The Society of the Plastics Industry, Inc., 1978.
91. Conley, David O., "Status of pigmented low shrink SMC or BMC," in *28th Annual Technical Conference, Reinforced Plastics/Composites Institute,* The Society of the Plastics Industry, Inc., 1973.
92. Gott, S. L., Suggs, J. L., and Blount, W. W., "New unsaturated polyester resins based on TMPD Glycol for FRP composites," in *28th Annual Technical Conference, Reinforced Plastics/Composites Institute,* The Society of the Plastics Industry, Inc., 1973.
93. Ward, D. Douglas, and Robbins, Greg, "Glass reinforced polyester and its impact on business machines," in *30th Anniversary Technical Conference, Reinforced Plastics/Composites Institute,* The Society of the Plastics Industry, Inc., 1975.
94. Maaghul, J., and Potkanowicz, E. J., "HMC—A high performance sheet molding com-

pound," in *31st Annual Technical Conference, Reinforced Plastics/Composites Institute,* The Society of the Plastics Industry, Inc., 1976.

95. Johansen, Kenneth, "The first XMC production part," in *32nd Annual Technical Conference, Reinforced Plastics/Composites Institute,* The Society of the Plastics Industry, Inc., 1977.
96. Kalnin, I. L., "Graphite fiber reinforced epoxy bulk molding compounds," in *28th Annual Technical Conference, Reinforced Plastics/Composites Institute,* The Society of the Plastics Industry, Inc., 1973.
97. Gruenewald, Rainer, and Walter, Oskar, "Fifteen years' experience with sheet molding compound," in *30th Anniversary Technical Conference, Reinforced Plastics/Composites Institute,* The Society of the Plastics Industry, Inc., 1975.
98. Finsel, James R., and Younge, K. Robert, "Hybrid BMC's—Fiberglas reinforced polyester thermosets displacing traditional molding materials," in *35th Annual Technical Conference, Reinforced Plastics/Composites Institute,* The Society of the Plastics Industry, Inc., 1980.
99. Best, James R., and Kelly, Thomas C., "The growth of the fiber glass reinforced thermoplastics market in the United States," in *29th Annual Technical Conference, Reinforced Plastics/Composites Institute,* The Society of the Plastics Industry, Inc., 1974.
100. Whicker, G. L., "Inorganic fiber reinforced thermoplastics," in *Fourth International reinforced Plastics Conference,* London, Nov., 1964, British Plastics Federation, London, 1964.
101. LNP, "Design properties profile of LNP Thermocomp fortified thermoplastics," Bulletin 250–570, LNP Corp., Pennsylvania.
102. Kjellmark, Jr., E. W., and Johannes, R. K., "Glass and mineral reinforced nylons; their advantages and uses," in *30th Anniversary Technical Conference, Reinforced Plastics/Composites Institute,* The Society of the Plastics Industry, Inc., 1975.
103. Orenski, P. J., Berger, S. E., and Ranney, M. W., "Silane coupling agents—Performance in engineering plastics," in *28th Annual Technical Conference, Reinforced Plastics/Composites Institute,* The Society of the Plastics Industry, Inc., 1973.
104. Maine, F. W., Shepherd, P. D., Golemba, F. J., Moskal, E. A., and Osborne, A. D., "A new family of reinforced thermoplastics," in *28th Annual Technical Conference, Reinforced Plastics/Composites Institute,* The Society of the Plastics Industry, Inc., 1973.
105. Cameron, A. B., Heron G. F., and Wicker, G. L., "Asbestos reinforced thermoplastics," in *28th Annual Technical Conference, Reinforced Plastics/Composites Institute,* The Society of the Plastics Industry, Inc., 1973.
106. Theberge, John, Arkles, Barry, Cloud, Pete, and Gerakaris, Steve, "Elevated temperature performance of thermoplastic composites," in *30th Anniversary Technical Conference, Reinforced Plastics/Composites Institute,* The Society of the Plastics Industry, Inc., 1975.
107. Wambach, A. D., and Kramer, M., "VALOX-420-SEO—A glass-reinforced thermoplastic polyester with built-in flame retardance," in *28th Annual Technical Conference, Reinforced Plastics/Composites Institute,* The Society of the Plastics Industry, Inc., 1973.
108. Theberge, John, Arkles, Barry, and Robinson, Rick, "Carbon fiber reinforced thermoplastics," in *29th Annual Technical Conference, Reinforced Plastics/Composites Institute,* The Society of the Plastics Industry, Inc., 1974.
109. Menges, G., Harnier, A. V., and Heisse, P., "Properties of carbon fiber reinforced polyamides in dependence of the manufacturing process," in *28th Annual Technical Conference, Reinforced Plastics/Composites Institute,* The Society of the Plastics Industry, Inc., 1973.
110. Theberge, John E., Arkles, Barry, Cloud, Peter J., and Goodhue, Richard, "Gear and bearing designs with lubricated and reinforced thermoplastics," in *28th Annual Technical Conference, Reinforced Plastics/Composites Institute,* The Society of the Plastics Industry, Inc., 1973.
111. Theberge, John E., "Recent product advances in thermoplastic composites," in *35th Annual*

Technical Conference, Reinforced Plastics/Composites Institute, The Society of the Plastics Industry, Inc., 1980.

112. Benjamin, B. S., "Expanded polystyrene and rigid polyurethane foam cores in structural sandwich construction," *Civil Engineering and Public Works Review* (Nov., 1965).
113. United Kingdom Atomic Energy Authority, "Plastics at Aldermaston," June, 1965.
114. The Celotex Corporation, "Technifoam—Structural Bureau Report 308," Building Officials Conference of America, Oct., 1969.
115. General Latex and Chemical Corporation, "Typical properties, 16-F-1502," Vultafoam Technical Bulletin, March 1970.
116. The Urethane Systems Manufacturer's Committee, "Urethane foam systems—Value and versatility," The Cellular Plastics Division, SPI.
117. Barber, Elaine, Nelson, John, and Beck, Warren, "Improving urethane foams with glass bubbles," in *32nd Annual Technical Conference, Reinforced Plastics/Composites Institute,* The Society of the Plastics Industry, Inc., 1977.
118. Traeger, R. K., *Journal of Cellular Plastics* (Sept., 1967).
119. Griffin, J. D., and Skochdopole, R. E., "Plastic foams," in *Engineering Design for Plastics,* edited by E. Baer, Reinhold Publishing Corporation, New York, 1964, Chap. 15.
120. Amspec, Inc., "STYROFOAM brand plastic foam," Building Insulation, 7.14, Sept., 1973.
121. Shell Plastics Advisory Service, "Cellular plastics in roof structures," London, 1965.
122. Hughes, B., "Structural plastics for sandwich construction," M.Sc. (Eng.) thesis, University of London, 1966.
123. Hughes, B., and Wajda, R. L., "Plastics sandwich panels with various foamed core materials, and their behavior under load," in *Conference on Plastics in Building Structures,* London, 1965, Pergamon Press, Oxford, 1966.
124. Gallagher, R. B., Harpell, G. A., Kamens, E. R., and Kamath, V. R., "Properties, processing and applications of polyester foams prepared from LUCEL AZO foaming agents," in *32nd Annual Technical Conference, Reinforced Plastics/Composites Institute,* The Society of the Plastics Industry, Inc., 1977.
125. Eakin, J. L., "Urethane structural foams: key properties give them a competitive edge," *Plastics Engineering,* Vol. 34, No. 7 (July, 1978).
126. Matonic, Delbert E., "Molding PC structural foam on standard injection machines," *Plastics Engineering,* Vol. 32, No. 9 (Sept., 1976).
127. O'Dell W. W., and Graham, D. L., "Structural behavior of sandwich panels with foamed-plastics cores," in *Conference on Plastics in Building Structures,* London, 1965, Pergamon Press, Oxford, 1966.
128. Gibbs and Cox, Inc., *Marine Design Manual for Fiberglas Reinforced Plastics,* McGraw Hill Book Co., Inc., New York, 1960.
129. Panshin, B. I., Popov, V. A., Fedorenko, A. G., Buyanov, G. I., Efimova, V. S., and Gorskii, K. P., "Mechanical properties of cellular plastics determining their durability as load-bearing cores. Durability under static load," *Soviet Plastics* (Dec., 1964).
130. Oberdick, W. A., "Application of minimum structure to cellular plastics," in *Conference on Plastics in Building Structures,* London, 1965, Pergamon Press, Oxford, 1966.
131. Hauser, Consuelo M., Hauser, Ray L., and Norling, James F., "Designing strength and stiffness into GRP panels," *Plastics Engineering,* Vol. 34, No. 11, (Nov., 1978).
132. Holland, K. M., and Forsberg, R. P., "Developments in reinforced plastics honeycomb core materials," in *Proc. 12th Technical and Management Conference, Reinforced Plastics Division,* SPI, Feb., 1957.
133. Holloway, M. W., and MacDonald, N. C., "Composite structures as high strength, light weight fillers," in *Fourth International Reinforced Plastics Conference,* London, Nov., 1964, British Plastics Federation, London.

134. Kray, Raymond J., "NCNS resins for structural and electrical applications," in *30th Anniversary Technical Conference, Reinforced Plastics/Composites Institute,* The Society of the Plastics Industry, Inc., 1975.
135. Shearer, D. C., "A one-step fabrication of honeycomb sandwich radomes," in "*Proc. 20th Technical and Management Conference, Reinforced Plastics Division,* SPI, Feb., 1965.
136. Kuenzi, E. W., "Structural sandwich design criteria," Forest Products Laboratory Report 2161, 1959.
137. Doherty, D. J., Ball, G. W., and Walker, M. G., "The physical properties of rigid urethane foam sandwich panels," in *Conference on Plastics in Building Structures,* London, 1965, Pergamon Press, Oxford, 1966.
138. Benjamin, B. S., "Folded plate structures in plastics," Ph.D. thesis, University of London, 1965.
139. Fenton, J. B., "Sandwich panels for vehicle bodies," *Automotive Design Engineering* (Nov., 1962).
140. Hoff, N. J., and Mautner, S. E., "Bending and buckling of sandwich beams," *J. Aero. Sciences,* Vol. 15, No. 12 (1948).
141. March, H. W., "Effects of shear deformation in the core of a flat rectangular sandwich panel," Forest Products Laboratory Report 1583, 1955.
142. Reissner, E., "Small bending and stretching of sandwich type shells," N.A.C.A. Technical Note No. 1832, 1949.
143. Ericksen, W. S., "Effects of shear deformation in the core of a flat rectangular sandwich panel," Forest Products Laboratory Report 1583-D, 1958.
144. Born, J., *Hipped Plate Structures,* Crosby, Lockwood and Son Ltd., London, 1962.
145. Benjamin, B. S., "Single skin folded plate construction in plastics," in *International Conference on Building Materials,* Warsaw, 1967.
146. Benjamin, B. S., "An account of some experimental work on a prototype folded plate barrel vault of plastics sandwich construction," in *International Conference on Space Structures,* London, 1966, Blackwell Scientific Publications Ltd., 1967.
147. Haas, A. M. *Design of Thin Concrete Shells,* John Wiley and Sons, Inc., 1967, Vol. 2.
148. Gilkie, R. C., and Robak, D., "Recent developments in plastics stressed-skin pyramidal roof systems," in *Conference on Plastics in Building Structures,* London, 1965, Pergamon Press, Oxford, 1966.
149. Gilkie, R. C. "A comparison between the theoretical and experimental analysis of a stressed skin system of construction in plastics and aluminum," in *International Conference on Space Structures,* London, 1966, Blackwell Scientific Publications Ltd., 1967.
150. Robak, D., "The structural use of plastics pyramids in doublelayer space grids," in *International Conference on Space Structures,* London, 1966, Blackwell Scientific Publications Ltd., 1967.
151. Gilkie, R. C., "Pyramids in lightweight roof systems, Ph.D. thesis, University of London, 1967.
152. Whitney, G. S., Anderson, B. G., and Birnbaum, N., "Reinforced concrete folded plate construction," *J. Struct. Division, Proc. Am. Soc. C.E.,* Vol. 85 (Oct., 1959).
153. Simpson, H., "Design of folded plate roofs," *J. Struct. Division, Proc. Am. Soc. C.E.,* Vol. 84 (Jan., 1958).
154. Parme, A. L. L., "Direct solution of folded plate concrete roofs," Bulletin of the International Association for Shell Structures, No. 6.
155. Yitzhaki, D., *The Design of Prismatic and Cylindrical Shell Roofs,* Haifa Science Publishers, Haifa, Israel, 1958.
156. Goldberg, J. E., and Leve, H. L., "Theory of prismatic folded plate structures," International Association of Bridge and Structural Engineering, Zurich, No. 17.

157. Venkatramiah, S., Benjamin, B. S., and Gandhi, N. A., "Edge disturbance effects on inner plates in multi-bay folded plate construction," *Indian Concrete Journal,* June, 1962.
158. The Concrete Association of India, "Design of deep girders," 1962.
159. Kehrer, K. K., "Prefabricated shell structure of glass reinforced plastics for a greenhouse," in *Conference on Plastics in Building Structures,* London, 1965, Pergamon Press, Oxford, 1966.
160. Zhidkov, V. D., "Some examples of space structures of aluminum sheets," in *Analysis of Space Structures,* edited by A. A. Umanski, Moscow, 1964, Vol. 9.
161. Timoshenko, S., and Woinowsky-Krieger, S., *Theory of Plates and Shells,* 2nd ed., McGraw-Hill Book Co. Inc., New York, 1959.
162. Benjamin, B. S., "Transverse behavior of the external unit of prismatic folded plate structures in plastics," in *International Conference on Space Structures,* London, 1966, Blackwell Scientific Publications Ltd., 1967.
163. Volmir, A. S., *Flexible Plates and Shells,* Moscow, 1956.
164. Szymczyk, J., *Arches. Tables for Statical Analysis,* Pergamon Press, Oxford, 1961.
165. Greszczuk, L. B., "Theoretical and experimental studies on properties and behavior of filamentary composites," in *Proc. 21st Technical and Management Conference, Reinforced Plastics Division,* SPI, Feb., 1966.
166. Girkmann, K., *Flächentragwerke,* Springer-Verlag, Vienna, 1959.
167. Billington, D. P., *Thin Shell Concrete Structures,* McGraw-Hill Book Co., Inc., New York, 1965.
168. Heger, F. J., "Design of reinforced plastics shell structures," in *Plastics in Building,* edited by I. Skeist, Reinhold Publishing Corporation, New York, 1966, Chap. 6.
169. Flügge, W. *Statik und Dynamik der Schalen,* Springer-Verlag, Berlin, 1962.
170. Pflüger, A., *Elementare Schalenstatik,* Springer-Verlag, Berlin, 1957.
171. Gondikas, P., and Salvadori, M. G., "Wind stresses in domes," *J. Engineering Mechanics Division, Proc. Am. Soc. C.E.,* Vol. 86, n. EM 5 (Oct., 1960).
172. Thorkildsen, R. L., "Mechanical Behavior," in *Engineering Design for Plastics,* edited by E. Baer, Reinhold Publishing Corporation, New York, 1964, Chap. 5.
173. Timoshenko, S., and Gere, J. M., *Theory of Elastic Stability,* McGraw-Hill Book Co., Inc., New York, 1961.
174. Hannoosh, M. M., and Muldoon, R. A., "Trends in radome design parameters," in *Proc. 18th Technical and Management Conference, Reinforced Plastics Division,* SPI, Feb., 1963.
175. Schmidt, H., "Ergebnisse von Beulversuchen mit doppelt gekrummten Schalenmodellen," in *Proc. of the Symposium on Shell Research,* North Holland Publishing Co., Amsterdam, 1962.
176. Davis, P., "Rigid reinforced plastics radomes for ground environments," in *Proc. 12th Technical and Management Conference, Reinforced Plastics Division,* SPI, Feb., 1957.
177. Benjamin, B. S., *Structures for Architects,* Ashnorjen Bezaleel Publishing Co., Kansas, 1975.
178. Gibson, J. E. *The Design of Cylindrical Shell Roofs,* E. & F. N. Spon, London, 1961.
179. Salvadori, M., and Levy, M., *Structural Design in Architecture,* Prentice-Hall, Inc., New Jersey, 1967.
180. Livesley, R. K., *Matrix Methods of Structural Analysis,* Pergamon Press, Oxford, 1964.
181. Zienkiewicz, O. C., and Cheung, Y. K., *The Finite Element Method in Structural and Continuum Mechanics,* McGraw-Hill Book Co., Ltd., London, 1967.
182. Rubinstein, M. F., *Matrix Computer Analysis of Structures,* Prentice-Hall, Inc., New Jersey, 1966.
183. Willems, Nicholas, and Lucas Jr., William M., *Matrix Analysis for Structural Engineers,* Prentice-Hall, New Jersey, 1968.

184. Von Karman, T., "The strength of thin plates in compression," *Trans. Am. Soc. M. E.* (June, 1932).
185. Needham, R. A., "The ultimate strength of aluminum alloy formed structural shapes in compression," *J. Aero. Sciences* (April, 1954).
186. Frost, R. J., "The ultimate strength of corrugated plates under normal and axial loads," M. Phil. thesis, University of London, 1968.
187. Baker, A. R., "Stiffness Method," Lecture notes in the course on 'Advanced Structural Analysis,' Hatfield Polytechnic, 1967.
188. Nooshin, H., "A technique for the analysis of structures having partially constrained joints," *Civil Engineering and Public Works Review,* (May, 1966).
189. Ramirez-Rivera, A., "Topological and geometrical considerations in the analysis of space structures by the electronic computer," in *International Conference on Space Structures,* London, 1966, Blackwell Scientific Publications, Ltd., 1967.
190. Prynne, P., "A computer investigation into the behavior of a long braced barrel vault roof," in *International Conference on Space Structures,* London, 1966, Blackwell Scientific Publications, Ltd., 1967.
191. Jennings, A., and Majid, K. I., "The computer analysis of space frames using sparse matrix techniques," in *International Conference on Space Structures,* London, 1966, Blackwell Scientific Publications, Ltd., 1967.
192. Zienkiewicz, O. C., "The finite element technique as a mathematical approximation and a civil engineering tool," in *Symposium on Finite Element Techniques and Applications,* University of Birmingham, April, 1967.
193. Clough, R. W., "The finite element method in structural mechanics," in *Stress Analysis,* edited by O. C. Zienkiewicz and G. S. Holister, John Wiley and Sons, London, 1965, Chap. 7.
194. Clough, R. W., "The finite element in plane stress analysis," in *Proc. 2nd Am. Soc. C.E. Conference on Electronic Computation,* Sept., 1960.
195. MIT, "ICES STRUDL-1. The Structural Design Language Engineering User's Manual," IBM United Kingdom, Ltd., Sept., 1967.
196. Villello, M., "Technology: The infinitely expandable future of air structures," *Architectural Forum,* Vol. 133 (Sept., 1970).
197. Murata, Y., "Osaka World's Fair: The United States Pavilion and Fuji Group Pavilion," *Japan Architect,* Vol. 45 (May–June, 1970).
198. Benjamin, B. S., *The Analysis of Braced Domes,* Asia Publishing House, 1963.
199. Benjamin, B. S., "A shell analogy for Schwedler domes," *Indian Concrete Journal,* (Nov., 1960).
200. Mitchell, L. H., "A shell analogy for framed domes," Note ARL/SM 208, Research and Development Branch, Aeronautical Research Labs., Australia, Dec., 1953.
201. Hossdorf, H., "Design construction and experience with post-tensioned polyester roof in the Swiss National Exhibition," in *Conference on Plastics in Building Structures,* London, 1965, Pergamon Press, Oxford, 1966.
202. Benjamin, B. S., *Building Construction for Architects and Engineers, Volume I,* Ashnorjen Bezaleel Publishing Co., Kansas, 1978.
203. Note on "Shelter for Migrant Workers" Interbuild, Sept., 1966.
204. Rial, E., "Reinforced plastics in Italy for building and technical applications—Developments and achievements 1964–66," in *Fifth International Reinforced Plastics Conference,* London, Nov., 1966, British Plastics Federation, London.
205. Note on "Lampshade roof for cheap housing," *New Scientist,* June 8, 1967.
206. Note on "Pleated roof in plastics," *Interbuild* (Oct., 1966). See also note on "Low cost plastics roofs," *Engineering* (Oct., 1966).

207. Benjamin, B. S., "A prefabricated plastics roof for low-cost housing," *Reinforced Plastics* (Nov., 1966).
208. Winfield, Armand G., and Winfield, Barbara L., "Major breakthrough in low cost plastic housing using jute reinforced polyesters: The CARE/Winfield/Bangladesh House," in *29th Annual Technical Conference, Reinforced Plastics/Composites Institute,* The Society of the Plastics Industry, Inc., 1974.
209. Van Giessen, J. H., "Air transportable emergency shelter," Report FPO-AFY, No. 114, Avio-Fokker, Ypenburg, Jan., 1976.
210. Benjamin, B. S., "Plastics shelters for slum clearance in poor countries—A problem statement," in *30th Anniversary Conference, Reinforced Plastics/Composites Institute,* The Society of the Plastics Industry, Inc., 1975.
211. Gadgil, Satish G., "Low cost shelters for the poor in underdeveloped countries with emphasis on India," M. Arch. thesis, University of Kansas, 1975.
212. Benjamin, B. S., "Plastics shelter for slum clearance in poor countries," in *Plastics in Building, Regional Technical Conference,* Society of Plastics Engineers, Inc., Eastern New England Section, Boston, Nov., 1976.

AUTHOR INDEX

Abbott, B. W., 46, 59, 369
Abdul Kalam, A. P. J., 368
Ainsworth, L., 65, 66, 370
Alvarez, R. T., 121, 371
Anderson, B. G., 221, 374
Anthony, F. M., 371
Antonacci, J., 102, 369
Arkles, B., 136, 137, 139, 372

Bacon, C. E., 32, 38, 369
Baer, E., 368
Baker, A. R., 376
Ball, G. W., 374
Barber, E., 154, 373
Bax, J., 66, 100, 370
Beck, W., 154, 373
Benjamin, B. S., 19, 22, 64, 67, 75, 81, 130, 143, 144, 145, 156, 190, 210, 214, 223, 238, 242, 255, 258, 315, 317, 332, 333, 334, 353, 355, 365, 367, 368, 370, 373, 374, 375, 376, 377
Berger, S. E., 132, 372
Best, J. R., 132, 372
Billington, D. P., 375
Birnbaum, N., 221, 374
Blount, W. W., 124, 371
Boller, K. H., 65, 370
Born, J., 374
Bowen, D. H., 85, 368
Brass, L. L., 105, 370
Breitigam, W. V., 105, 370
Brink, N. O., 55, 369
Broutman, L. J., 46, 59, 369
Browning, C. E., 368
Buyanov, G. I., 179, 373

Cameron, A. B., 133, 372
Chamis, C. C., 109, 110, 111, 112, 371
Cheung, Y. K., 375
Chiao, T. T., 106, 122, 371
Cloud, P. J., 136, 139, 372
Clough, R. W., 376
Collister, J., 129, 368
Conley, D. O., 124, 371
Crowder, J. R., 367

Daniel, I. M., 110, 111, 112, 371
Darmory, F. P., 121, 371
Davenport, A. J., 367
Davis, P., 294, 375
Desai, R. R., 44, 369
Dietz, A. G. H., 369
Dixon, B. G., 55, 369
Doherty, D. J., 374

Eakin, J. L., 178, 373
Efimova, V. S., 179, 373
Enos, J. H., 105, 370
Ericksen, W. S., 200, 374
Ewchuk, W., 368

Faughnan, K. A., 368
Fedorenko, A. G., 179, 373
Fenton, J. B., 374
Feodosjew, B., 293
Ferrarini, J., 55, 369
Findley, W. N., 75, 100, 103, 370
Finsel, J. R., 130, 372
Flugge, W., 375
Forsberg, R. P., 182, 373
Frost, R. J., 310, 311, 313, 376

Gadgil, S. G., 377
Gallagher, R. B., 171, 367, 373
Gandhi, N. A., 375
Gerakaris, S., 136, 372
Gere, J. M., 375
Gibson, J. E., 375
Gilbu, A., 81, 370
Gilkie, R. C., 317, 374
Girkmann, K., 375
Goldberg, J. E., 221, 222, 374
Goldfein, S., 66, 67, 370
Golemba, F. J., 132, 372
Gondikas, P., 375
Goodhue, R., 139, 372
Gorskii, K. P., 179, 373
Gott, S. L., 124, 371
Graham, D. L., 179, 373
Green, D. H., 47, 369
Greszcsuk, L. B., 274, 375
Griffin, J. D., 373
Gruenewald, R., 130, 372
Guevin, P. R., 47, 369
Gumerman, C., 8, 367

Haas, A. M., 374
Haddad, G. N., 80, 370
Hanley, D. P., 371
Hannoosh, M. M., 375
Hanson, M. P., 109, 110, 371
Harnier, A. V., 137, 139, 372
Harpell, G. A., 171, 367, 373
Hart, G. L., 370
Hauser, C. M., 182, 373
Hauser, R. L., 182, 373
Hawley, G. C., 102, 369
Hays, C. W., 129, 368
Heisse, P., 137, 139, 372
Hensel, J. D., 115, 371
Heger, F. J., 375
Heron, G. F., 133, 372
Hertz, J., 55, 369
Hill, J. E., 368
Hilton, H. H., 368
Hoff, N. J., 374
Holland, K. M., 182, 373
Holloway, M. W., 182, 373
Hossdorf, H., 376
Hughes, B., 159, 179, 200, 373

Illing, G., 133, 368

Jaspers, H., 105, 371
Jennings, A., 376
Johannes, R. K., 132, 372
Johansen, K., 127, 372
Jutte, R. B., 124, 371

Kabelka, J., 46, 56, 62, 68, 77, 369
Kalnin, I. L., 127, 372
Kamath, V. R., 171, 367, 373
Kamens, E. R., 171, 367, 373
Kehrer, K. K., 375
Kelly, T. C., 132, 372
Kjellmark, E. W., 132, 372
Knox, J. R., 368
Kramer, M., 136, 372
Kray, R. J., 185, 374
Kuenzi, E. W., 374

Lant, T. P. R., 100, 370
Leve, H. L., 221, 222, 374
Levy, M., 227
Liber, T. M., 110, 111, 112, 371
Linton, T. J., 368
Livesley, R. K., 375
Longenecker, D. M., 55, 369
Lucas, W. M., 375
Lucy, M. H., 102, 370

Maaghul, J., 125, 128, 371
MacDonald, N. C., 185, 373
Macphail, I., 85, 368
Maier, R. E., 368
Maine, F. W., 132, 372
Majid, K. I., 376
Makowski, Z. S., 367
March, H. W., 197, 198, 200, 201, 374
Matonic, D. E., 179, 373
Mautner, S. E., 374
McCammond, D., 368
McCorsley, C. C., 79, 370
McKay, G. R., 8, 367
Menges, G., 137, 139, 372
Milewski, J. V., 29, 30, 369
Mitchell, L. H., 376
Montella, R., 80, 370
Morgan, P., 367
Moskal, E. A., 132, 372
Muldoon, R. A., 375
Murata, Y., 376
Mushtari, 293

Needham, R. A., 308, 311, 376
Nelson, J., 154, 373
Nooshin, H., 331, 376
Norling, J. F., 182, 373

Oberdick, W. A., 179, 373
O'Dell, W. W., 179, 373
Orenski, P. J., 132, 372
Osborne, A. D., 132, 372
O'Toole, J. L., 367

Pal, P. K., 85, 368
Panshin, B. I., 179, 373
Paramasivam, T., 368
Parme, A. L. L., 221, 374
Passaglia, E., 368
Pearson, E. A., 368
Penn, L., 106, 122, 371
Pfluger, A., 375
Philip, A. R., 84, 370
Piano, R., 367
Popov, V. A., 179, 373
Potkanowicz, E. J., 125, 371
Poveromo, L. M., 115, 371
Powell, D., 367
Price, H. L., 102, 370
Pritchard, G., 47, 369, 370
Prynne, P., 376

Ramirez-Rivera, A., 376
Rangel, G., 369
Ranney, M. W., 132, 372
Rassell, 47, 55, 86
Ratcliffe, W. F., 368
Reis, J. F., 101, 368
Reissner, E., 200, 374
Rial, E., 376
Riddell, M. N., 367
Riewald, P. G., 112, 368
Robak, D., 374
Robbins, G., 124, 371
Robinson, R., 137, 139, 372
Rolston, J. A., 81, 370
Ronkka, Y., 367
Rosato, D. V., 38, 369
Rubinstein, M. F., 375

Salvadori, M., 375
Sandorff, 294
Schmidt, H., 294, 375
Schrock, B. J., 369
Scott, R. J., 43, 369
Serafini, T. T., 115, 371
Shaffer, B. W., 369
Shearer, D. C., 186, 374
Shepherd, P. D., 132, 372
Simpson, H., 221, 374
Skochdopole, R. E., 373
Sleddon, G. J., 367
Smarook, W. H., 172, 367
Sommella, J. H., 43, 369
Sonneborn, R. H., 369
Sturgeon, J. B., 121, 371
Suggs, J. L., 124, 371
Sundstron, G. A., 129, 368
Surkin, 293
Szymanski, W. A., 105, 370
Szymczyk, J., 375

Taneja, N., 47, 369
Theberge, J. E., 136, 137, 139, 140, 372
Thomas, R. E., 105, 370
Thompson, A. W., 64, 90, 102, 370
Thorkildsen, R. L., 375
Timoshenko, S., 227, 253, 375
Traeger, R. K., 156, 158, 373
Trampenau, R. H., 43, 369
Tsien, H. S., 293
Turner, S., 368

Umanski, A. A., 375

Vaccarella, P. W., 105, 370
Van Giessen, J. H., 377
Vannucci, R. D., 115, 371
Venkatramiah, S., 375
Villello, M., 376
Volmir, A. S., 253, 255, 375
Von Karman, T., 293, 306, 310, 311, 376

Wajda, R. L., 373
Walker, M. G., 374
Walter, O., 130, 372
Wambach, A. D., 136, 372
Ward, D., 124, 371
Wells, H., 85, 368
Whitney, G. S., 221, 374
Wicker, G. L., 132, 133, 372

Willems, N., 375
Wilson, T. B., 43, 369
Winfield, A. G., 358, 377
Winfield, B. L., 358, 377
Woinowsky-Krieger, S., 227, 253, 375
Wolkowitz, W., 115, 371
Woodhams, R. T., 127, 369
Wynes, T., 105, 371

Xanthos, M., 102, 127, 369

Yitzhaki, D., 221, 374
Younge, K. R., 130, 372

Zhidkov, V. D., 258, 375
Zienkiewicz, O. C., 375, 376
Zweben, C., 112, 368

SUBJECT INDEX

Abrasion resistance, 32, 79
ABS, 3, 31, 132, 133, 174
Acetals, 132
Acrylics, 2, 5, 6, 7, 79, 80, 132, 317. *See also* Polymethylmethacrylate
Additives, 9, 144
Aluminum, 1, 218, 219, 259, 294, 308, 310, 355
Amosite, 26, 27, 87. *See also* Asbestos
Anchorage, 346–348, 360. *See also* Foundations
Angle plies, 56, 109–112, 129
Anthophyllite, 26, 27. *See also* Asbestos
Aramid, 3, 24, 28, 29, 105, 106, 110, 112, 115, 122, 124
Arch. *See* Barrel vault
Architectural forms, 209, 364
ARP, 86, 87
Asbestos, 2, 24, 26, 27, 75, 86, 101, 102, 127, 132, 133, 182
Autoclave, 31, 33
Axial force, 3, 187, 301, 302, 312, 313, 342

Barrel vault (or shell), 214, 215, 220, 224, 247, 248, 256–258, 262, 264, 266, 271, 275, 295–297, 299, 300, 315, 331
Beam (including Beam action), 209, 213, 222, 224, 233–236, 247, 248, 296, 300, 301, 303, 304
Bending moment, 3, 187, 188, 190, 200, 218, 222, 227, 229, 231, 233, 238, 240, 242, 243, 250–252, 255, 256, 264, 266, 271, 277, 279, 280, 284, 288, 291, 297, 299–301, 303, 313, 342
Bending stress, 188. *See also* Flexural stress
Berms, 346, 347
Beryllium oxide, 30
Blow molding, 38
Boltzmann superposition, 16
Boron, 3, 24, 28, 30, 105, 110, 111, 115
Buckling, 195, 201–203, 207–210, 215, 216, 220, 225, 235, 258, 264, 293, 294, 301, 304, 305, 314, 317, 332, 341, 353
Building construction, 1, 9
Building regulations, 9

Calendering, 38
Carbon fiber. *See* Graphite
Casting, 38
Centrifugal casting, 31, 35
Chemical resistance, 28, 47, 55, 79, 82, 87, 101, 103, 105
Chopped strand mat, 5, 24, 25, 35, 38, 42–44, 47, 62, 79, 82, 86, 87, 124, 259
Chrysotile, 26, 27, 101, 133. *See also* Asbestos
Clay, 24, 31, 127, 140
Cloth, 2, 5, 24–26, 38, 42, 43, 46, 55–57, 62, 64, 75, 84, 89, 100, 102, 103, 182, 190, 347. *See also* Fabric
Columns, 213, 355
Composite folded plates, 210, 213, 214, 222, 247, 258
Compression modulus, 4, 42, 46, 143, 145, 154, 156, 158, 160, 171
Compression molding, 102, 122, 124. *See also* Press molding
Compression ring, 348
Compressive strength, 42, 62, 143–145, 154, 158–160, 171, 172, 182, 185, 276
Compressive stress, 194, 225, 292, 297, 300, 304
Concrete, 1, 7, 235, 304, 332, 348

Cone, 210, 215
Conical dome. *See* Dome
Connections, 226, 248, 259, 261, 275, 295, 302, 355. *See also* Joints
Conoid, 216
Contact molding, 31–33, 44, 64, 84, 85, 90, 106, 192, 259, 295, 358
Continuous laminating, 31, 37
Continuous pultrusion, 31, 35
Core, 3, 84, 142, 145, 159, 173, 174, 178, 179, 182, 185–190, 192, 193, 195–197, 199, 201, 206, 219, 224, 236, 249, 259, 263, 264, 266, 351, 355, 364
Core failure. *See* Core
Corrosion resistance, 1, 80
Corrugations, 210, 215, 220
Cotton, 182, 347
Creep, 4, 7, 12, 14–16, 19, 30, 68, 75–77, 81, 87, 100, 103, 105, 115, 122, 137, 139, 158, 159, 179, 354
Creep modulus. *See* Creep
Creep rupture, 4, 7, 67, 68, 115, 347
Crocidolite, 26, 27, 86, 87. *See also* Asbestos
Cross-sectional area, 224, 247, 256, 264, 296, 297, 300, 314, 341
Cryogenic temperatures, 55

Deflected surfaces. *See* Deflections
Deflections, 5, 23, 68, 132, 189, 197–202, 206, 207, 210, 223–229, 231, 233, 235, 236, 238–243, 248–253, 261, 263, 264, 266, 276, 284, 300
Diaphragms, 210, 214, 222, 226, 234, 235, 296, 297, 299, 300
Directed fiber, 31, 35, 36
Disadvantages of plastics, 1
Dome, 130, 210, 215, 258, 275–280, 284, 288, 291, 293–295, 304, 317, 331–334, 341, 342, 352, 353, 355, 358, 363, 364
Dry condition. *See* Environmental conditions
Dynamic modulus, 6

Ellipsoid, 216
Elliptic paraboloid, 216
Encapsulation, 31, 34
Environmental conditions, 4, 8, 27, 65, 81, 84, 100, 259, 276
Epoxy, 2, 6, 25, 28, 88–90, 100, 101, 103, 106, 110–113, 121, 122, 127, 133, 182
Expanded polystyrene, 3, 5, 158, 159, 190
Expanded PVC, 3, 159, 179
Extruded polystyrene, 179
Extrusion molding, 38

Fabric, 29, 37, 42, 46, 56, 57, 64, 77, 85, 103, 186, 346, 354. *See also* Cloth
Facing failure. *See* Facings
Facings, 3, 187–189, 191–193, 195–197, 199, 201, 297, 300, 351
Felt, 26, 29, 85, 182
Filament winding, 5, 25, 31, 34, 35, 79, 80, 82, 84, 88, 100, 106
Fillers, 9, 24
Finite element method, 224, 331, 332, 352
Fire resistance, 9, 10, 363
Flexible polyurethane foam, 179
Flexural modulus, 5, 7, 42, 43, 46, 47, 84, 86, 100, 105, 115, 128, 133, 136, 137, 142, 172
Flexural rigidity, 187, 190, 218, 223, 228, 249
Flexural strength, 42, 47, 55, 84–87, 100, 105, 109, 115, 124, 128, 130, 133, 137, 142, 154, 172, 276
Flexural stress, 292. *See also* Bending stress
Folded plates, 5, 209, 210, 221–226, 228, 235, 236, 295, 296, 306, 311–315, 317, 331–334, 341, 342, 353, 355, 363
Foundations, 216, 301, 302, 304, 346, 347
FRP, 5, 9, 11, 32, 38, 42–44, 46, 47, 55–57, 62, 64, 66–68, 75–77, 79–81, 87, 130, 187, 190, 215, 259, 275–277, 294, 297, 303, 304, 311, 317, 332, 351, 353, 355

Gel-coat, 32, 84, 85, 295, 355
Geodesic dome. *See* Dome
Glass, 2, 24, 25, 29, 31, 33, 34, 36, 42, 46, 57, 61, 62, 66, 75, 79, 81, 82, 86, 90, 100, 102, 103, 105, 106, 112, 122, 124, 125, 127, 132, 133, 136, 137, 139, 140, 172, 174, 182, 190, 192, 274, 295
Glass spheres (or bubbles), 25, 129, 130, 133, 154, 156
Graphite (or Carbon), 2, 3, 24, 27–30, 100, 105, 106, 109–113, 115, 121, 124, 127, 132, 136, 137, 139, 140, 172
Grids, 219, 220, 327, 355

Hand lay-up. *See* Contact molding
Hessian, 26, 82, 84

High performance composites, 27, 28, 29, 88, 100, 101, 105, 106, 110, 111, 122, 125. *See also* specific materials
Honeycomb cores, 179, 182, 185, 186, 351
Hybrid composites, 28, 111–113, 115, 133, 137
Hybrid structure, 346, 349
Hyperbolic paraboloid, 217, 218, 275, 300–305, 353

Impact, 5, 29, 113, 133, 137
Inflatable membranes (or structures), 346, 348–350, 358
Injection molding, 31, 37, 132, 173
Inserts, 34
Insulation, 10, 185, 192, 355, 363

Joints, 210, 261, 318, 319, 331, 334, 335, 340, 341, 364, 365. *See also* Connections
JRP, 82, 84–86, 358
Jute, 2, 24–26, 82, 84, 85, 358

Kraft paper, 364

Laplace transforms, 17, 18
Light transmission, 8
Long term behavior. *See* specific mechanical property
Low-cost housing, 2, 333, 355, 360

Matched-die molding, 38
Matrix methods, 224, 225, 228, 306, 317, 353
Maxwell model, 14, 15
Melamine, 103
Membranes. *See* Tension membranes, Inflatable membranes
Membrane forces, 253, 255, 256, 277, 278, 284, 288, 291, 301–304
Membrane stress, 217, 243, 244, 246, 251, 292
Membrane theory, 277, 280, 282, 288
Mica, 24, 30, 31, 102, 124, 127, 132, 133, 140
Microfibers, 30
Milled fiber, 28, 29, 34
Mineral fibers, 24, 31
Modulus of elasticity, 4, 5, 12, 26, 28, 30, 46, 179, 189, 199, 201, 209, 271, 276, 280, 293, 294, 307, 308, 322, 341
Modulus of rigidity. *See* Shear modulus
Moment of inertia, 224, 247, 256–258, 264, 296, 306, 314, 341

NCNS resins, 182, 185
Needleloom mat, 25
Neoprene, 347, 351
Nylon, 6, 24, 25, 28, 31, 132, 133, 136, 139, 174, 347

Overlay mat, 25

Paper, 2, 182, 183, 185, 186, 364
Phenolic, 2, 6, 101–103, 115, 182, 185
Phenolic foam, 3, 159, 160, 171
Phenoxy, 133
Phenylene-oxide, 174, 179
Plate (including Plate action), 209, 210, 212, 222–228, 230, 231, 233–236, 238, 242, 243, 246–249, 251–253, 257, 258, 262–264, 266, 296, 306–308, 310, 311, 314, 315, 317
Plate theory, 189, 200, 224
Plenum chamber, 31, 36
Pneumatic structures, 346. *See also* Inflatable membranes
Poisson's ratio, 42, 56, 59, 61, 179, 228, 238, 249, 253, 271, 280, 307
Polyamides, 137, 139, 185, 218, 219, 306, 317
Polycarbonate, 28, 132, 174, 179
Polyester, 2, 3, 24, 25, 47, 62, 66, 79, 85–88, 100, 103, 106, 124, 127, 130, 132, 136, 171, 174, 179, 182, 258, 295, 358
Polyester foam, 3, 171
Polyethylene, 132, 347, 365
Polyimide, 28, 106, 109, 111, 115, 121, 136
Polymethylmethacrylate, 2, 19, 315. *See also* Acrylics
Polypropylene, 28, 132, 133
Polystyrene, 133
Polysulfone foam, 3, 28, 172, 174
Polyurethane, 47, 174, 178. *See also* Rigid polyurethane foam
Portal frames. *See* Rigid frames
Prefabrication, 1, 210, 226, 235, 247, 355
Preform, 35, 36, 38, 44
Premix molding, 31, 36, 37, 44. *See also* Press molding
Prepreg, 28, 127
Press molding, 44, 86, 90, 102, 106, 186, 295, 358. *See also* Premix molding
Pressure bag, 31, 33
Prismoidal folded plates, 210, 212
Proof stress, 308, 313
Pultrusion, 29. *See also* Continuous pultrusion

PVC, 2, 5, 6, 79, 80, 132, 347
PVF, 8
Pyramids (including Pyramidal folded plates), 210, 212, 215, 220

Radomes, 186, 294, 351, 352
Rankine criterion, 292, 293
Rate of strain, 4, 5, 142–144, 158, 159
Reinforced thermoplastics. *See* Thermoplastics
Reinforcement. *See* specific types of reinforcement
Ribbing, 210, 220, 294
Rigid frames (or frameworks), 213, 224, 247, 314, 326, 330
Rigid polyurethane foam, 3, 5, 143–145, 154, 156, 158–160, 171, 190, 259, 355, 364
Roof (or Roofing), 10, 19, 88, 130, 210, 212, 220–222, 247, 258, 333, 346, 350, 355, 356, 364
Roving, 5, 24, 25, 28, 29, 33–36, 38, 62, 66, 89, 124

Sand, 31, 81, 127
Sandwich construction, 3, 33, 42, 142, 173, 174, 179, 186–190, 192, 193, 195, 197, 201, 205, 207, 209, 215, 216, 219, 223–225, 236, 247, 248, 258, 271, 297, 300, 352, 355, 364
Sapphire whiskers, 3
Shear crimping, 196
Shear deformations, 187, 189, 190, 198, 200–202, 206–208, 224, 248, 263, 266
Shear deflections. *See* Shear deformations
Shear force, 187, 189, 190, 200, 219, 224, 229, 280, 288
Shear modulus, 42, 59, 144, 159, 160, 171, 182, 188, 199, 264, 341
Shear rigidity, 188, 190
Shear strength, 42, 77, 115, 144, 159, 160, 171, 172, 182
Shear stress, 188, 292
Shells, 5, 209, 214–216, 275, 293–297, 300, 301, 303, 304, 331, 332, 352, 353
Silicon carbide, 30
Silicone, 2, 103, 182
Sisal, 25, 26
Skeletal structure, 258, 306, 317, 331–333, 353. *See also* Space frame
Skylights, 9
Slab, 209
SMC/BMC, 24, 103, 105, 122, 124, 125, 127–130
Smoke, 11
Space frame (or frameworks), 186, 225, 306, 328, 330, 331, 334, 340, 341, 352
Spray-up, 31, 33, 34
Static fatigue. *See* Creep rupture
Steel, 1
Stiffness method, 225, 258, 317, 331, 332, 340, 341
Strength/weight ratio, 1, 35
Stress crazing, 8, 91
Stress relaxation, 12, 14–16, 19
Stress/strain curves, 4, 12, 18, 19, 57, 110
Stressed skin structures, 5, 209, 212, 218, 220, 306
Structural foam, 3, 173, 174, 178, 179
STRUDL, 225, 258, 306, 314, 332–334, 341, 353
Surface spread of flame, 10, 11
Surface tissue, 25, 295

Tape, 34, 109, 115
Temperature, 4, 6, 7–9, 12, 14, 19, 20, 23, 27, 34, 37, 47, 55, 59, 62, 67, 68, 75, 88, 100, 101, 103, 105, 106, 115, 121, 124, 132, 136, 137, 142, 143, 156, 172, 174, 182, 294
Tensile modulus, 4, 42, 43, 46, 68, 86, 87, 105, 106, 115, 143, 158–160, 171, 172
Tensile strength, 27, 42, 64, 84, 86, 90, 91, 100, 102, 105, 106, 109, 113, 115, 121, 124, 130, 136, 137, 143, 158, 160, 171, 172, 192, 276
Tensile stress, 292, 293, 297, 300, 304
Tension structures, 209, 216–218, 354
Tensioned structures, 353
Terylene, 347
Thermal insulation, 2. *See also* Insulation
Thermoplastics, 2, 4, 5, 7, 8, 24, 25, 37, 77, 79, 130, 132, 133, 136, 137, 140, 174, 179. *See also* specific materials
Thermosets (or Thermosetting materials), 2, 3, 6, 24, 115, 130, 160, 174, 185. *See also* specific materials
Timber, 304
Time, 4, 7, 12, 14–16, 19, 20, 64–68, 75, 76, 90, 130, 142, 294, 354
Toroid, 216
Tow, 28

Toxic gases, 11
Transfer molding, 37, 122, 124
Twill, 26, 84

Ultimate strength, 4, 7, 20, 22, 65, 67, 68, 276, 293. *See also* specific types
UV radiation, 8. *See also* Environmental conditions

Vacuum forming (including Vacuum bag), 2, 31, 32
Vinyl esters, 103, 105, 124
Viscoelastic behavior, 11, 12, 18, 19
Voigt model, 15, 16
Von Mises criterion, 292, 293

Walls (including Wall panels), 9, 10, 364
Warp, 5, 25, 55, 57, 59, 62, 68
Water slurry, 31, 32, 36
Waterproofing, 360, 365
Weathering, 8, 9, 32, 43, 47, 81, 82, 84, 85, 130, 218, 259, 293–295, 356, 358
Weft, 5, 25, 55, 57, 59, 62, 84
Wet condition. *See* Environmental conditions
Whiskers, 24, 30
Wollastonite, 24, 30, 31, 127, 132, 140
Woven roving, 5, 24, 29, 38, 42, 43, 55, 79, 105. *See also* Roving
Wrinkling, 195, 196

Yarns, 29, 82
Yield stress (or Point), 4, 115, 145, 158, 311